GIS and Archaeological Site Location Modeling

GIS and Archaeological Site Location Modeling

EDITED BY

Mark W. Mehrer
Konnie L. Wescott

CRC Press
Taylor & Francis Group
Boca Raton London New York

CRC Press is an imprint of the
Taylor & Francis Group, an **informa** business
A TAYLOR & FRANCIS BOOK

Cover art courtesy of Malcolm Ridges.

CRC Press
Taylor & Francis Group
6000 Broken Sound Parkway NW, Suite 300
Boca Raton, FL 33487-2742

First issued in paperback 2019

© 2006 by Taylor & Francis Group, LLC
CRC Press is an imprint of Taylor & Francis Group, an Informa business

No claim to original U.S. Government works

ISBN-13: 978-0-415-31548-7 (hbk)
ISBN-13: 978-0-367-39143-0 (pbk)
Library of Congress Card Number 2005044884

Library of Congress Cataloging-in-Publication Data

GIS and archaeological site location modeling / edited by Mark W. Mehrer, Konnie L. Wescott.
       p. cm.
   Includes bibliographical references and index.
   ISBN 0-415-31548-4 (alk. paper)
   1. Archaeology--Mathematical models--Congresses. 2. Geographic information
systems--Congresses. 3. Archaeology--Data processing--Congresses. I. Mehrer, Mark. II. Wescott, Konnie. III. Title.

CC80.4.G57 2001
930.10285--dc22   2005044884

**Visit the Taylor & Francis Web site at
http://www.taylorandfrancis.com**

**and the CRC Press Web site at
http://www.crcpress.com**

Dedication

To Jim (glad you're home safe from Iraq), Jackie, and Allison

–KLW

For Denise, Paige, and Alexander who make it all worthwhile

–MWM

Preface

This volume began as an idea for a conference. The idea was that the goals of archaeological predictive modeling needed to be reexamined in light of then-current criticisms, such as: site location cannot be modeled because ancient cultures cannot be modeled; site location cannot be modeled on the basis of known site locations because the population of known sites is biased by sampling errors; and, site models based on environmental factors are environmentally deterministic and therefore fatally flawed.

At the same time, advances in GIS (geographic information systems) software and personal-computing power had put sophisticated tools in the hands of archaeologists with an interest in predictive modeling. In other disciplines, GIS was being employed to model species diversity in forests, predict wetland dynamics, model health-care availability, and a host of other useful tasks. Surely these other disciplines had provoked profound criticism of their "fatal flaws." Yet they persisted and were doing something useful nevertheless. Perhaps there might be something useful to be done in archaeology by using GIS in service of predictive modeling if only we could see our way through the criticisms and neutralize our flaws.

True, it is unlikely that archaeologists will successfully model ancient society because it is too remote and too many mysteries remain. Also true, we are unlikely to predict the location of the next important site in the region of our choice because computers cannot be expected to perform a task that we are unable to formulate with our minds. And yet, other disciplines were using GIS methods and data to make useful models.

Our conference was inspired by the knowledge that some archaeologists were actually producing useful models — models that helped land managers and resource planners make better informed and more reasonable decisions. Other archaeological scholars were developing new methods and improving old ones. Decision support was a tacit if not proclaimed goal, the fatal flaws of societal modeling or site prospection notwithstanding. It seemed like a good time to assemble a broad range of experts to establish a baseline for site-location models.

And it was a good time, too. The response to our call for papers was gratifying. Responses came from Australia, Austria, Belgium, France, Greece, The Netherlands, Slovenia, the United Kingdom, and throughout the U.S. The international enthusiasm was especially welcome because the Wescott and Brandon (2000) edited volume was due out and had primarily emphasized the Western Hemisphere.

The conference center at Argonne National Laboratory's Advanced Photon Source was an outstanding venue. Its seclusion, security, and excellent facilities no doubt added to the freedom allowing ideas to readily flow. Our schedule offered plenty of time for discussion in addition to paper delivery. The discussion time was well used by the audience, who offered generous responses with lively give and take. After the papers were delivered, most of the participants were able to stay for an extended discussion about the immediate future of the modeling endeavor. This, too, was a lively exchange with an overtone not unlike what you might expect in the first full meeting of a newly formed organization. Well, we did not actually create a new organization, but I have noticed that some of the new notions we kicked around are now, 4 years later, mentioned more often in the literature — notions like "decision-support" and "baseline establishment."

Almost all of the presenters followed through by submitting papers for publication. That is why this is such a hefty volume. Each contribution was well conceived and professionally written, as I am sure you will agree.

MWM
DeKalb, Illinois

Acknowledgments

We thank all of the contributors to this volume as well as the additional presenters and participants at the GIS and Archaeological Predictive Modeling Conference (pictured here). Everyone's cooperation, support, and persistence are much appreciated. We thank Northern Illinois University for its financial support and Argonne National Laboratory for the use of its facility.

Attendees of the GIS and Archaeological Predictive Modeling Conference, Argonne National Laboratory, March 2001. (Courtesy of Argonne Photo Library.)

Mark W. Mehrer
Department of Anthropology
Northern Illinois University
De Kalb, Illinois

Konnie L. Wescott
Argonne National Laboratory
Argonne, Illinois

Editors

Mark W. Mehrer is an archaeologist with research interests in North American prehistory, settlement studies, household archaeology, remote sensing, and GIS. He has conducted research in midwestern North America. He is an associate professor in the Department of Anthropology, Northern Illinois University, where he teaches archaeology and directs NIU's Contract Archaeology Program.

Konnie L. Wescott is an archaeologist with Argonne National Laboratory. Her work for Argonne centers on the environmental assessment process, specifically the evaluation of potential impacts of proposed federal actions on cultural resources. She is also involved in activities in support of environmental assessments at various federal facilities throughout the U.S., such as conducting archaeological surveys, developing cultural resource management plans, and performing historic-building inventories. Her research interests include the use of GIS for modeling site locations and performing impact analyses, as well as Mesoamerican archaeology and museum studies. She is lead editor of a related book entitled *Practical Applications of GIS for Archaeologists*, published by Taylor and Francis in 2000.

Contributors

Matthew Cole   Environmental Services, Inc., Raleigh, North Carolina

Christopher D. Dore   Cartography and Geospatial Technologies Department, Statistical Research, Inc., Tucson, Arizona, and Department of Anthropology, University of Arizona, Tucson, Arizona

Michiel Gazenbeek   Centre d'Etudes Préhistoire, Antiquité, Moyen Age, Sophia-Antipolis (Valbonne), France

Steve Gould   GAI Consultants, Inc., Monroeville, Pennsylvania

Trevor Harris   West Virginia University, Morgantown, West Virginia

Eugenia G. Hatzinikolaou   Department of Geography and Regional Planning, National Technical University of Athens, Athens, Greece

Carrie Ann Hritz   Oriental Institute, University of Chicago, Chicago, Illinois

Hans Kamermans   Faculty of Archaeology, Leiden University, Leiden, The Netherlands

Kira E. Kaufmann   Department of Anthropology, University of Wisconsin, Milwaukee, Wisconsin

Frank J. Krist, Jr.   USDA Forest Service, Forest Health Technology Enterprise Team (FHTET), Fort Collins, Colorado

James Kuiper   Environmental Assessment Division, Argonne National Laboratory, Argonne, Illinois

Kenneth L. Kvamme   University of Arkansas, Fayetteville, Arkansas

James Levenson   Environmental Assessment Division, Argonne National Laboratory, Argonne, Illinois

Gary Lock   Institute of Archaeology, University of Oxford, Oxford, United Kingdom.

Scott Madry   Informatics International, Chapel Hill, North Carolina

Christian Mayer   Federal Commission on Historical Monuments, Department of Archäology, Vienna, Austria

Philip B. Mink, II   Kentucky Archaeological Survey, Lexington, Kentucky

Jerry Mount   Department of Geography, Southern Illinois University, Carbondale, Illinois

Linda S. Naunapper   Archaeological Research Laboratory, University of Wisconsin, Milwaukee, Wisconsin

David Pollack   Kentucky Heritage Council, Kentucky Archaeological Survey, Frankfort, Kentucky

Ben Resnick   GAI Consultants, Inc., Monroeville, Pennsylvania

Malcolm Ridges   Department of Environment and Conservation, Armidale, NSW, Australia

Kevin R. Schwarz   Department of Anthropology, Southern Illinois University, Carbondale, Illinois

Scott Seibel   Environmental Services, Inc., Raleigh, North Carolina

Zoran Stančič   Institute of Anthropological and Spatial Studies, Scientific Research Centre of the Slovenian Academy of Sciences and Arts, Ljubljana, Slovenia

B. Jo Stokes   Westchester Community College, State University of New York, Valhalla, New York

Tatjana Veljanovski   Institute of Anthropological and Spatial Studies, Scientific Research Centre of the Slovenian Academy of Sciences and Arts, Ljubljana, Slovenia

Bruce Verhaaren   Environmental Assessment Division, Argonne National Laboratory, Argonne, Illinois

Philip Verhagen   RAAP Archeologisch Adviesbureau BV, Amsterdam, The Netherlands

Frank Vermeulen   Department of Archaeology and Ancient History of Europe, Ghent University, Belgium

LuAnn Wandsnider Department of Anthropology and Geography, University of Nebraska, Lincoln, Nebraska and Statistical Research, Inc., Tucson, Arizona

Konnie L. Wescott Argonne National Laboratory, Argonne, Illinois

Thomas G. Whitley Brockington and Associates, Inc., Norcross, Georgia

Matt Wilkerson Office of Human Environment, North Carolina Department of Transportation, Raleigh, North Carolina

Contents

Section 4:   Quantitative and Methodological Issues

Section 5:   Large Databases and CRM

Section 6:   Modeling Applications in Progress

Section 1:

Introduction

1

There and Back Again: Revisiting Archaeological Locational Modeling

Kenneth L. Kvamme

CONTENTS

1.1   Introduction

Predictive modeling — the practice of building models that in some way indicate the likelihood of archaeological sites, cultural resources, or past landscape use across a region — has its roots in the 1960s and earlier. Such models were implicit in the earliest expressions of settlement archaeology (e.g., Willey 1953) and in later work that actually formulated explicit statements about prehistoric location (e.g., Williams et al. 1973). The First Age of Modeling, in the early to mid-1980s, saw many stumbling blocks to be overcome: ways of thinking that concentrated more on difficulties and sources of variation that seemed to dictate why archaeological models could *not* be developed, the "processualist school" that advocated deductive or "lawlike" behavioral statements as a basis for modeling and decried uses of statistical methodologies based on simple correlations, and a lack of effective computer technology for the application of models across regions. Yet, despite these disadvantages, real progress was made, largely in university research settings made possible by cultural resource management (CRM)-funded projects. Some of these advances included recognition of sampling biases in archaeological databases, procedures for characterization of background environments, applications of univariate and multivariate statistical tests and models, the use of independent test samples for model performance assessments, and the pioneering applications of geographic information system (GIS) technology in the discipline (see Judge and Sebastian 1988; for historical overviews see Kvamme 1995; Wheatley and Gillings 2002: 165–181).

The Second Age of Modeling, now ongoing, is very different in form and orientation. Readily available digital data and ease of GIS software application facilitate the entire modeling process, and ample funding has created incentive. There is now a multimillion dollar archaeological modeling industry, but based almost entirely within CRM settings. One key benefit of this work has been the collation and standardization of archaeological knowledge within modeling regions into computer databases; another has been the building of diverse GIS layers for those regions (Mink et al., Chapter 10, this volume). Both are of great use to the archaeological community. Yet, given the volume of work and its scope — archaeological models have been developed for entire states and large segments of Canadian provinces (e.g., Dalla Bona and Larcombe 1996; Hobbs 1996; Madry et al., Chapter 15, this volume) — shortcomings exist. Funding agencies may be willing to support development of modeling *applications,* but not new research into methods or more-anthropological interests revolving around the interpretation of results and the incorporation of findings into the knowledge base of archaeology. Moreover, much of this work does not get published, and there has been a sameness to approaches that suggests a lack of innovation beyond basic procedures established during the First Age. In other words, advances in archaeological location modeling have not generally kept pace with new methodologies developed in such diverse fields as GIS, satellite remote sensing, economic geography, and wildlife biology. Fortunately, the chapters that follow in this volume serve to correct many of these deficiencies.

In this chapter I examine some of the key issues in the First Age of Modeling that yet impact and impinge on the conduct of modeling today. I hope to clear up several sticky issues. Being somewhat of a fossil from the First Age, I necessarily digress and offer some historical background from my own experience in the growth of modeling. Beyond this, I present a theoretical justification for the practice of archaeological location modeling, review several important new methodologies that have arisen in the past decade, and discuss how they might be incorporated within our modeling tool kits.

1.2   Not So in Bongo-Bongo: Cultural Variation and Modeling

Most North American archaeologists are trained within departments of anthropology. The province of that field claims the full range of variability among all peoples, in all places, in all times (Hoebel 1966). Such tremendous variation in cultures and behaviors is mind-boggling to contemplate, and I believe it structures how the anthropologically trained view the world and approach their research. Focus tends to be placed on variation or *differences* between cultures, and in archaeology, the unique artifacts, sites, or dates; the spectacular find; the oldest; the richest, and the extraordinary tend to receive focus.

In stark contrast, scientific practice in most disciplines focuses on regularities or *patterns*, on commonalities, on recognizing order in the chaos of the natural world by formulating generalizations or rules (laws, principles) of increasing specificity. The anthropological tendency to concentrate on differences and contrasts among phenomena stifles such progress, resulting in little more than a compendium of variation. In spite of this, a large anthropological movement did arise in the mid-20th century that examined systematic cultural patterns, hoping to elucidate regularities underlying human behaviors. Known as "cross-cultural methodology" and culminating in such endeavors as the *Cross-Cultural Survey* and the *Human Relations Area Files*, countless cultural patterns and causal and functional relationships were investigated between such phenomena as types of social organization and warfare, or form of residence, or environmental type, or religious practices, and other factors (e.g., see Murdock 1949, 1967). As is always the case with anthropological data, exceptions to general rules were frequent: a culture or cultures could be found that did not "fit the pattern." I am reliably informed that when George Peter Murdock, a central figure in cross-cultural methodology, was confronted with the unique society once too often, he exclaimed in exasperation "not so in Bongo-bongo," a theme relevant here.

About 15 years ago I decided to investigate this penchant for the unique, this focus on chaos rather than pattern, by having students in my anthropological statistics class at the University of Arizona (where I was then employed) undertake an experiment with the help of the larger student body. Each student interviewed ten individuals — upper class undergraduates, graduate students, or faculty — who would have well-inculcated modes of thought according to their fields of study. Each interviewed five from anthropology and five from physical sciences like physics, engineering, chemistry, or astronomy. The interviewees were asked to write a descriptive statement about two similar objects, in this case a common wooden pencil and a Bic pen.[1] The results in no way constitute a random sample, but I think they are enlightening. About two-thirds of the anthropologists asserted contrasts or *differences* in their responses, with statements like "one is green, the other white," "one has a metal tip, the other a graphite one," "one cross-section is octagonal, the other is circular," and so on. In the more science-based group, nearly the opposite occurred, with almost three-quarters seeing *commonalities* like "both are roughly cylindrical," "both have about the same mass," or "both have a conical tip."

These perspectives on anthropological thinking are relevant to many of the difficulties that I and others faced in developing approaches to archaeological locational modeling nearly a quarter-century ago, and they may even apply today. Instead of focusing on problem-oriented solutions to modeling human locational behavior, much energy gets diverted to complaining about the many problems, difficulties, and "deficiencies" of the archaeological record, or to variations in human behavioral practices, or to insufficient digital representations or algorithms in GIS, or to the inadequacies of contemporary maps, and on and on. A list of some of the sources of variation that have been used as arguments against modeling is given in Table 1.1. (Ironically, most of it comes from the pioneering collection on archaeological predictive modeling edited by Judge and Sebastian 1988; more on this volume below.) These many difficulties and dimensions of variation have served to deflect our attentions away from pathways that might lead to successful models; they also emphasize the many challenges one is faced with in modeling past human locational behaviors.

To give a sense of balance, I formulated a similar list containing reasons why we *can* pursue models of archaeological location, but it came down to only three simple points.

1. Human behavior is patterned with respect to the natural environment and to social environments created by humanity itself.
2. We know or can learn something about how people interacted with these environments by observing relationships between human residues (i.e., the archaeological record) and environmental features.
3. GIS provides a tool for mapping what we know.

TABLE 1.1

A Few Sources of Variation Posing Difficulties in the Archaeological Modeling Process

Archaeological

- Many archaeological sites are buried, and we cannot model them because we do not and cannot know about their distributions
- Known site distributions in extant government files and databases are biased because of (a) the haphazard way in which many were discovered and (b) variations in obtrusiveness, visibility, and preservation
- Many known sites are inaccurately located on maps and in databases
- One cannot model archaeological site distributions because "site" is a meaningless concept; human behavior did not occur in discrete bounded areas but formed a continuum over the landscape
- Functional, temporal, or cultural site types cannot be readily determined for most sites in an archaeological database, yet profound locational differences must exist between the types
- We must be able to model and understand the archaeological formation process, both natural and cultural, before we can model where sites might be found

Environmental

- Past environments were very different from present ones, so we cannot model the past based on the present
- Models based on landscape variables are meaningless
- We do not know the locations of resources important in past times, such as water sources, springs, edible-species distributions, lithic raw material sources, and the like

Behavioral

- Human behavior is too idiosyncratic to be modeled; one cannot model the unique
- One must understand and model complete behavioral systems before archaeological models can be built
- Site location is more a function of unknown (and frequently unknowable) social environments representing dimensions that we cannot map
- The most interesting sites are the (idiosyncratic) ones that do *not* fit the pattern
- Environmental variables shown to be important to site locations may only be proxies for variables that were actually important

Technical

- Blue-line features on topographic maps are frequently arbitrary and unreliable indicators of water
- Modern soil types are meaningless because they are changed from the past and, in any case, are frequently irrelevant to past farming practices
- GIS data have insufficient resolution and poorly represent the real world
- GIS data are inaccurate
- Linear distances computable in GIS are meaningless
- Models based on statistics cannot meet random-sampling assumptions because most extant data were not obtained by random sampling
- Models derived from random cluster sampling are misspecified because they do not adjust for underestimated variances
- Grouping sites of many types into a single, site-present class creates too much variability to be modeled
- Models based on site presence–absence criteria are misspecified because one cannot assume site absence

1.3   The First Age of Modeling: A Personal Narrative

As a master's candidate in the mid-1970s, I was excited by the possibilities of the New Archaeology and assumed that knowledge of statistical methods would go a long way toward solving the problems of archaeology, as were many of my fellow students. (Fortunately, we were blessed with a rather good agricultural statistics department at Colorado State University, in which many of us took classes.) I remember working with discriminant functions on a lithics problem when a fellow student, Jim Chase (now with the U.S. Bureau of Land Management, Wyoming), asked me if I thought they might be applied to environmental variables map-measured at known site-present and site-absent samples to develop a model that might ultimately be employed to make predictions about archaeological locations. I replied that I thought it a splendid idea, and fortunately remembered it.

A few years later, while working for a small archaeological company, a proposal request by the U.S. Bureau of Land Management (BLM) called for (1) a large Class 2 survey (a random sample survey) in the central Rocky Mountains and (2) the mapping of likely locations of archaeological sites for management purposes, based on patterns in the sample data. At that time, such maps were typically composed of giant polygons corresponding to broad environmental tracts like valley bottoms, open grasslands, or juniper forests, and "predictive" guidance was typically given by estimates of site density per zone based on survey sample data (Figure 1.1a). Environmental types with high estimated archaeological densities were deemed "more sensitive" for management and planning purposes than zones of low densities (e.g., Camilli 1984; Ebert 1978; Plog 1983).

Our successful proposal, for what became known as the Glenwood project, employed canonical discriminant functions and, without belaboring details, we generated a model that appeared on the basis of jackknifed validation tests on the sample data, plus a second independent data set (also a random sample survey), to offer good performance in the range of 80 to 85% correct (Kvamme 1980). It is emphasized that all this occurred before GIS or even personal computers (PCs) were available, so mapping results in the form of a probability surface was not easily undertaken.

Instead, I programmed the discriminant functions into a Texas Instruments TI-59 calculator, an amazing gizmo that read from or wrote to tiny magnetic strips, and this program was given to the BLM in lieu of predictive maps. To assess a property about the potential for archaeological resources, the land manager would go to the proper map, hand-measure the six relevant environmental variables (e.g., slope, elevation, local relief, height above river), enter them into the calculator by pressing preprogrammed function keys, and it would spit out a p-value for that locality (i.e., an estimated probability of archaeological site-presence conditional on the environmental measurements). This methodology was actually employed by the BLM to

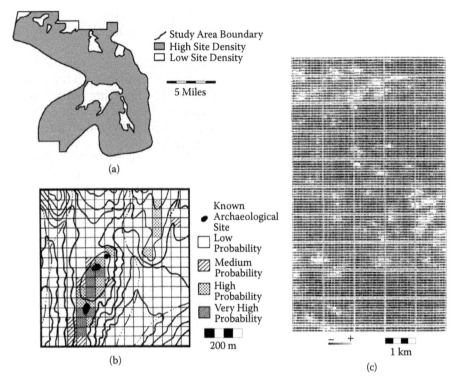

FIGURE 1.1
Early archaeological "prediction" maps. (a) A map based on gross environmental zones with estimates of likely site density (after Plog 1983). (b) An early, precomputer age, hand-drafted archaeological probability surface (after Kvamme 1980). (c) The first archaeological probability surface derived completely through computer measurement of map variables (after Kvamme 1983).

assist in property assessments for a number of years. It was of some significance that this model could be applied to characteristics associated with a *point* on a map, because previously the resolution of most models was at the level of the environmental zone or community, typically many hectares (or even square kilometers) in area (Figure 1.1a).

The idea of producing a mappable archaeological probability surface was in my mind, however, and our team was bent on including one in our final report (Kvamme 1980). We did so by enlarging a single quarter-section (quarter of a square mile; about 800 × 800 m) that contained three archaeological sites independently discovered by another project. We superimposed a 50 × 50-m grid, producing a 16 × 16 matrix of 256 cells, and in each we hand-measured the six predictor variables of the model, for 1536 measurements. Lacking a PC, and with the nearest mainframe computer more than 150 km away, we elected to spend an afternoon with our TI-59 and entered the six measurements for each cell by hand to generate the 256 *p*-values. Finally, we hand-drafted a probability surface that, pleasingly, corresponded well with our notions about the most likely places where archaeological sites would be found (based on years of experience in the area), and with the known site locations (Figure 1.1b).

This exercise convinced me that (1) our archaeological modeling methodology was far superior to any other approach then available; (2) the model itself, the statistical analyses leading up to it, and its mapping offered great potential for understanding human-environmental interactions; and (3) an automated mechanism was absolutely necessary to map the model functions over broad regions. In 1980 I returned to graduate school for a doctorate at the University of California at Santa Barbara to study with Michael Jochim, whose then-recent book *Hunter-Gatherer Subsistence and Settlement: A Predictive Model* (1976) was of prime relevance (this work offers a rare example of an archaeological model derived through mathematical deduction); with Albert Spaulding, the founder of statistical reasoning in archaeology; and to study in the university's Department of Geography, then leading the country in quantitative geography and computerized map handling, later to grow into GIS.

Without GIS software in 1980–1983, we employed a computer system for handling satellite data known as VICAR (video imaging communication and retrieval) that was connected with something called IBIS (image-based information system), which allowed special-purpose FORTRAN subroutines to be linked through horrendous IBM JCL (job control language), all on punch cards. I became a programmer. Without scanning technology and no ready-made digital maps, I first learned to communicate with digitizers so that digital representations of elevation contour lines could be produced, and ultimately a DEM (digital elevation model) after interpolation. FORTRAN routines were written for computing slope, aspect, local relief, ridge and drainage lines, terrain variance; generating distance surfaces from stream vectors; and other operations. While working on this embryonic GIS, I was able to computer-generate my first archaeological probability surface, presented at the annual meeting of the Society for American Archaeology (SAA) in 1981, with improvements in subsequent meetings (published in Kvamme 1983; Figure 1.1c).

At those SAA meetings, I made two important contacts. One was Sandra Scholtz (now Parker) of the Arkansas Archeological Survey, who had been independently developing a nearly identical modeling methodology in their Sparta Mine project, in Arkansas. Their big stumbling block was also the lack of an automated means to measure map variables, but with a circumscribed area, they were able to hand-measure a suite of relevant environmental variables within grid cells 200 m in size (to reduce the number of measurements), from which they generated prehistoric- and historic-site probability surfaces using SAS statistical software (Scholtz 1981; Parker 1985). They were using a relatively new and more robust classification algorithm known as logistic regression (based on the recommendation of James Dunn, Department of Mathematics, University of Arkansas), which proved fortuitous, because Alan Strahler, who pioneered applications of logistic regression in remote sensing (Maynard and Strahler 1981), held a visiting professorship at UC-Santa Barbara the following year. My second important contact was Bob Hasenstab, then a student at the University of Massachusetts (now at the University of Illinois-Chicago), who had also programmed a GIS from scratch, in FORTRAN, to enable cultural resource modeling studies of high resolution in the Passaic River area of New Jersey (Hasenstab 1983). These and other associations led to the first GIS and archaeology symposium, held at the 1985 SAA meeting and quaintly titled: "Computer-Based Geographic Information Systems and Archaeology: A Tool of the Future for Solving Problems of the Past."

Post-doctorate employment took me to the University of Denver, where I became involved in the volume *Quantifying the Present and Predicting the Past: Theory, Method, and Application of Archaeological Predictive Modeling* (edited by Judge and Sebastian and completed by 1985, but not published until 1988). Government-sponsored project authors (often part of consulting firms) had to be part of successful proposals in a national competition, a fact that was certainly one ingredient that contributed to ensuing problems, because individuals who should have been part of it either did not bid (they did not know about it) or did not submit competitive proposals. Ironically, many that ultimately joined the project had little or no previous experience in archaeological location modeling. The result was considerable chaos, leading to its many-year delay to publication and to issues still influencing contemporary work that warranted closer scrutiny.

Several editors and authors of *Quantifying the Present and Predicting the Past* were ardent followers of the processualist school of archaeology (devoted to understanding elements of culture process or change), who informed us that only models generated through deductive reasoning were "good" and potentially "explanatory," while models utilizing statistical methods were not only "inductive" (a bad word at the time), but "merely correlative" and incapable of explanatory insight. Furthermore, it was asserted that "models must span the entire explanatory framework rather than simply concentrating on those things we want to predict .... It is human organizational systems that must be modeled, as well as all those complicating factors between the highest level of human behavior and the archaeological record" (Ebert and Kohler 1988: 105). This seemed a tall (and naïve) order to fill that, if followed, left archaeological modeling dead in the water before it could even leave port. I was stunned because not only had I and others already developed "successful" models by 1984 (e.g., Kvamme 1980, 1983; Parker 1985; Scholtz 1981) (Figure 1.1b and Figure 1.1c), but I believed (and still do) that (1) the type of lawlike or rule-based statements that were advanced as "deductive" models are practical for understanding only relatively trivial cultural processes, (2) such simple models are unsuitable for applications owing to their comparatively low power (and in any case none existed that could be applied), and (3) that there was a complete misunderstanding of the role of statistical methods in applied research settings, points that I tried to convey in my principal chapter (Kvamme 1988a). Since its publication, along with several papers a few years later (Kvamme 1990a, 1990b, 1992), my interests in modeling have only recently been rekindled by an unlikely source: working with students of biology, I have become aware of a tremendous renaissance in modeling approaches made possible by the GIS revolution, as the following sections will demonstrate.

1.4   Perspectives on "Correlative" and "Deductive" Models

Critics of archaeological models derived through statistical methods such as discriminant functions or regression are wrong in assuming they are based solely on "mere correlations." This can rarely be the case because even the simple act of selecting variables for analysis demands an *a priori* theoretical perspective that comes from previous work, training, and exposure to the theoretical currents of a discipline. Statistical models should most properly be viewed as a means of estimating appropriate weights for theoretically derived variables. Without such a mechanism it is unlikely that robust weights can be derived, resulting in models of lower power. A deductive model based on anthropological theory, previous findings, or ethnographic analogs might define variables relevant to past location behaviors. But without recourse to statistical calibration based on sample data, how those variables might be combined, weighted, or thresholded to achieve a GIS mapping becomes something of an art. A simple Boolean combination, for example, means that each variable receives an equal and arbitrary weight of unity; altering those weights in a more complex model implies a level of theoretical knowledge not generally possible. Moreover, such models must perform suboptimally compared with those with weights derived from statistical theory and suitably constructed random samples. Making claims about the superiority of the former is therefore ironic. Dalla Bona and Larcombe (1996) deduced an excellent suite of variables through ethnohistoric and contemporary native informant accounts concerning prehistoric settlement in northwestern Ontario, for example, but their GIS mapping was only possible after close calibration of model weights against empirical archaeological distributions.

Wildlife biologists utilize GIS to map models of species distributions and habitat (analogous to archaeological sites) with the advantage of a more mature view of the modeling process (being firmly wedded to empirical data and possessing a statistical tradition that goes back to the 19th century). Most biological models begin with theory, usually meaning a list of variables relevant to the locations or habitat of the species of interest derived from prior knowledge and work. Based on species locations observed in field data, the resultant models, including discriminant and logistic regression functions, give insights into interactions between variables, identify significant relationships, confirm or refute hypothetical associations, and expose relative strengths of relationships. Additionally, GIS mappings in the form of species probability or abundance surfaces prove insightful because relationships between species and environment become graphically clear, revealing the relative clumping, dispersion, or patchiness of the result (e.g., see Bian and West 1997; Clark et al. 1993; various papers in Scott et al. 2002). Khaemba and Stein (2000: 836), for example, state outright that their models are deductively derived because they begin with the prior knowledge that "elephants generally prefer tall grassland and shrubby vegetation."

How is the foregoing different from deducing that settlements of a farming culture should be situated in well-watered valley bottomlands, near level fields with good soils? Wheatley and Gillings (2000: 166) observe that

> A distinction between data and theory [driven models] is not universally recognized, and most archaeologists accept that the two are not independent — data is collected within a theoretical context, and so may be regarded as theory-laden, while theories are generally based to some extent on empirical observations.... It is impossible in any practical sense to implement a predictive modeling method that is based entirely on either of these tactics.

The archaeological dichotomy that has arisen claiming distinct correlative and deductive approaches to modeling is an unfortunate historical accident; they need not be different but can and should be one and the same.

1.5 Theoretical Justification of Archaeological Location Modeling

1.5.1 Background Concepts

In developing models, we need to be clear whether we are trying to model the *systemic context* or the *archaeological context*, as originally codified by Schiffer (1972). The former refers to the living, behavioral state of a human group or society. The latter refers to the static, nonbehavioral state of archaeological materials, the physical record that archaeologists study. Even a perusal of the literature on modeling suggests that it is frequently unclear which context is being modeled, despite critical differences in assumptions, approaches, likely difficulties, and possible outcomes. Explanatory or deductive models appear to be generally concerned with the systemic context, but from a cultural resource management standpoint the goal clearly seems to be the modeling of the archaeological context.

In approaching the latter, we must first recognize that if a goal of modeling is the mapping of locations where archaeological resources are likely to occur, then logically the equivalent is the mapping of locations where such resources are *unlikely* to exist. The elimination of portions of a region that are unlikely to contain archaeological resources becomes a useful way of approaching the modeling problem.

The definition of the *niche* of a species, as defined in quantitative ecology, provides a second vital perspective. The niche can be defined as the total range of conditions in the environment under which a population lives and replaces itself (Pianka 1974: 186). In a landmark paper, Hutchinson (1957) emphasized that the niche can be determined empirically by measuring the location of individuals of the population along multiple dimensions of environment, with the range defining a niche space in a "hypervolume" of measurements. That space can be visualized as a variable probability density function (PDF), with certain locations within it more ideal for the species than others. In fact, the "ideal habitat" of a species can be represented by the mean vector of measurements on each variable, as indicated by the locations of the species itself. Less desirable habitat is then inferred as any deviation from the mean vector. (This perspective forms the basis of an important modeling approach discussed in Section 1.7.3.) The obvious application of this perspective to ideas of human niche spaces and settlement distributions was first extended to the field of geography by Hudson (1969). Of more importance are the implications it offers as a logical basis for archaeological modeling (see Kvamme, 1985 for an early attempt).

1.5.2 A Deductive Model

Let us begin with three observations:

> *Observation 1*: The human organism lives within the natural environment.

> *Observation 2*: The environmental variation within any circumscribed region is large.

> *Observation 3*: The niche of the human organism is that portion of the environment that it utilizes or to which it has access.

The human niche, N, may correspond with the entire environmental range, E, of a region, where the niche space equals the environmental space, $N = E$. The niche space, however, may typically include only a subset of the total environmental space, $N < E$, because areas of steep slopes, cliff faces, water bodies, lava fields, glaciers, wetlands, high altitudes, dense vegetation, and other contexts may be inaccessible or unutilized (Figure 1.2). The level of niche space accessibility may also be partially dependent on technological level, other cultural circumstances, and resource distributions. For example, cliff faces might harbor an important food resource (bird eggs) that become accessible only with appropriate technology (sturdy ropes).

> *Corollary 1*. If $N < E$, then locational modeling *must* be productive if N can be defined.

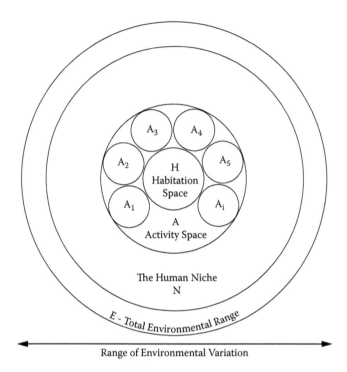

FIGURE 1.2
The human niche, activity, and habitation spaces can be viewed as subsets of increasing specificity within the total environmental range of a region.

That portion of E larger than N may presumably be eliminated as a region unlikely to contain archaeological resources (Figure 1.2). As with other species, N may be determined empirically by environmental measurements made at species locations, in this case, archaeological sites.

>*Observation 4*: Human activity is nonuniformly distributed within its niche space. Some portions of the environment are more heavily utilized or accessed than others.

This corresponds with the variable PDF implied by Hutchinson (1957) and the concept of ideal habitat within the niche space. It depends on cultural practices, resource distributions, and the nature of activity sets.

>*Corollary 2.* Even when $N = E$, it can be assumed that N will follow a variable PDF (because human activity is nonuniformly distributed). Consequently, locational modeling must still be productive if "ideal" and "poor" locations (habitat) within the human niche space can be distinguished.

The analysis of empirical archaeological distributions through a host of statistical or other means (e.g., Kellogg 1987; Kvamme 1990c) can potentially indicate favorable and unfavorable localities. Favorable places generally correspond with specific classes of activity, fitting well with the idea that the PDF is multimodal.

Observation 5: If we do not include simple travel between locations, then activities are tied to *places*. Various types of human activity are frequently associated with particular environmental circumstances.

Fishing occurs in or adjacent to streams or lakes and nut gathering where nuts grow, for example.

Corollary 3. The association between particular activity classes with specific environmental contexts dictates that modeling specific site-type distributions must be productive. These activity-specific "niches" comprise small subsets, A_i, within the human niche space, N (Figure 1.2). Each are determined by relatively few environmental dimensions according to specific activity needs.

The creation of distinct models for individual types of archaeological sites (functional, temporal, or other) is something rarely undertaken in the literature of modeling, yet it forms an area of certain improvement.

Corollary 4. The sum or union of all places where people concentrate activity forms an activity space, $A = \Sigma A_i$, that is smaller than the human niche space (i.e., $A < N$) if simple travel between places is disallowed (Figure 1.2). This forms the logical basis of Corollary 2.

Observation 6: Places of human habitation or settlement, with long-term needs and variable activities represented, must be sited according to *many* dimensions of environment dictated by the many needs of community.

A long-term settlement might be located according to its needs for defensibility *and* proximity to water *and* quality of agricultural soils *and* level slope *and* fuel resources, etc.

Corollary 5. Habitation activities with many location requirements may be more predictable than sites of specialized activities with relatively few, defining a comparatively small "niche." Owing to this fundamental difference, a habitation space, H, is defined, where $H < A$ (Figure 1.2).

If a simple activity, A_i, requires only environmental condition e_1 to occur, but H requires $e_1 \cap e_2 \cap e_3 \cap e_4$, then generally $H < A_i$.

Observation 7: People also construct a social environment that influences locational behavior. If the natural environment defines a possible range of conditions for the placement of activities, then the social environment imposes further restrictions and order. Road networks and the necessity of intersettlement spacing, for example, further alter the PDF within the human niche space.

Corollary 6. Consideration of variables of the social environment must be productive because the range of variation within suitable areas or favorable spaces in the natural environment becomes further reduced.

1.5.3 Summary

The foregoing suggests that if we can view human uses of space in terms of subsets of environmental variation, and identify those subsets as a basis for modeling, then archaeologically useful results must be possible if we consider "useful" to mean the elimination of regions unlikely to contain archaeological resources. Consideration of individual site types and variables of the social environment will allow models to focus on narrower ranges of variation, improving performance, and long-term habitations or settlements should be highly predictable owing to their more restrictive environmental requirements. To achieve the full potential of this perspective, a number of continuing issues and methodological improvements must first be addressed.

1.6 The Second Age of Modeling: Continuing Issues

With nearly a quarter-century of serious work in archaeological location modeling, it is clear that several issues, some of which may be insurmountable, remain at the forefront of difficulties. They include the problems of modeling multiple site types, paleoenvironmental reconstructions, and sampling issues. Lack of resolution in these areas continues to affect the power and specificity of models and, indeed, what we are able to model.

1.6.1 Archaeological Site Types

A handful of lithics, a couple of sherds, or a few tools gained through a limited surface reconnaissance does not usually allow reliable reconstruction of the kinds of activities that occurred at an archaeological site, identification of the culture(s) that used or produced the artifacts, or accurate estimates of the amount of activity that occurred, length of occupation, or dates of use.

This dilemma, typical of the vast majority of sites in any region, has forced modelers to throw them into one large "pot" that can only be labeled "human activity present," or to ignore them, relying solely on well-dated and understood sites. The latter tactic is undesirable because the sample size of known archaeological locations becomes so reduced that meaningful statistical analyses become untenable. The former is the principal reason why most models remain dichotomous (i.e., site-present versus site-absent).

Ethnography and common sense indicate that sites associated with various functions are located differently: a fishing spot, a plant-gathering location, a hunter's kill and butchering site, and a long-term settlement will generally be located in unlike places. Moreover, different cultural groups may have unique responses to the same environment, with large variations in locational behavior. Finally, temporal differences between sites may correlate with vastly changed environmental circumstances. It seems obvious that when placing all sites in a single group for modeling, the enormous variation associated with all human activity yields models of lower power and specificity. What we end up modeling is the sum total of the human "activity space" of Figure 1.2. In defense, it must be noted that models of surprising power have nevertheless been created following this simple site-presence–absence approach. Brandt et al. (1992), for example, lumped sites of all types and periods in the Netherlands into a single class (representing a remarkable breadth of functions, cultures, and chronology) and achieved models that performed surprisingly well (suggesting some sort of commonality to locational behaviors or perhaps site visibility).

Defining meaningful site types and modeling each as a separate class is probably the greatest potential improvement to the quality of archaeological models (see Stančič and Veljanovski 2000 for an excellent example). Aside from better recording, more field time, increased funding, retrieval of larger samples of artifacts, better analysis methods, and improved theory that might point to site function, there appears to be few ways out of this quandary. Larger, more permanent settlements are sometimes more visible, enabling models of settlement location (i.e., the "habitation space" of Figure 1.2) as opposed to all site locations (the "activity space"), a useful undertaking.

One certain area of improvement lies in removing rock shelters or cave sites from consideration in the modeling equation. These kinds of sites were invariably utilized for a range of activities, yet unlike all other archaeological sites, their placement in the landscape is *not* dictated by human choice. Rather, the loci of rock shelters and caves are determined by a peculiar and idiosyncratic set of *geological* variables, including rock type, exposure, hydrology, terrain shape, and other factors. We can model human choices that placed other kinds of sites in the landscape, but to model the use of rock shelters and caves, complex geological models and variables must be pursued that undoubtedly open up other problems. These sites should therefore not be considered with other site types in combined modeling operations because the larger range of locational variance introduced will upset model performance. The best approach for handling them may simply lie in improved mapping that locates caves and rock shelters.

1.6.2 The Paleoenvironment

Research has demonstrated significant empirical and theoretical relationships between environment and archaeological distributions. Yet, in nearly all cases, it has been modern environmental conditions instead of past circumstances that have been investigated. In most regions, contemporary conditions are very different from those of the past, especially the distant past, raising the question of just how well models based on the present situation can predict the past. After all, it was then-contemporaneous conditions that were relevant to locational decisions and choices made by past peoples. While it can be argued that terrain form (and its many derivative measures) is relatively stable through time, it is well-known that plant communities migrate up and down altitudinal clines with climatic change and that rivers and streams wildly meander within valleys over relatively short periods, for example.

It would seem that reconstruction of paleoenvironments is a necessary first step in the archaeological modeling enterprise (see Kamermans, Chapter 5, this volume). Paleoclimatic data from tree rings and other sources might be employed to model life-zone altitudinal changes, pollen data could point to former environmental compositions, and erosion and hydrological models could be used to reconstruct past landforms and stream channels, for example.

When one considers that paleoenvironmental reconstructions are potentially necessary for each time period relevant to the archaeological sites in a region, however, such a task becomes daunting and has rarely been undertaken (for exceptions, see Boaz and Uleberg 2000; Gillings 1995; van Leusen 1993; Nunez et al. 1995). Moreover, paleoenvironmental and paleoclimatic reconstructions are difficult and capable of only very broad generalizations, with little specificity in terms of the point-by-point requirements of GIS-based models. (We ideally want representations of the paleoenvironment on a pixel-by-pixel basis.) Also raised is the question of error budgets in GIS models based on such data. Recent work has shown significant levels of error, even in present-day maps (see Goodchild and Gopal 1989). Past reconstructions of an environment will likely contain huge errors owing to their imprecision. Because error is cumulative in a multidimensional model, it is quite likely that results could be unusable. For example, even assuming an unrealistic 90% level of accuracy (however accuracy might be defined), with only five reconstructed environmental layers, the overall accuracy becomes $.9^5 = .59$, dismal indeed. Unless reliable paleoenvironmental reconstructions can be generated, it is clear that we must proceed with caution. At the same time, it might also be argued that any paleoenvironmental reconstruction, however poor it might be, must be better than using present-day data.

Most practitioners will continue to employ present-day maps and digital data sets as a basis for modeling, if only because of ready availability. One benefit is that map error is at least known and quantifiable. Focus should be given to variables less sensitive to change, such as landform characteristics. Other tactics might also be employed to mitigate the effects of change. For example, instead of using distance measures to current rivers or streams (assuming proximity to water is a meaningful criterion), distance to the edge of the floodplain might instead be considered to eliminate the effects of meandering.

1.6.3 Sampling

Most regional models necessarily employ extant records of archaeological sites from the region of interest. Sampling biases that exist in these data sets are well-known and arise from such circumstances as (1) the tendency of archaeologists to discover sites where they believe they should be or in places with easier access (near roads, towns), (2) the arbitrary but nonrandom locations of development projects that have required cultural resource surveys, or (3) the greater obtrusiveness of larger sites and settlements (e.g., sites with mounds or earthworks). Models based on these kinds of databases are biased, and entire archaeological contexts may not be well represented (see Kvamme 1988b for ways to reduce or mitigate such biases; Wescott, Chapter 3, this volume, discusses sampling issues).

Some archaeological projects have had the luxury and budgets to employ random sampling designs and pedestrian surveys to procure unbiased samples for model development (e.g., Thomas 1975; Warren and Asch 2000). Most have employed some form of cluster sampling, conducting surveys within randomly selected blocks of large size (e.g., 500-m squares, quarter-sections). One reason is convenience: it is easy to locate a relatively small number of large quadrats on a map and on the ground. Yet, sampling elements like archaeological sites within clusters creates negative consequences, such as (1) reduced estimates of variability (because places sampled occur in a relatively small number of clusters), and (2) a lack of independence between data elements owing to their spatial proximity or the autocorrelation effect (Kvamme 1988a).

In the past decade we have moved into a very different world where we can now accurately locate ourselves through GPS (global positioning system) technology. Let us throw out large cluster blocks and utilize small (subhectare) parcels (or even points) for survey and random-element sampling designs (Scheaffer et al. 1979). With preselected coordinates, it is simply a matter of pressing "go to" on the GPS to reach a new locality, and a survey of nearby randomly selected places can be preplanned to minimize travel requirements. In so doing, we can attempt to attain "ideal" sampling designs that allow meeting of statistical assumptions, permit more-representative sampling of environmental and archaeological variability, and increase the likelihood of independence between observations.

1.7 The Second Age of Modeling: Possible Improvements

Although much contemporary modeling work is of high quality and innovation is apparent, other refinements seem possible in such areas as developing new variables through GIS, utilizing new modeling approaches and algorithms, and in methods for evaluating model performance. The following subsections offer a number of ideas, suggestions, and new methods that might be utilized in archaeological locational modeling.

1.7.1 Independent Variables

Acquiring better data and variables that might bear on archaeological locations is one domain that can improve our ability to model. As technology improves, our potential in this area is increased. High spatial resolutions for digital elevation and satellite data mean that we can capture more detail of the landscape that could be relevant to certain classes of past activities, for example. Moreover, a lack of consideration of the social environment has been justly criticized by European scholars (Gaffney and van Leusen 1995), pointing to other dimensions for improvement.

1.7.1.1 The Natural Environment

Variables that quantify aspects of the natural environment will generally remain most important in archaeological location modeling owing to their ready availability in digital or map form and their importance to human locational behavior. In general, we need to move beyond simple terrain variables like slope and aspect, or linear distances to blue-line water features on maps. We now have access to powerful GIS tools that offer potentially more. We should pursue quantification of subtle variations in terrain shape and complexity, identify local high points and saddles (Duncan and Beckman 2000), quantify solar insolation, terrain texture, and local relief changes above and below locations for possible relationships with past activities, particularly in hunter-gatherer contexts. Llobera (2000) and Bell and Lock (2000) reveal great improvements in modeling movement over the landscape; perhaps it is now time to move beyond simple linear proximity measures.

One particular area of promise lies in drainage runoff algorithms that objectively define flow based on landform shape in DEM (Burrough and McDonnell 1998: 193–198), allowing movement away from the frequently subjective and arbitrary blue-line features on topographic maps. They allow

quantification of accumulated flow to any pixel in a region; simple reclassification methods can then define drainage networks of any rank or complexity for proximity-based analyses. The continuous scale of accumulated flow
itself might also be of interest.

Vegetation and biomass-biocomplexity diversity indices derived from satellite imagery are yielding much insight into patterns of regional plant productivity and health (Sabins 1997: 404). They have been largely ignored in
archaeological modeling (see Gisiger 1996, however), despite their apparent
potential, particularly in the large tracts of land in the Americas and elsewhere little changed from recent prehistory.

1.7.1.2   The Social Environment

Social variables typically refer to characteristics of the human-created environment. In complex societies it is markets, central places, intervillage spacing, road networks, political boundaries, and the like that drive uses of space.
The relative importance of the natural versus social environments to locational behavior strikes some sort of balance, with one or the other more
important, depending on needs and the nature of activity requirements. In
general, we might imagine a continuum where the relative influence of these
domains is a function of cultural-technological complexity (Figure 1.3, while
realizing that such a generalization may not apply to Bongo-bongo). While
both are important to any society, in hunter-gatherer contexts the social
environment probably is less so, if only because there frequently are no

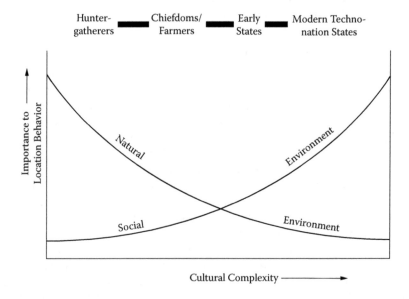

FIGURE 1.3
The relative importance of natural versus social environments to locational behavior is related
to cultural complexity.

markets, central places, road networks, and related phenomena that characterize settings of greater cultural complexity.

Social variables have rarely been employed in archaeological location modeling (Gaffney and van Leusen 1995). One reason lies in data availability. Maps of the natural environment (albeit the present environment) are easy to obtain, and frequently can be instantly downloaded through the Internet. This is not true of social variables, where the loci of contemporary villages, markets, religious centers, or roads are frequently difficult to obtain for past times. In general, it is only in well-studied archaeological regions where, after decades of work, a reasonable semblance of past social landscapes can be reconstructed. Yet, even in these ideal contexts, such reconstructions are likely only partial: missing villages, road segments, or unknown political boundaries are likely (see Vermeulen, Chapter 14, this volume).

A somewhat more subtle issue lies in the need to establish contemporaneity between features in the social landscape. Measuring proximity to a road or political center is only relevant if those features are coeval with the social milieu being modeled. This requirement further restricts consideration of many social variables to well-studied archaeological regions with good chronological control. Madry and Rakos (1996) were able to model prehistoric travel routes based on the arrangement and viewsheds of a series of contemporaneous Celtic hill forts in France. Chadwick (1978, 1979) was even more restrictive by modeling Late Helladic settlement distributions based on the distribution of settlements in preceding periods in Mycenae.

Obviously, archaeological location models that fail to address the social dimension owing to a lack of data or effort only get at a portion of the variability in past site-selection behaviors (Corollary 6, above): that portion related solely to the natural environment, which can be small (Figure 1.3). Recent debate and applications in this area, particularly by Europeans (with generally better knowledge of archaeological regions), are therefore encouraging (Gaffney and van Leusen 1995; Gaffney et al. 1995; Stančič and Kvamme 1999; Wheatley 1996).

1.7.2 Other Modeling Algorithms

With the growth of GIS technology and its ready acceptance by government, industry, and academia, together with intense focus on regional modeling in other disciplines like biology, medical science, and economics, there has been a remarkable explosion in modeling methods and algorithms in the past decade. Approaches in this literature range from simple Boolean intersections, to additive binary layers, weighted additive layers, fuzzy versions of the foregoing, Dempster-Shafer models, log-linear and logit models, dominant-category clustering models, neural-network algorithms, Mahalanobis D^2 statistics, suites of classifiers from image-classification methodologies like maximum likelihood, and the ever-popular discriminant functions, including logistic regression, to name a few (e.g., Bian and West 1997; Clark et al.

1993; Eastman et al. 1995; Gabler et al. 2000; van Manen et al. 2002; Vila et al. 1999; Wang 1990).

Despite this great variety of available approaches for modeling many types of spatially distributed phenomena, there has been relatively little variation in the archaeological literature in the methods that have been employed. Logistic regression, a robust nonparametric classifier, has been particularly popular in archaeological model development, as has discriminant-function analysis, the parametric alternative (Parker 1985; Scholtz 1981; Kvamme 1983, 1988a; Warren and Asch 2000; Wheatley and Gillings 2002: 172). Both are examples of linear statistical models, and even here, recent improvements exist. Generalized additive models (GAM) appear to offer a significant advance over the generalized linear model, for example, because they replace the linear component of the model with an additive one that identifies and describes nonlinear trends and threshold effects, which are far more common in nature than linear ones (Hastie and Tibshirani 1990).

1.7.3   Forget Those Nonsites: Single-Class Approaches

As a means of modeling the archaeological context, the two-class approach can be justified because there are places that contain material evidence of past activities (archaeological sites) and others that do not (nonsites). Yet, even if thorough field investigation fails to encounter archaeological evidence at some locus, there is a nonzero probability that archaeological remains may actually be present; for example, they might be buried, be lying under vegetation, or simply have been overlooked.

A similar perspective arises in the modeling of biological species occurrence. Much like the archaeological site present-absent dichotomy, such studies employ sightings or radiotelemetry on tagged animals to compare their presence-absence against mappable habitat variables in GIS settings. Logistic regression-based and other probability surfaces are then developed for species presence (Bian and West 1997; Dettmers et al. 2002). Argument has recently been vigorous against use of a species-absent class for model calibration, however, because the lack of a den or nest at the time of a field investigation does not imply its absence in times past or future (Clark et al. 1993; Dettmers and Bart 1999).[2] This quandary has led to alternative modeling approaches of great power that fit well within long-held theoretical perspectives stemming from perspectives on niche (as in Figure 1.2).

These approaches focus on a *single* species-present class (analogous to an archaeology-present class). Calibrating to a species-present sample, the mean on any one environmental variable represents an estimate of "ideal habitat" for that species on that variable; in a multivariate context, it is the mean vector μ that represents ideal habitat across a series of variables. Less desirable habitat is inferred by any deviation from μ, agreeing well with the classic species niche model developed by Hutchinson (1957) that emphasizes an ideal "niche-space" within an *n*-dimensional hypervolume of relevant environmental parameters. The most common metric for evaluating locations in this perspective is the Mahalanobis distance statistic (in matrix notation)

$$D^2 = (x - \mu)' \, \Sigma^{-1} \, (x - \mu)$$

which is interpreted as a squared normalized distance between a location's measurements (x) and μ (Σ is the variance-covariance matrix). While D^2 is a valid metric on its own, it tends to be highly skewed, and a χ^2 transformation allows a 0-to-1 rescaling that, if multivariate normality is assumed, can be interpreted as a p-value analogous to a posterior probability obtained with more-conventional discriminant or logistic regression functions. These D^2 or p-values are then mapped by GIS on a pixel-by-pixel basis, yielding a "deviation from ideal habitat" or a species-probability surface, respectively (e.g., Clark et al. 1993; van Manen et al. 2002).

While offering an alternative to more-conventional and -accepted methodologies, this approach presents its own series of problems. One cannot undertake a stepwise F-to-enter solution, for example. One has to *know* which variables are relevant and go with them, but this does not appear to be a problem in the biological sciences. As alluded to earlier, variables selected are typically derived from *a priori* theoretical ideas.

1.7.4 Models of Greater Specificity

Environmental variation in large project areas can be enormous, and past human adaptations and uses were undoubtedly numerous. Given the size of some projects (e.g., whole states and significant proportions of Canadian provinces), gradients or differences in cultural practices, or even cultural types, might occur, and variables relevant in one subarea might not even apply to another. A model fine-tuned to the more limited variation of a small region should theoretically better "fit" that region's cultural and environmental variability compared with a global model that can only "average" relationships over huge areas. To illustrate, I built one logistic regression model using all data from a 600-km² region, and then a second model using data only from an 8.5 × 5.5-km subarea. The model for the latter, because it dealt only with the archaeological and environmental variation in the subarea, offered a much better fit with the data, and all performance indicators were markedly higher (Figure 1.4; Kvamme 1988a).

One might therefore consider partitioning a large project area into a series of small blocks and building a fine-tuned model for each. Such distinct and independent models would undoubtedly perform better, but arbitrary "seams" or discontinuities in model results would likely occur at borders between the individual blocks. Such effects arise from environmental and archaeological differences between the blocks, resulting in reduced interpretability and quality of presentation. (Defects like massive jumps in estimated archaeological probabilities can only be explained by the arbitrary locations

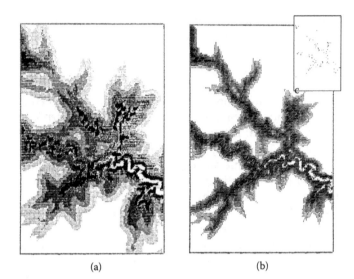

(a)   (b)

FIGURE 1.4
Archaeological probability surfaces obtained through logistic regression analyses of open-air lithic scatters in an 8.5 × 5.5-km (46.75 km²) study block in southeastern Colorado (after Kvamme 1988a). (a) Model derived from all open-air lithic scatters (n = 269) and environmental variation in a larger 600-km² project area. (b) Model derived only from data occurring within the smaller study block (n = 95). (c) Distribution of known open-air lithic scatters in study block. Black signifies high archaeological probabilities.

of boundaries between the study blocks.) With the significant computing power at hand today, an alternative approach may be workable. A moving window, kilometers in diameter, could potentially be employed to build a model utilizing data that only occur within it. The window would then be centered location-by-location throughout the project area, causing model results at each locus to be based on environmental and archaeological characteristics that are the most relevant, resulting in models of greater specificity. Other benefits could potentially accrue from such an approach. Mean vectors, confidence coefficients, and model parameter estimates within each window could be mapped and examined over space. Variations in the relative sizes and signs of regression coefficients, for example, could point to the relative importance of particular variables as environmental and archaeological circumstances change across a region.

1.7.5 Measures of Model Performance

It is uniformly agreed that a model *must* be tested before one can place reliance in it, and this stricture should apply to any model regardless of its means of derivation. Various methods of resampling (e.g., cross-validation, jackknifing, bootstrapping) have been developed that can provide robust estimates of performance (Verbyla and Litvaitis 1989). The ultimate test, however, is against samples independent of those used to develop a model.

While these points have been well belabored before (Kvamme 1988a; Warren and Asch 2000), a number of alternative performance statistics can greatly enhance interpretation of various model qualities.

Our goal is the modeling of archaeological phenomena across space, yet our focus is not on the archaeological site but on the *location* and whether or not a site is likely to be present. (The location should be regarded as a point on the landscape.) Let event S signify the actual presence of an archaeological site (or whatever archaeological phenomenon is of interest) of a type we wish to model at a location. S' then indicates the absence of such a site at a location. An archaeological model can be regarded as a collection of irregular polygons that are mapped onto the landscape that indicate locations that are "favorable," "likely," or "probable" to contain an archaeological site of the type(s) being modeled. Let M denote the event that a model, however derived, assigns a location as "likely" for the site type of interest. M' is its complement, meaning that the site type is unlikely according to the model. M and M' therefore represent the GIS mapping of model predictions, but it must be discrete for this formulation. If a model mapping is continuous (as in a probability surface) or ranked (e.g., polygons indicating variations in archaeological likelihood), a GIS reclassification must be made at some threshold to achieve M and M'. Models with continuous or ranked outcomes therefore have the advantage that statistics may be generated under a variety of thresholds and graphed to yield richer and more insightful performance indications (e.g., see Kvamme 1992; Warren and Asch 2000).

Most modelers focus on percent correct statistics for known archaeological site classes, or $100\,p(M\,|\,S)$ (the probability that a model specifies a site when one is known to actually be present; the "$|\,S$" means "given" that a site is present). It is obtained simply by working out the percentage of known sites that a model gets right. The $p(M\,|\,S)$ can be referred to as "model accuracy" for the archaeological class. Other statistics yield additional insights about performance.

The probability of S, the archaeological phenomenon of interest, can be estimated by $p(S) =$ (total area of all known sites in class)/(total area field surveyed).[3] Its value is extremely important: it indicates the base rate or *a priori* chance that the archaeological phenomenon of interest will occur at some location (Kvamme 1990a). Its meaning can be grasped if one considers throwing darts haphazardly at a map; the chance of a dart falling on a site of the type of interest is $p(S)$. It is frequently difficult to estimate when examining regional literature because relevant data are seldom published. Values I have been able to compute in North American surveys range from about $.001 \leq p(S) \leq .04$, signifying that archaeological phenomena may be regarded as *rare events* in most regions (forming a principal reason why robust archaeological models are challenging). The $p(S)$ should be regarded as a base model that our efforts must beat. Stated differently, any model must yield $p(S\,|\,M) > p(S)$; otherwise, it is worthless (where $p[S\,|\,M]$ indicates the probability of a site *given* that a model specifies a site; see Kvamme [1990a]). Obviously, for rigorous statistical treatment, the sites used to estimate $p(S)$ should ideally have been discovered by a program of random sampling, but a ballpark estimate can also offer useful insights. The $p(S')$ = 1 − $p(S)$ is the probability of the site class being absent at a location, typically large in value.

The $p(M)$ is the base probability that a model indicates a site, determined by the area of its mapping: $p(M)$ = (total area mapped by model to event M)/(total study area), a trivial computation with GIS. In other words, if $p(M)$ = .3, then we would expect even a worthless model without predictive capacity to correctly specify about 30% of the sites in a region by chance alone, simply because it covers 30% of the region's area. The $p(M)$ is related to the precision of a model because it represents the proportion of a region mapped to event M, and one goal is to *minimize* that area to produce models of greater specificity. We might therefore define *model precision* as $p(M')$ = 1 − $p(M)$; the higher its value, the more precise (or smaller the region) is a model's mapping. The ratio $p(M)/p(S)$, also related to precision, can be regarded as an index of *model fit*. It indicates how many times larger the area of a model's mapping is for a site class (M) compared with the actual estimated area of the site class (S). Although we wish this ratio to be small, values might typically range between 20 and 100, pointing to the imprecision of most models, but also to the fact that although many locations possess characteristics typical of sites, they do not contain one owing to low site densities or small $p(S)$.

It is emphasized that one can make $p(M)$ = 1 by mapping *every* location in a region to M to achieve a perfectly accurate model where $p(M|S)$ = 1. (Because every location is classified as "site likely," all archaeological sites are correctly indicated.) In this case, however, model precision $p(M')$ = 0, and we have a worthless model. A useful statistic is therefore $p(M|S) - p(M)$, which signifies the *improvement over chance* a model offers, after correcting for its area. For example, a model that correctly indicates 75% of known sites ($p[M|S]$ = .75) in a mapping that covers only 30% of a study region ($p[M]$ = .3) represents an improvement over chance of 45% ($p[M|S] - p[M]$ = .45).

Other statistics pertain more to archaeological-class discovery probabilities. The importance of $p(S|M) > p(S)$ was previously emphasized; the ratio $p(S|M)/p(S)$ is therefore meaningful. It should be greater than unity and indicates how many *times* better than chance the probability is of an archaeological site when the model indicates one. The $p(S|M')$, the probability of an archaeological site given that the model indicates its *absence*, is also a noteworthy statistic. It should be less than $p(S)$ at any locus, and the ratio $p(S)/p(S|M') > 1$ should occur, indicating that when a model does *not* indicate an archaeological site, the probability of one occurring is less than the base-rate probability, $p(S)$. Finally, the ratio $p(S|M)/p(S|M')$ is one of the most significant computable statistics because it indicates how many *times* more likely an archaeological site is when a model (M) indicates one compared with when it does not (M'). A summary of these statistics is given in Table 1.2, and a worked example with typical application numbers is given in Table 1.3.

TABLE 1.2

Derivation and Interpretation of Model Performance and Related Statistics

Statistic	Derivation	Interpretation
$p(S)$	From survey data: (total area of site class)/ (total area field surveyed)	Base rate or chance probability of archaeological site class in study region
$p(M)$	Determined exactly by GIS: (total area of model)/(total study area)	Base rate or chance probability that a model will indicate a site; proportion of study region mapped to M
$p(M')$	$1 - p(M)$	Model precision; high values indicate high precision
$p(M)/p(S)$	Ratio	Model fit; indicates how many times larger a model mapping is than the total site-class area
$p(M\|S)$	Estimated by proportion of known archaeological sites correctly specified by model	Model accuracy; probability that a model will correctly indicate a site: $100 \times p(M\|S) =$ percent correct
$p(S\|M)$	Estimated by proportion of locations in M that contain archaeological sites	Probability of archaeological site presence when model specifies a site
$p(S\|M')$	Estimated by proportion of locations in M' that contain archaeological sites	Probability of archaeological site presence when model does *not* specify a site
$p(M\|S) - p(M)$	Subtraction	Improvement that model offers over chance in specifying known archaeological sites
$p(S\|M)/p(S)$	Ratio	Model improvement ratio; indicates how many times more likely a site is in M than the base-rate site probability
$p(S)/p(S\|M')$	Ratio	Model improvement ratio; indicates how many times less likely a site is in M' than the base-rate site probability
$p(S\|M)/p(S\|M')$	Ratio	Model improvement ratio; indicates how many times more likely a site is in M versus M'

1.7.6 Significance Tests

Assuming a random sample for the archaeological site class, one-sample (and one-tailed) tests for differences in proportion (Conover 1999) may be applied to test H_0: $p(M\|S) \le p(M)$ or H_0: $p(S\|M) \le p(S)$. The latter hypothesis is a more difficult one to test, however, because the typically low values of $p(S)$ and $p(S\|M)$ and the relatively small difference between them demand large samples to minimize the probability of a Type II error. Values of $p(M\|S)$ and $p(M)$ are typically much more central, and large differences in values usually are achieved in even mediocre models. Archaeological samples used in testing should be independent of the model (i.e., not used in model development).

TABLE 1.3

Example of Derivation of Model Performance Statistics

		Actual Circumstances		
		S	S'	
	M	$p(S \cap M)$ = .0085	$p(S' \cap M)$ = .3915	$p(M)$ = .4
	M'	$p(S \cap M')$ = .0015	$p(S' \cap M')$ = .5985	$p(M')$ = .6
Model Predictions		$p(S)$ = .01	$p(S')$ = .99	1.00

Derivation of cell data:

If model accuracy is $p(M|S)$ = .85, then $p(S \cap M) = p(M|S)\, p(S)$ = (.85)(.01) = .0085

The remainder of the table is completed by subtraction because:
- $p(S \cap M') = p(S) - p(S \cap M)$
- $p(S' \cap M) = p(M) - p(S \cap M)$
- $p(S' \cap M') = p(S') - p(S' \cap M) = p(M') - p(S \cap M')$

Derivation of other statistics:

- $p(M') = 1 - p(M) = 1 - .4 = .6$
  (model precision: 60% of region eliminated as unlikely to contain sites)
- $p(M)/p(S) = .4/.01 = 40$
  (model fit: its mapping is 40 times larger than the likely total site area being modeled)
- $p(M|S) - p(M) = .85 - .4 = .45$
  (model gives .45 improvement over chance)
- $p(S|M) = p(S \cap M)/p(M) = .0085/.4 = .0213$
  (probability of site when model indicates one)
- $p(S|M') = p(S \cap M')/p(M') = .0015/.6 = .0025$
  (probability of site when model *does not* indicate one)
- $p(S|M)/p(S) = .0213/.01 = 2.13$
  (site is 2.13 times more likely in M than base-rate chance of site)
- $p(S)/p(S|M') = .01/.0025 = 4$
  (site is four times less likely in M' than base-rate chance of site)
- $p(S|M)/p(S|M') = .0213/.0025 = 8.52$
  (site is 8.52 times more likely in M than in M')

Note: Model performance statistics are derived assuming that the regional *a priori* probability of an archaeological site is $p(S)$ = .01; that the model maps 40% of the region to the site class, $p(M)$ = .4; and that the model correctly indicates 85% of known sites, $p(M|S)$ = .85.

1.7.7 Confidence Intervals

Percent correct statistics (e.g., 100 $p[M|S]$, 100 $p[M|S']$) and estimated site class probabilities ($p[S]$) used in model evaluations should routinely be associated with binomial confidence limits, provided that the samples can be regarded as random samples. This is rarely done despite early example applications (Kvamme 1988a, 1992). Such data gives an idea of the variability associated with each estimate.

1.7.8 *p(S)* Is Not Constant

Foregoing discussions assume $p(S)$, the base or *a priori* probability of an archaeological site class, is constant in a study region. Yet, it undoubtedly also varies, like everything else, within the confines of a project area. There are many examples in the literature where high archaeological densities might occur in one area, while other regions tend to be relatively devoid of sites. Instead of computing $p(S)$ *once* for an entire region and treating it as a constant, we now have the ability through GIS to compute it continuously within an extended neighborhood to produce a $p(S)$ *surface*, allowing its treatment as a model parameter ($p[S]$ could be computed much like a local site-density surface within a moving window of, say, a 5-km radius). The result could then be combined through Bayesian methods with model outcomes based on environmental or other relationships, acting much like a weighting effect, improving the overall model.

Unfortunately, the $p(S)$ is a function of *known* archaeological densities and therefore subject to wide errors in its estimation stemming from small samples and sampling biases. Reasonable sample sizes and distributions of extant archaeological sites across regions must be present to explore this approach.

1.7.9   Issues of Scale: Near and Far Perspectives

The importance of scale when considering human land use and settlement choice has been emphasized by several authors, including Allen (1996) in a study of Iroquoian settlement. Variations in climate, soils, and food resource densities are seen to influence use and habitation at global and regional levels, while local on-site conditions such as slope and proximity to water dictate immediate site placements. Jochim's (1976) overview of hunter-gatherer settlement choice suggests much the same thing: high resource densities in a region might initially attract human groups, but the specific characteristics of individual locations dictate actual camp or settlement selections. Other examples can be found in ethnography (e.g., Western and Dunne 1979).

It might therefore be argued that there are at least "near" and "far" scales of phenomena that influence human settlement choice. It is not just variation at the immediate on-site level that accounts for regional archaeological patterning, but the distribution of resources, soil type, climate, and other factors at a wider scale. For example, we can imagine that high densities of game or ripening nuts, good soils for crops, or ready availability of water might draw people to a particular valley, but immediate settlement choice is dictated by the characteristics of a particular place: its accessibility, slope, shelter quality, proximity to water and fuel, view, defensibility, and so on. Most archaeological models have focused only on variation at immediate locations, typically on a cell-by-cell basis in a raster GIS of high spatial resolution (e.g., Brandt et al. 1992; Custer et al. 1986; Dalla Bona and Larcombe 1996; Hobbs 1996; Kvamme 1988a, 1992; Parker 1985; Scholtz 1981; van Leusen 1993; Warren and Asch 2000).

The variable *a priori* probabilities within large-radius windows of the previous section might be one way to incorporate regional variations within site-specific locational models. Alternatively, "far" perspectives could be explicitly incorporated within modeling efforts. One tactic might utilize variables that quantify variation within large areas (e.g., temperature or rainfall data, biomass/bioproductivity, soil quality with a 10-km radius) simultaneously with site-specific or "near" variables. Another might be the development of *distinct* near and far models that are later combined through Bayesian or other methods. With the former of high resolution and focusing on variation at immediate locations (e.g., 30-m pixels), the latter might employ pixels 5-km in size or larger and more global variables that can get at modeling gross variations in site densities and the question of why certain regions seem to have been more preferred in the past than others.

1.8   Conclusions: Direct Discovery Methods

Perhaps we are going about the process of archaeological site discovery all wrong. Instead of trying to *model* past human locational behavior based on sometimes questionable theoretical assumptions or empirical relationships found in often-poor samples, why not try to *directly locate* archaeological sites through other means? After all, technology is marching forward faster than we can either imagine or grasp, and there are a host of new remote sensing methods within easy reach.

Suppose we could just fly through the air and simply record information that leads to the identification of archaeological sites? We have actually been able to do that for nearly a century through conventional aerial photography where the regular geometric shapes common to human settlements (square, circular, oval, rectangular, and linear features) are easily recognized. In Europe, aerial archaeology is a commonplace tool and is probably the most productive site-discovery method employed by archaeologists (see Wilson 2000). Yet, it is relatively unutilized and rarely recognized as such in North America, despite pioneering efforts in the Southwest and elsewhere by Charles Lindberg in the 1920s (Avery and Berlin 1992: 226–227). In addition, we now have access to similar imagery from space, in the form of IKONOS and Quickbird satellite data, for example, with global coverage at spatial resolutions at 1 m or better, yielding results nearly as detailed as aerial photography (see Fowler 2002; Hritz, Chapter 19, this volume).

Whether from the air or space, high-resolution panchromatic or multi-spectral data offer enormous potential for detecting and locating archaeological resources over broad areas. Besides discovery through direct visualization of settlement components, past work has demonstrated that human occupations leave characteristic spectral signatures detectable in air and space imagery (e.g., Custer et al. 1986). In other words, predictive archaeological models of high power can potentially be developed from air or space remote-sensing data alone. Furthermore, the very methods that form the core of current approaches to archaeological modeling — various statistically based discriminant functions — are central to the methods of digital image classification that are employed to identify such features in remotely sensed imagery. Sadly, almost no effort has been invested in this line of research. Yet, owing to the frequently lackluster performance of many of our current archaeological models, it is incumbent on us to utilize every means, and every angle possible, to develop robust models for locating sites of our cultural heritage. From a statistical standpoint, there is little difference whether one associates site locations with an infrared wavelength band or a terrain slope layer, and there can be significant information in spectral data that can lead to more powerful models. We might ultimately imagine a hybrid approach, where models based on remote-sensing data are combined with models based on more-conventional environmental and other relationships to develop powerful tools for this task.

I believe we should and will turn increasingly to direct discovery methods for locating and mapping archaeological sites, for they can provide useful information in many contexts, and the various technologies are only going to improve. But where is the "theory" here; where are the "deductions"? In the end we must ask ourselves: "what are we trying to do?" Are we trying to find, map, and manage our planet's cultural heritage, or are we trying to develop cultural theories of location choice? The answer can be either one or both. We must recognize both to be valid in sometimes complementary pursuits.

Acknowledgments

I owe Paul Nickens a debt of gratitude for allowing me to try a new modeling methodology in a contract situation in 1978–1979. Mike Jochim, Albert Spaulding, and Mike Glassow at UC-Santa Barbara gave me great flexibility in letting me follow my nose in doctoral work, and the BLM supported it. Dan Martin (BLM) gave great encouragement in my involvement in *Quantifying the Present and Predicting the Past*. Bert Voorrips was kind enough to request a summary chapter on the method and theory of modeling that allowed me to consolidate ideas in a 1990 publication. Dawn Browning, now a Ph.D. student in wildlife biology at the University of Arizona, pointed me to the literature in biological habitat modeling. Eileen Ernenwein, environmental dynamics Ph.D. student at the University of Arkansas, gave advice that helped smooth out rough edges to this paper. Jo Ann Christein Kvamme polished it up too, from a perspective that includes decades of involvement in modeling, ranging from digitizing, to measurement, fieldwork, and write-up.

Notes

1. This experiment was dreamed up by my good friend Michael G. Spitzer, now at Washington State University.
2. Because the archaeological record is static, our situation is somewhat different. Archaeological evidence is either present or absent at a location; it is only unreliability in our detection of the evidence that gives rise to uncertainty.
3. If site-area data are unavailable, a typical or average site area might be employed for each site of the type of interest.

References

Allen, K.M.S., Iroquoian landscapes: people, environments, and the GIS context, in *New Methods, Old Problems: Geographic Information Systems in Modern Archaeological Research*, Maschner, H.D.G., Ed., Occasional Paper 23, Center for Archaeological Investigations, Southern Illinois University, Carbondale, 1996, pp. 198–222.

Avery, T.E. and Berlin, G.L., *Fundamentals of Remote Sensing and Airphoto Interpretation*, 5th ed., Macmillan, New York, 1992.

Bell, T. and Lock, G., Topographic and cultural influences on walking the Ridgeway in later prehistoric times, in *Beyond the Map: Archaeology and Spatial Technologies*, Lock, G., Ed., IOS Press, Amsterdam, 2000, pp. 85–100.

Bian, L. and West, E., GIS modeling of elk calving habitat in a prairie environment with statistics, *Photogrammetric Engineering and Remote Sensing*, 63, 161–167, 1997.

Boaz, J. and Uleberg, E., Quantifying the non-quantifiable: studying hunter-gatherer landscapes, in *Beyond the Map: Archaeology and Spatial Technologies*, Lock, G., Ed., IOS Press, Amsterdam, 2000, pp. 101–115.

Brandt, R., Groenewoudt, B.J., and Kvamme, K.L., An experiment in archaeological site location: modeling in the Netherlands using GIS techniques, *World Archaeology*, 24, 268–282, 1992.

Burrough, P.A. and McDonnell, R.A., *Principles of Geographical Information Systems*, Oxford University Press, Oxford, 1998.

Camilli, E., An ecological cover-type map of the San Juan Basin, northwestern New Mexico, in *Remote Sensing in Cultural Resource Management: The San Juan Basin Project*, Drager, D.L. and Lyons, T.R., Eds., Cultural Resource Management Division, National Park Service, Washington, DC, 1984, pp. 39–56.

Chadwick, A.J., A computer simulation of Mycenaean settlement, in *Simulation Studies in Archaeology*, Hodder, I., Ed., Cambridge University Press, Cambridge, 1978, pp. 47–58.

Chadwick, A.J., Settlement simulation, in *Transformations: Mathematical Approaches to Culture Change*, Renfrew, C. and Cooke, K., Eds., Academic Press, New York, 1979, pp. 237–255.

Clark, J.D., Dunn, J.E., and Smith, K.G., A multivariate model of female black bear habitat use for a geographic information system, *Journal of Wildlife Management*, 57, 519–526, 1993.

Conover, W.J., *Practical Nonparametric Statistics*, 3rd ed., Wiley, New York, 1999.

Custer, J.F., Eveleigh, T., Klemas, V., and Wells, I., Application of Landsat data and synoptic remote sensing to predictive models for prehistoric archaeological sites: an example from the Delaware coastal plain, *American Antiquity*, 51, 572–588, 1986.

Dalla Bona, L. and Larcombe, L., Modeling prehistoric land use in northern Ontario, in *New Methods, Old Problems: Geographic Information Systems in Modern Archaeological Research*, Maschner, H.D.G., Ed., Occasional Paper 23, Center for Archaeological Investigations, Southern Illinois University, Carbondale, 1996, pp. 252–274.

Dettmers, R. and Bart, J., A GIS modeling method applied to predicting forest songbird habitat, *Ecological Applications*, 9, 152–163, 1999.

Dettmers, R., Buehler, D.A., and Bartlett, J.B., A test and comparison of wildlife modeling techniques for predicting bird occurrence at a regional scale, in *Predicting Species Occurrences: Issues of Accuracy and Scale*, Scott, J.M., Heglund, P.J., Morrison, M.L., Haufler, J.B., Raphael, M.G., Wall, W.A., and Samson, F.B., Eds., Island Press, Washington, DC, 2002, pp. 607–615.

Duncan, R.B. and Beckman, K.A., The application of GIS predictive site location models within Pennsylvania and West Virginia, in *Practical Applications of GIS for Archaeologists: A Predictive Modeling Toolkit*, Wescott, K.L. and Brandon, R.J., Eds., Taylor and Francis, London, 2000, pp. 33–58.

Eastman, J.R., Jin, W., Kyem, P.A.K., and Toledano, J., Raster procedures for multi-criteria/multi-objective decisions, *Photogrammetric Engineering and Remote Sensing*, 61, 539–547, 1995.

Ebert, J.I., Remote sensing and large-scale cultural resources management, in *Remote Sensing and Nondestructive Archeology*, Lyons, T.R. and Ebert, J.I., Eds., National Park Service, Washington, DC, 1978, pp. 21–34.

Ebert, J.I. and Kohler, T.A., The theoretical basis of archaeological predictive modeling and a consideration of appropriate data-collection methods, in *Quantifying the Present and Predicting the Past: Theory, Method, and Application of Archaeological Predictive Modeling*, Judge, W.J. and Sebastian, L., Eds., U.S. Department of the Interior, Bureau of Land Management, Washington, DC, 1988, pp. 97–171.

Fowler, M.J.F., Satellite remote sensing and archaeology: a comparative study of satellite imagery of the environs of Figsbury Ring, Wiltshire, *Archaeological Prospection*, 9, 55–69, 2002.

Gabler, K.I., Laundre, J.W., and Heady, L.T., Predicting the suitability of habitat in southeast Idaho for pygmy rabbits, *Journal of Wildlife Management*, 64, 759–764, 2000.

Gaffney, V. and van Leusen, P.M., Postscript: GIS, environmental determinism and archaeology, in *Archaeology and Geographical Information Systems: A European Perspective*, Lock, G. and Stančič, Z., Eds., Taylor and Francis, London, 1995, pp. 367–382.

Gaffney, V., Stančič, Z., and Watson, H., The impact of GIS in archaeology: a personal perspective, in *Archaeology and Geographical Information Systems: A European Perspective*, Lock, G. and Stančič, Z., Eds., Taylor and Francis, London, 1995, pp. 211–229.

Gillings, M., Flood dynamics and settlement in the Tisza valley of northeast Hungary: GIS and the Upper Tisza project, in *Archaeology and Geographical Information Systems: A European Perspective*, Lock, G. and Stančič, Z., Eds., Taylor and Francis, London, 1995, pp. 67–84.

Gisiger, A., A Spatial Analysis of Human Adaptation Patterns Using Continental-Scale Data, unpublished M.A. thesis, Department of Anthropology, University of Arkansas, Fayetteville, 1996.

Goodchild, M. and Gopal, S., Eds., *The Accuracy of Spatial Databases*, Taylor and Francis, London, 1989.

Hasenstab, R.J., A Preliminary Cultural Resource Sensitivity Analysis for the Proposed Flood Control Facilities Construction in the Passaic River Basin of New Jersey, report submitted to the Passaic River Basin Special Studies Branch, Department of the Army, New York District Army Corps of Engineers, Soil Systems, Inc., Marietta, GA, 1983.

Hastie, T.J. and Tibshirani, R.J., *Generalized Additive Models*, Chapman and Hall, New York, 1990.

Hobbs, E., Preliminary results from Mn/Model, BRW, Inc., Minneapolis, *Mn/Model Newsletter*, 2, 1–3, 1996.

Hoebel, E.A., *Anthropology: The Study of Man*, 3rd ed., McGraw-Hill, New York, 1966.

Hudson, J.C., A location theory for rural settlement, *Annals of the Association of American Geographers*, 59, 365–381, 1969.

Hutchinson, G.E., Concluding remarks, *Cold Spring Harbor Symposium on Quantitative Biology*, 22, 415–427, 1957.

Jochim, M.A., *Hunter-Gatherer Subsistence and Settlement: A Predictive Model*, Academic Press, New York, 1976.

Judge, W.J. and Sebastian, L., Eds., *Quantifying the Present and Predicting the Past: Theory, Method, and Application of Archaeological Predictive Modeling*, U.S. Department of the Interior, Bureau of Land Management, Washington, DC, 1988.

Kellogg, D.C., Statistical relevance and site locational data, *American Antiquity*, 52, 143–150, 1987.

Khaemba, W.M. and Stein, A., Use of GIS for a spatial and temporal analysis of Kenyan wildlife with generalized linear modeling, *International Journal of Geographical Information Science*, 8, 833–853, 2000.

Kvamme, K.L., Predictive model of site location in the Glenwood Springs resource area, in A Class II Cultural Resource Inventory of the Bureau of Land Management's Glenwood Springs Resource Area, Nickens, P.R., Ed., U.S. Department of the Interior, Bureau of Land Management, Grand Junction District, CO, Nickens and Associates, Montrose, CO, 1980.

Kvamme, K.L., Computer processing techniques for regional modeling of archaeological site locations, *Advances in Computer Archaeology*, 1, 26–52, 1983.

Kvamme, K.L., Determining empirical relationships between the natural environment and prehistoric site locations: a hunter-gatherer example, in *For Concordance in Archaeological Analysis: Bridging Data Structure, Quantitative Technique, and Theory*, Carr, C., Ed., Westport Publishers, Kansas City, MO, 1985, pp. 208–238.

Kvamme, K.L., Development and testing of quantitative models, in *Quantifying the Present and Predicting the Past: Theory, Method, and Application of Archaeological Predictive Modeling*, Judge, W.J. and Sebastian, L., Eds., U.S. Department of the Interior, Bureau of Land Management, Washington, DC, 1988a, pp. 325–428.

Kvamme, K.L., Using existing archaeological survey data for model building, in *Quantifying the Present and Predicting the Past: Theory, Method, and Application of Archaeological Predictive Modeling*, Judge, W.J. and Sebastian, L., Eds., U.S. Department of the Interior, Bureau of Land Management, Washington, DC, 1988b, pp. 301–323.

Kvamme, K.L., The fundamental principles and practice of predictive archaeological modeling, in *Mathematics and Information Science in Archaeology: A Flexible Framework*, Voorrips, A., Ed., Holos-Verlag, Bonn, 1990a, pp. 257–295.

Kvamme, K.L., GIS algorithms and their effects on regional archaeological analysis, in *Interpreting Space: GIS and Archaeology*, Allen, K.M.S., Green, S.W., and Zubrow, E.B.W., Eds., Taylor and Francis, London, 1990b, pp. 112–126.

Kvamme, K.L., One-sample tests in regional archaeological analysis: new possibilities through computer technology, *American Antiquity*, 55, 367–381, 1990c.

Kvamme, K.L., A predictive site location model on the High Plains: an example with an independent test, *Plains Anthropologist*, 37, 19–40, 1992.

Kvamme, K.L., A view from across the water: the North American experience in archaeological GIS, in *Archaeology and Geographical Information Systems: A European Perspective*, Lock, G. and Stančič, Z., Eds., Taylor and Francis, London, 1995, pp. 1–14.

Llobera, M., Understanding movement: a pilot model towards the sociology of movement, in *Beyond the Map: Archaeology and Spatial Technologies*, Lock, G., Ed., IOS Press, Amsterdam, 2000, pp. 65–84.

Madry, S.L.H. and Rakos, L., Line-of-sight and cost-surface techniques for regional research in the Arroux River Valley, in *New Methods, Old Problems: Geographic Information Systems in Modern Archaeological Research*, Maschner, H.D.G., Ed., Occasional Paper 23, Center for Archaeological Investigations, Southern Illinois University, Carbondale, 1996, pp. 104–126.

Maynard, P.F. and Strahler, A.H., Logit classifier, a general maximum likelihood discriminant for remote sensing applications, in *Proceedings of the Fifteenth International Symposium on Remote Sensing of Environment*, American Society for Photogrammetry and Remote Sensing, Ann Arbor, MI, 1981, pp. 213–222.

Murdock, G.P., *Social Structure*, Macmillan, New York, 1949.

Murdock, G.P., *Ethnographic Atlas*, University of Pittsburgh Press, Pittsburgh, 1967.

Nunez, M., Vikkula, A., and Kirkinen, T., Perceiving time and space in an isostatically rising region, in *Archaeology and Geographical Information Systems: A European Perspective*, Lock, G. and Stančič, Z., Eds., Taylor and Francis, London, 1995, pp. 141–152.

Parker, S., Predictive modeling of site settlement systems using multivariate logistics, in *For Concordance in Archaeological Analysis: Bridging Data Structure, Quantitative Technique, and Theory*, Carr, C., Ed., Westport Publishers, Kansas City, MO, 1985, pp. 173–207.

Pianka, E.R., *Evolutionary Ecology*, Harper and Row, New York, 1974.

Plog, F.T., Study Area 1: Kaibab National Forest, in *Theory and Model Building: Refining Survey Strategies for Locating Prehistoric Heritage Resources*, Gumerman, G.J., Ed., Anthropological Reports 1, Prescott College, Prescott, AZ, 1983, pp. 7–36.

Sabins, F.F., *Remote Sensing*, 3rd ed., W.H. Freeman, New York, 1997.

Scheaffer, R.L., Mendenhall, W., and Ott, L., *Elementary Survey Sampling*, 2nd ed., Duxbury Press, North Scituate, MA, 1979.

Schiffer, M.B., Archaeological context and systemic context, *American Antiquity*, 37, 156–165, 1972.

Scholtz, S.C., Location choice models in Sparta, in *Settlement Predictions in Sparta: A Locational Analysis and Cultural Resource Assessment on the Uplands of Calhoun County, Arkansas*, Lafferty, R.H., III, Otinger, J.L., Scholtz, S.C., Limp, W.F., Watkins, B., and Jones, R.D., Arkansas Archaeological Survey Research Series 14, Arkansas Archeological Survey, Fayetteville, 1981, pp. 207–222.

Scott, J.M., Heglund, P.J., Morrison, M.L., Haufler, J.B., Raphael, M.G., Wall, W.A., and Samson, F.B., *Predicting Species Occurrences: Issues of Accuracy and Scale*, Island Press, Washington, DC, 2002.

Stančič, Z. and Kvamme, K.L., Settlement pattern modelling through Boolean overlays of social and environmental variables, in *New Techniques for Old Times, CAA98*, Barcelo, J.A., Briz, I., and Vila, A., Eds., BAR International Series 757, Tempus Reparatum, Oxford, 1999, pp. 231–237.

Stančič, Z. and Veljanovski, T., Understanding Roman settlement patterns through multivariate statistics and predictive modeling, in *Beyond the Map: Archaeology and Spatial Technologies*, Lock, G., Ed., IOS Press, Amsterdam, 2000, pp. 147–156.

Thomas, D.H., Nonsite sampling in archaeology: up the creek without a site? in *Sampling in Archaeology*, Mueller, J.W., Ed., University of Arizona Press, Tucson, 1975, pp. 61–81.

Van Leusen, P.M., Cartographic modelling in a cell-based GIS, in *Computing the Past: Computer Applications and Quantitative Methods in Archaeology, CAA92*, Andresen, J., Madsen, T., and Scollar, I., Eds., Aarhus University Press, Aarhus, Denmark, 1993, pp. 105–123.

Van Manen, F.T., Clark, J.D., Schlarbaum, S.E., Johnson, K., and Taylor, G., A model to predict occurrence of surviving butternut trees in the southern Blue Ridge Mountains, in *Predicting Species Occurrences: Issues of Accuracy and Scale*, Scott, J.M., Heglund, P.J., Morrison, M.L., Haufler, J.B., Raphael, M.G., Wall, W.A., and Samson, F.B., Eds., Island Press, Washington, DC, 2002, pp. 491–497.

Verbyla, D.L. and Litvaitis, J.A., Resampling methods for evaluating classification accuracy of wildlife habitat models, *Environmental Management*, 13, 783–787, 1989.

Vila, J., Wagner, V., Neveu, P., Voltz, M., and Lagacherie, P., Neural network architecture selection: new Bayesian perspectives in predictive modelling application to a soil hydrology problem, *Ecological Modelling*, 120, 119–130, 1999.

Wang, F., Fuzzy supervised classification of remote sensing images, *IEEE Transactions on Geoscience and Remote Sensing*, 28, 194–201, 1990.

Warren, R.E. and Asch, D.L., A predictive model of archaeological site location in the Eastern Prairie Peninsula, in *Practical Applications of GIS for Archaeologists: A Predictive Modeling Toolkit*, Wescott, K.L. and Brandon, R.J., Eds., Taylor and Francis, London, 2000, pp. 5–32.

Western, D. and Dunne, T., Environmental aspects of settlement site decisions among Pastoral Maasai, *Human Ecology*, 7, 75–81, 1979.

Wheatley, D., The use of GIS to understand regional variation in earlier Neolithic Wessex, in *New Methods, Old Problems: Geographic Information Systems in Modern Archaeological Research*, Maschner, H.D.G., Ed., Occasional Paper 23, Center for Archaeological Investigations, Southern Illinois University, Carbondale, 1996, pp. 75–103.

Wheatley, D. and Gillings, M., *Spatial Technology and Archaeology: The Archaeological Applications of GIS*, Taylor and Francis, London, 2002.

Willey, G.R., *Prehistoric Settlement Patterns In the Virú Valley, Peru*, Bulletin 155, Bureau of American Ethnology, Smithsonian Institution, Washington, DC, 1953.

Williams, L., Thomas, D.H., and Bettinger, R.L., Notions to numbers: Great Basin settlements as polythetic sets, in *Research and Theory in Current Archaeology*, Redman, C.L., Ed., John Wiley, New York, 1973, pp. 215–237.

Wilson, D.R., *Air Photo Interpretation for Archaeologists*, Arcadia Publishing, Charleston, SC, 2000.

Section 2:

Theoretical and Methodological Issues

2

Enhancing Predictive Archaeological Modeling: Integrating Location, Landscape, and Culture

Gary Lock and Trevor Harris

CONTENTS

2.1   Introduction

There is a beguiling aura surrounding predictive archaeological modeling that is compelling. This allure has been reinforced in recent years by the scientific and technological legitimation provided by geographic information systems (GIS). Using knowledge of the environmental variables that "first influenced the activities of original inhabitants," GIS layers are produced that identify locations where combinations of environmental variables match the patterns observed at known prehistoric sites (Kuiper and Wescott 1999: 1). Thus, high-potential areas of prehistoric sites can be identified using environmental data from known archaeological sites in a region that corresponds well with known sites (Kuiper and Wescott 1999: 2). While most developers of such predictive models acknowledge limitations in their analysis, and of predictive archaeological modeling in general, there can be no doubt that in North America these models are seen to "provide planners with a guide showing areas that would likely require less time, effort, and money to develop from a cultural resource compliance standpoint" and to augment and prioritize areas for "evaluation, monitoring, or mitigation" (Kuiper and Wescott 1999: 2; Wescott and Kuiper 2000). In other cultures, however, landscape analysis has taken an alternative route that reflects differing epistemological approaches to landscape archaeology. This chapter seeks to explore such differences in the approaches to landscape archaeology and predictive modeling. The aim is to identify strengths and weaknesses in archaeological predictive-modeling approaches and to suggest enhancements to modeling methods that incorporate cultural and humanistic archaeology as well as environmental factors in the modeling process.

The history, theory, method, and application of archaeological predictive models have been well publicized in several valuable texts, and it is not our intent to review such studies in this chapter (Kohler and Parker 1986; Judge and Sebastian 1988; Wescott and Brandon 2000). Nor is it our intent to denigrate such models, for there has clearly been a demand for such tools from planners and cultural resource managers faced with the daunting task of identifying prehistoric cultural sites while constrained by limited resources and tight timelines. Faced with these realities, the end is seen to justify the means. There are, however, certain tensions and issues associated with these approaches that we seek to clarify and use to point toward an archaeologically more sensitive approach to GIS-based archaeological predictive modeling.

2.2 Identifying the Tensions: Predictive Modeling and Landscape Archaeologies

In seeking to identify the tensions that exist within the wider context of landscape archaeology and archaeological predictive modeling, we draw attention to the differences that exist between landscape archaeology in the U.S. and the U.K. When reviewing a group of papers on landscape archaeology, the British archaeologist Barbara Bender went so far as to identify an "Atlantic divide" between approaches to these topics in Britain and America (Bender 1999). We suggest there is value in exploring the nature of this divide and the tensions between the alternative approaches taken in the two countries as a basis to proposing approaches that may begin to close not only the cultural gap, but contribute to next generation GIS predictive models.

Central to these tensions and the cultural "divide" are the demands of cultural resource management (CRM) to which predictive modeling is, and always has been, a handmaiden. Driven by the constraints of CRM, predictive modeling was developed to answer very specific questions that view the landscape as a current economic resource. The political, economic, and administrative context within which CRM archaeologists work defines their concerns based on *landscape as now*, that is, the recording and management of archaeological sites, usually within a legislative framework based on contemporary administrative perceptions of space, as well as "what exists where." In sharp contrast to this is the development of landscape archaeology, which incorporates a diversity of approaches that, in essence, explore the *landscape as then*, where the focus is on explanation and interpretations of past landscape understandings.

Interwoven within these arguments are other strands, of which two are relevant here. The first such strand concerns the role of technology and, in particular, the increasing reliance on, and dominance of, GIS. The development of predictive modeling and its application were a major area of interest predating GIS, as shown by a survey in 1986 that cites over 70 papers on the topic (Kohler and Parker 1986). Importantly then, the use of GIS has not determined the underlying philosophy of predictive modeling, but there can be no doubt that it has now become so embedded within predictive archaeological modeling that the technology itself must now be considered an essential part of the predictive methodology and the framework within which CRM now operates. The social practice of CRM and predictive modeling is now structured around the use of GIS and must acknowledge the epistemological issues that surround the use of this technology.

The second strand involves the development and use of archaeological theory and all of the complexity that this involves. To use a very simplistic caricature as a vehicle for the argument: the tensions are created by the differences between the applied scientism of processual archaeology and the attempted humanism of postprocessual approaches. Bender suggests that "the need to retain a strong scientific methodology" is evident in American landscape archaeology and is largely responsible for the "divide" (or at least the papers she was reviewing [Bender 1999: 632]) and has reinforced the move toward predictive modeling. Processual archaeology heavily utilizes formal models and modeling philosophy, such as site-catchment analysis and systems theory, and predictive modeling is one continuation of this intellectual tradition. The adoption of GIS in the mid to late 1980s, and especially the suitability of the raster data structure that facilitated the application of predictive modeling at a regional scale (Kvamme 1990), reinforced the methodology so strongly that theoretical development in other areas of landscape archaeology were largely overridden.

The emphasis on a quantitative methodology was at odds with humanistic approaches to landscape that were emerging contemporaneously, as exemplified in Tilley's influential *A Phenomenology of Landscape* (1994). The essence of this dichotomy is often crystallized as one of space versus place. Space is characterized as a void in which human activities are carried out. It is treated as the same void everywhere and at any point through time, a neutral backdrop for processual spatial modeling. Place, on the other hand, is a culturally defined locale that acts as a medium for action and is part of human experience and activity. Places are fluid and capable of taking on different meanings at different times, but they are always formative within personal and social activities. This juxtaposition of traditions forces us to confront the qualitative complexity of social landscapes and the quantitative reductionism of formally modeled space.

There is a large literature on humanized approaches to landscape, albeit with a strong theoretical theme (see, for example, Tilley 1994; Hirsch and O'Hanlon 1995; Ashmore and Knapp 1999; Thomas 2001). It is because these approaches are so explicitly theoretical that they create such a challenge for GIS applications generally and predictive modeling in particular. Whereas processual models are relatively methodologically concise and reproducible within a GIS context (through buffering, overlaying, and statistical tests of association, for example), the text-rich description that forms the basis of postprocessual landscape work is focused on descriptive theory rather than methodology. These concerns fed into the now well-rehearsed arguments about GIS heralding the return to environmental determinism (Gaffney and van Leusen 1995), a critique that landscape archaeology has tried to address but that seems to have left predictive modeling largely unmoved. This dichotomy between theory and practice in landscape archaeology has been confronted since at least 1993 (Wheatley 1993) and has produced a growing literature concerned with the theorizing of archaeological GIS (encapsulated within Wheatley 2000; Wise 2000). It must be said, however, that progress is slow and there are actually very few innovative applications that have moved very far beyond the theorizing (Lock 2001).

Of course, GIS is a technology that operates within wider arenas, and the elaboration of tensions concerning its epistemology is not unique to archaeology. The GIS and society debate, which addresses the role of GIS in society, seeks to identify the assumptions, principles, and practices affecting the way in which analysis and the acquisition of knowledge is pursued through GIS. Two general themes within this debate form an important background to the more specific interests of predictive modeling that is now almost entirely GIS-dependent (Lock and Harris 2000). The first theme concerns the nature of the data and argues that data do not exist but are created. Data are a social construction, and the "for whom, by whom, and for what purpose" is based within a mix of social, political, and economic contexts and interests (Taylor and Overton 1991). The second theme concerns the potential exclusion of much information from GIS because it is qualitative in nature and not capable of being measured and represented by the spatial primitives of point, line, or polygon. The GIS and society discourse recognizes that alternative forms of knowledge representation are crucial to understanding the nature of place and are largely excluded from GIS, resulting in a single "capturing" of an official view of reality that is heavily biased toward a scientific data-driven representation (Mark 1993). Indeed, it has been argued that one reason why GIS has been so spectacularly successful is because it represents a single noncontradictory view of the world (Harris et al. 1995). This is of particular interest to predictive modeling within national and regional CRM systems and their ability, or inability, to incorporate alternative views of the past within a seemingly inflexible quantitative GIS.

This dichotomy, between a GIS-based predictive modeling and landscape theory, in essence focuses attention on how people and nature are represented within GIS. It is of interest here to consider the ideas of Michael Curry in his book *Digital Places: Living with Geographical Technologies* (Curry 1998). Curry classifies GIS into PaleoGIS and GIS_2. PaleoGIS represents most current GIS applications, and certainly most archaeological GIS, which are defined by their underrepresentation of the basic elements of human experience that give meaning to the world. The challenge is to move toward GIS_2, and this aspect is further explored below.

2.3 The West Virginia Predictive Model

To illustrate and develop these points further, we briefly recount here a predictive-modeling study undertaken by one of the authors in West Virginia as part of an environmental impact assessment project associated with a proposed high-power 765-kV transmission line crossing West Virginia into Virginia. To comply with Section 106 legislation and to support CRM efforts, a predictive model was developed (Gozdzik 1997). The model drew upon similar predictive models developed elsewhere that were adapted for use in southern West Virginia (see, for example, the Fort Drum, NY study by Hasenstaab and Resnick [1990]). Known historical and archaeological sites were identified from state records, and two models were constructed to identify the spatial probability of prehistoric and historic sites in the region. Similar approaches were used in both the prehistoric and historic studies, but for the sake of brevity only the former is discussed here. Some 588 known prehistoric sites were located on GIS maps. For each site, four environmental parameters related to distance of site to water, site slope, site elevation, and site soil type and drainage (based on http://www.nrcs.usda.gov/technical/techtools/ststsgo_db.pdf STATSGO data) were identified and the details extracted from the GIS (Figure 2.1 [State Soil Geographic Data Base]). Using exploratory data analysis, the environmental factors associated with the known archaeological sites in the region were explored through the analysis of univariate, bivariate, and hypervariate distributions. Based upon a graphical intuitive approach (GIA), a classification of three groups was devised, as detailed in Table 2.1. The GIA draws heavily on a review of parameter distributions and relationships, but it also enables an intuitive understanding of the archaeology of the region to influence the selection of threshold boundaries, as in Table 2.1. The environmental parameters reflect the known archaeology of the region in that, put simply, most sites can be found close to sources of fresh

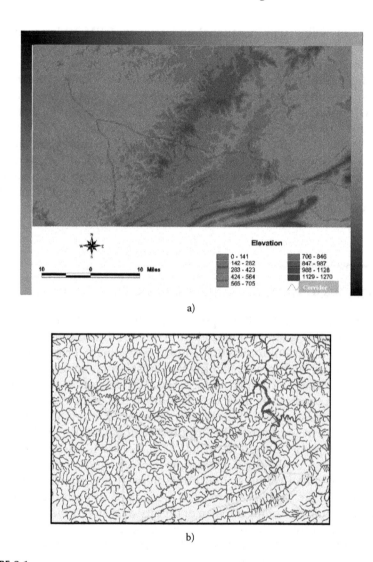

a)

b)

FIGURE 2.1
Constructing the West Virginia predictive model based on environmental variables (a, b), resulting in a site-sensitivity model for prehistoric sites (c, d).

water, on land of shallow slope, on well-drained and fertile soil, and at low elevations. Three site-density categories were identified comprising locations of high, moderate, and low probability of locations containing archaeological sites. Some 10% of the study area fell within the high-likelihood category, 49% in the moderate category, and 41% in the low-likelihood category.

This approach is necessarily totally dependent on the knowledge gained from known and recorded archaeological sites. Thus, as with

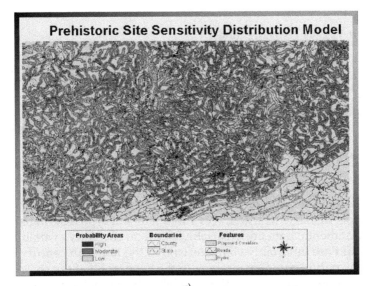

c)

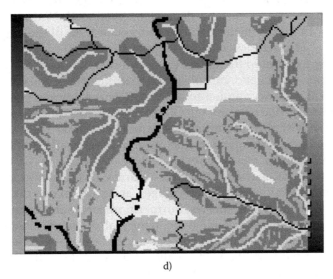

d)

FIGURE 2.1
(continued)

similar predictive-modeling projects, the study is heavily dependent on the vagaries that brought about the identification of these known sites and not on a systematic and methodological survey of the sites. The resulting silences in the constructed data set are not acknowledged, are unquantifiable, and could result in a distorted knowledge database upon which the predictive model is dependent. Furthermore, the actual model is entirely environmentally driven and assumes that the behavior patterns

TABLE 2.1

The West Virginia Predictive Model

	High Probability of Sites	Moderate Probability of Sites	Low Probability of Sites
Distance to water	0–230 m	231–500 m	>500 m
Slope	0–18°	19–30°	>31°
Elevation	400–560 m	561–840 m	>840 m
Soils	Well drained	Moderately well drained	Poorly drained
Roads (historic only)	0–150 m	151–500 m	>500 m

associated with the known sites are influenced by, and highly correlated with, slope, distance to water, elevation, and soil type. Having said that, the environmental data are invariably based on currently available digital sources and are invariably characteristic of the late 20th century and not of the time period commensurate with the geography of the society under study. As with other predictive models, the model does not seek to identify individual archaeological sites, but to target areas that contained archaeological sites to varying degrees of likelihood. Without question, the use of GIS greatly facilitated this process because of the emphasis on the physical characteristics of landscape. However, it should be noted that seeking to develop the model to incorporate nonquantitative aspects of landscape would generate considerable difficulty for the GIS analysis. No statistical test was applied to determine optimal group classification, and reliance was placed upon the GIA to blend archaeological knowledge of the region with the environmental parameters. Thus, there were many subjective elements involved in this "quantitative" methodology, including the selection of environmental factors, the scale of GIS data and analysis, the locational and attribute accuracy of the recorded archaeological sites and environmental coverages, and particularly, the subjective selection of category boundaries. Furthermore, and somewhat typical of similar studies, there is almost no temporal distinction between the archaeological sites because of the paucity of time data. As a result, all sites, for all time periods, were analyzed as a single cadre in the same analysis. Once again, we stress the intent here is not to denigrate the predictive-modeling process, but to realistically appraise the strengths and weaknesses of that process.

The resulting product was an impressive and persuasive map that displayed areas of high, moderate, and low site probability. The map was ostensibly based on a logical Boolean approach, was supported by the technology of GIS, could be replicated, and, importantly, provided a substantive product that would contribute toward the needs of the CRM community and meet compliance requirements. Measures of what constituted a "good" result, of course, remain unspecified, and without a comprehensive archaeological survey of the entire region (the avoidance of which was one of the purposes behind this model generation) the accuracy of the model will remain unknown.

2.4   Another West Virginia Predictive Model

In the context of the previous model, now consider the following predictive model. In similar fashion to the above approach, a new study area was designated that again emphasized distance to water (buffered from 100, 500, and >500 m), low slopes and valley bottomlands (<18°), fertile alluvial soils, favored old-field sites, and at a preferred elevation from just above sea level to 300 m and up to 770 m (Figure 2.2). Without stretching the point too far, these parameters are not greatly dissimilar to those specified above for the prehistoric site model. However, these latter parameters correspond not to the location of archaeological sites, but to the habitat of the common sycamore tree (*Platanus occidentalis L.*), a fast-growing, long-lived tree and one of the most common trees in eastern U.S. deciduous forests (Wells and Schmidtling 2001). The tree especially favors alluvial soils along streams and bottomlands and is very tolerant of wet soil conditions and grows well in proximity to water (though it is relatively intolerant to flooding). We hope you will excuse the ruse employed here, but the question arises as to what are the models actually predicting — nature or culture? Bearing in mind the minor differences between the prehistoric and the sycamore models, does this imply that human behavior can be modeled and predicted by locating the habitat of sycamore trees in the eastern U.S.? No such correlation is mentioned in the archaeology of the region. But of course this misses the point, for the prehistoric predictive model does not suggest to understand human behavior so much as to be able to predict, by whatever means, the likelihood of sites being located in a particular region. Indeed, based on this finding, that the distribution of sycamore habitat is highly correlated with the distribution of prehistoric sites in West Virginia, we may now have the basis of a new, single-variable, predictive model. The end in predictive modeling justifies the means. If the modeling of human behavior can be approximated by the distribution of the sycamore, or fertile soils, or a streambed, then however environmentally deterministic this may be, the model is a success.

It is clear from these two predictive examples that we consider the treatment of archaeological sites as simple data points, set solely in an environmentally determined space, as being problematic. Essentially, the archaeological site is reduced to a uniform, undifferentiated point in space — the McArchaeo site of predictive modeling — easily predictable because all points are the same and their location is entirely due to external variables. In contrast, we suggest that a continuum must exist that ranges from the extreme reductionism of archaeological sites as uniform points in space, to the full complexity that comes from consideration of the archaeological

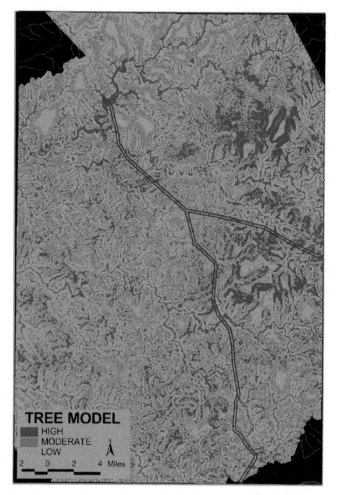

FIGURE 2.2
Predicting the common sycamore tree. The resulting model is based on distance from water, slope, soil type, and elevation.

site as a cultural entity. Seeking to model human–landscape interaction using uniform data points concentrates on the world to the exclusion of the subject. At the other end of the continuum lies Tilley's phenomenology *in extremis* — the subject to the exclusion of the world — which requires a landscape of cultural entities, each differentiated in endlessly complex variations of subjectivity. The theory, methodology, and practice are not yet developed to enable the integration of such complexity into GIS and predictive modeling, but the way forward is to start moving along the continuum away from sole reliance on simplistic reductionism toward an augmented cultural entity.

2.5 Humanizing the Landscape

We suggest that in determining site locations there is a need to move beyond a sole reliance on environmental factors (Ebert 2000) and to draw upon aspects of landscape archaeology that could provide additional determinants of site location. Both "landscape archaeology" and "postprocessualism" are umbrella terms that encompass a broad range of understandings, and while it is not possible to go into great detail here, we proffer some ideas that are pertinent to the aims of predictive modeling. Before doing so, however, there is an even more fundamental problem that underlies the whole tension between predictive modeling and landscape archaeology — that of the concept of "site." It has long been argued that the "site" as a unit of analysis is theoretically redundant and that individual and social interaction with the material world is a continuum across the landscape (Gaffney and Tingle 1984). Early concepts of "off-site archaeology" (Foley 1981) have matured into a general acceptance that an understanding of a site can only be gained through its contextual connections within its landscape, which comprises a complex web of spatial, temporal, and cultural links. Part of that understanding is necessarily concerned with location and why a site is where it is. Of course environmental factors contribute to this understanding, but they are unlikely to provide a satisfactory explanation, for the whole edifice of predictive modeling is site-based rather than landscape-based.

While landscape archaeology is diverse in its interests and applications, it is in essence concerned with *landscape as then* as accessed through what it means to be a socialized human being living within a particular landscape. Exploring such notions of landscape is not confined to archaeologists, but is a rich vein of thought spanning many subjects including geography, social theory, philosophy, anthropology, and history, often with an emphasis on a multidisciplinary approach (Muir 1999; Thomas 2001). We see some of the challenges facing predictive modeling being encapsulated in postprocessual writings. As Bender (1993: 1) suggests, "Landscapes are created by people — through their experience and engagement with the world around them. They may be close-grained, worked upon, lived in places, or they may be distant and half-fantasized." The differing forms of engagement with the physical world, and how those relationships can be used to make sense of an individual and a group's place in the wider scheme of things, have given rise to many different types of landscape that, because of their relevance to the concept of location, are worth exploring briefly here.

Much of this postprocessual research is concerned with landscapes, or more specifically with places and locales (whether natural features sites or monuments) that are imbued with cultural meaning through symbolism, often as part of cosmological, religious, ritual, or ideological beliefs. These traditional cultural properties or sacred landscapes involve sacred geographies and are founded on many different notions of what is sacred (Carmichael et al. 1994).

There is considerable overlap here with humanistic geography (Cosgrove and Daniels 1988) and with anthropology (Hirsch and O'Hanlon 1995), although these ideas have been widely applied to British prehistory in particular. Often incorporated into these approaches is a notion of time depth or historical narrative that is inscribed on a landscape through social memory (Ingold 1993) and reproduced over long periods of time through structured social practices (Barrett 1994). Time depth may involve ancestors, perhaps through genealogies, and mythology that charge certain places with power and significance through association. For nonliterate peoples, history can be written through landscape features and reproduced through social action and events that take place at these significant locales (Gosden and Lock 1998; Edmonds 1999). Aspects of landscape can also represent individual and group power and identity informally through some of the ideas above or through more structured mechanisms that convert political and economic hierarchies into forms of spatial control and access.

It is clear from the body of work briefly summarized above that many different social and cultural factors can influence site location. Predicting site location based on rational Western logic is to ignore a considerable amount of anthropological and archaeological evidence and theory for the convenience of methodological simplicity. There is much more to location than the interplay of environmental variables, and the modeling challenge to be confronted is to integrate both environmental and cultural concerns with the concomitant demands of both data and theory. This approach demands recognition of the context of individual sites and the importance of spatial and temporal contingency. Not only what surrounds a site influences its position, but also what came before it and the cultural development of the landscape. Landscapes are thus a web of spatial and temporal connections, and the implications of this are only just beginning to be explored by GIS-based landscape archaeologists. The few existing examples of research in humanizing landscapes are mainly based on what can be described as landscapes of perception and have involved attempts to model visibility and accessibility based on an embodied subject situated within the landscape. Wheatley and Gillings (2000) have shown the complexity of visibility studies and the shortcomings of a simplistic binary viewshed, together with ideas for developing the technique. The work of Llobera (1996, 2000) includes an accessibility index that models topographic and cultural influences on human movement across a landscape. Both visibility and accessibility can have an influence on site location, although trying to incorporate perception into a model raises its own philosophical tensions. Through notions of "sensuous geographies" (Rodaway 1994), Witcher (1999) differentiates between perception as the simple reception of information (i.e., a viewshed of what can be seen) and perception as mental insight (i.e., making sense of the view via socially constituted meaning). Despite the limitations of this work, these are serious attempts to move beyond the confines of PaleoGIS and the reliance on environmental data alone, and it remains to be seen how these can be incorporated into predictive modeling.

2.6   Moving from Data Points to Cultural Entities

To illustrate our thinking, we suggest one approach that would support ongoing predictive-modeling initiatives and build on the contribution of GIS, and yet redress some of the imbalance we have alluded to in the context of differing interpretations of landscape, whether based on sacred, ideological, or cosmological meaning, on perception, or on symbolisms of power. In this respect, we suggest treating sites not as uniform data points in space but as cultural entities set against an environmental backdrop (see Rouse 2000). Such an approach entails applying any knowledge about the archaeology of the site to the model itself. In Figure 2.3, the data point is interpreted not solely in terms of its proximity to water and the stream network, but is postulated to have an additional attractor value. Thus a burial mound becomes more than just another undifferentiated point intersecting environmental planes, but has a cultural meaning in its own right. In our West Virginia example we postulate that a burial mound may have attractor status in that it pulls groups toward it or retains groups in the vicinity through ancestor worship. Thus one could suggest a greater likelihood of sites being present in the area because of the nature of the archaeological site itself, the burial mound, having a sphere of influence over the surrounding landscape and that extended beyond the strict confines of the "site" itself (Figure 2.3). This influence could easily be modeled within GIS through the use of distance buffers or, preferably, friction surfaces that incorporated concepts of relative distance rather than the more limiting use of absolute distance. Alternatively, the mound might be interpreted as being a repulsor, where fear and taboo acted to deter access to the site and thereby created a vacant annulus surrounding the site devoid of prehistoric activity. Again, this is easily modeled in a GIS. A further situation might be one where the mound acted as a territorial symbol or marker and was thus placed in a dominant, highly visible location. While these examples are conjecture, in all three instances the interpretation of the site archaeology itself can both influence and refine the resulting site-probability map.

Cultural site interpretation can thus be used to augment the environmental point-data-driven models currently practiced to create an envirocultural probability model. Thus sites are not treated as homogeneous point data, but, where the archaeology allows, are treated individually and interpreted for their cultural significance. By augmenting an environmentally driven analysis of site probability with the addition of a cultural interpretation of the site entity, we suggest the basis for a midpoint on Bender's Atlantic divide. Thus we acknowledge that symbolic, cognitive, perceptual, and other qualitative reasoning can also influence behavior patterns. In short, we are arguing that the archaeology be put back into the archaeological predictor models. Furthermore, the approach outlined above could relatively easily be adapted within a GIS methodology. This approach does

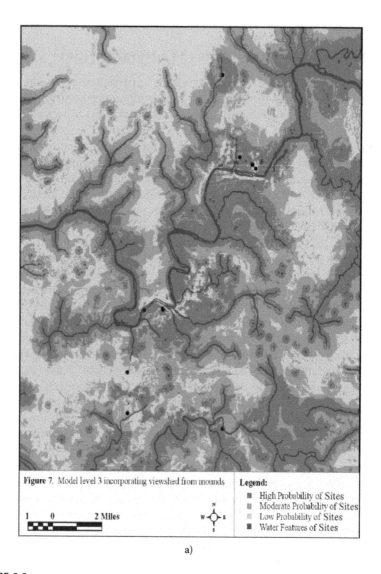

Figure 7. Model level 3 incorporating viewshed from mounds

Legend:
■ High Probability of Sites
■ Moderate Probability of Sites
■ Low Probability of Sites
■ Water Features of Sites

1   0   2 Miles

a)

FIGURE 2.3
Attempting to incorporate the social into predictive modeling. The probability models include
(a) viewsheds from a group of mounds and (b) attractor buffers.

entail greater effort because each site must be examined to determine its
role in the landscape. In most predictive models, the number of such known
sites has been limited and should not pose a significant resource problem.
Furthermore, it draws upon the skills of the archaeologist to interpret the
data rather than the existing heavy reliance on the data-processing skills
of the GIS analyst.

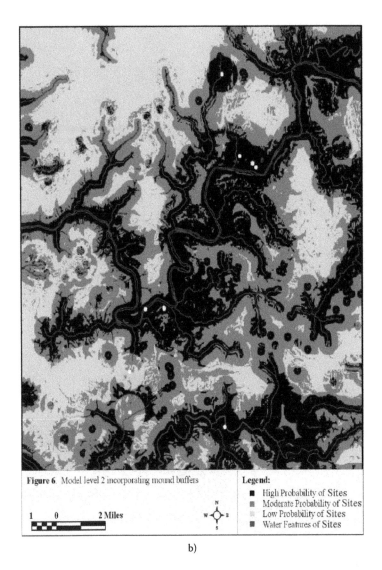

Figure 6. Model level 2 incorporating mound buffers

Legend:
■ High Probability of Sites
■ Moderate Probability of Sites
 Low Probability of Sites
■ Water Features of Sites

1   0   2 Miles

W ◇ E
N
S

b)

FIGURE 2.3
(continued)

2.7 Increasing the Envirocultural Complexity

The model outlined above should be seen as a first step toward some of the ideas outlined in this chapter. The challenge remains, however, of how to incorporate even greater complexity into the model. As already argued, this complexity must derive from both landscape theory and the specifics of

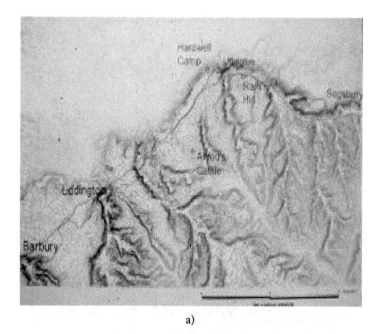

a)

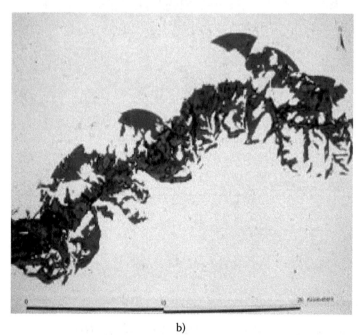

b)

FIGURE 2.4
Modeling location based on movement and visibility: the Ridgeway ancient trackway in Oxfordshire, England. A cost-path model (a) and visibility index (b) are both used to inform
locational decisions for a series of Iron Age hill forts along the route.

archaeological evidence from the region in question. We use two further examples here to illustrate how meaning and explanation based on the contextual detail of the archaeology of a region must feed into an understanding of site location. Both of these show that location can be independent of environmental variables in complex and subtle ways that are both spatially and temporally contingent and therefore could not, in these cases, be predicted by environmental variables alone.

The first example is based on the work of the Hillforts of the Ridgeway Project that is based on fieldwork conducted in an area surrounding the ancient Ridgeway track in Oxfordshire, England (Daly and Lock 2004). Excavation of three Iron Age hill forts located within 20 km of each other — Uffington Castle, Segsbury Camp, and Alfred's Castle — has emphasized marked differences between these sites, despite several surface similarities. The detailed studies have shown how each site developed from varying earlier activity, much of which is not evident from the surface. The Ridgeway was a crucial access route that antedates these sites and, together with the visibility characteristics of the area's distinctive topography, has emphasized the nuances of location (Bell and Lock 2000). In this area, site prediction from environmental variables would be somewhat meaningless because there is very little variation, and the story is in the detail of the archaeology that shows location based on the subtleties of vision and movement as well as historical context (Figure 2.4).

A second example focuses on Mesolithic and Neolithic sites along the River Danube in the area of the Iron Gates Gorge (Figure 2.5). Central to understanding this site is the transformation of traditional environmental variables into notions of affordances: how aspects of the landscape can offer potential for individual and social action to be played out. Distributions of flora and fauna can be converted into resource-scapes that offer opportunities for subsistence strategies. The occurrence of archaeological evidence can feed into notions of task-scapes, for example lithic scatters may represent locales of tool production that will have a relationship with resource-scapes that require lithic tools. The interplay between different types of affordances across the landscape creates lifeways for individuals and groups that challenge our understanding of their practices at a variety of scales from individual actions to seasonal movements. The location of sites is in a reflexive relationship with these activities, each influencing the other to produce social practice that not only reproduces cultural values, but also provides a mechanism for social change through deviant practices, however slight, being absorbed and continued.

The Iron Gates Gorge example also illustrates the importance of understanding the meaning of sites through the contextual detail that derives from a combination of theory and data (in this case, the link between symbolism, visibility, landscape features, and structural elements of settlements or sites). Excavation at the site of Lepenski Vir, for example, has revealed graves in alignment with a nearby and very distinctive trapezoidal mountain, Treskavac. House plans also mimic the mountain shape. Using Higuchi viewsheds,

a)

b)

FIGURE 2.5

Modeling location based on symbolism and context: the Iron Gates Gorge in Serbia. Note the alignments of excavated details such as houses and graves at the site of Lepenski Vir (a) with pronounced landscape features (b).

which incorporate near, medium, and distant banding, has shown that the site is located to include a variety of alignments on natural features at different distances in the surrounding landscape. The combination of landscape affordances and understanding each site as a cultural entity has enabled an understanding of location that goes beyond environmental determinism.

2.8 Conclusion: The Challenge

The intent of this chapter is to explore issues and approaches to archaeological predictive modeling. We suggest that many of the substantive issues become most dominant when we contrast the views of many American CRM-based predictive modelers and those of many British academic landscape archaeologists. These differences are framed within a classic modernist–postmodernist tension characterized on one hand by reductionism and on the other by the complexity of human existence. We acknowledge that these two positions are largely the result of different historical trajectories within archaeology, but they also operate within very different contexts today. While acknowledging the political, economic, and institutional constraints of CRM, we have also raised issues concerning the role of predictive modeling. At the same time, we suggest one example of how data points and environmentally driven predictive models may be augmented by the concept of cultural entity, which brings with it a richer understanding of site and landscape archaeology. While we are not claiming that landscape archaeology has the answers to these complex issues, it certainly provides an alternative way of thinking about sites and their locational characteristics. Drawing upon Bender's words, "I'm not suggesting that we have cracked these questions ... but ... because we have moved into a more reflexive relationship with the past and to the process of doing archaeology, we are uneasily aware that they need addressing" (Bender 1999: 632).

References

Ashmore, W. and Knapp, A.B., *Archaeologies of Landscape: Contemporary Perspectives*, Blackwell, Oxford, 1999.

Barret, J.C., *Fragments from Antiquity: An Archaeology of Social Life in Britain, 2900–1200 BC*, Blackwell, Oxford, 1994.

Bell, T. and Lock, G., Topographic and cultural influences on walking the Ridgeway in later prehistoric times, in *Beyond the Map: Archaeology and Spatial Technologies*, G. Lock, Ed., IOS Press, Amsterdam, 2000, pp. 85–100.

Bender, B., Ed., *Landscape: Politics and Perspectives*, Berg, Oxford, 1993.

Bender, B., Introductory comments, special section on dynamic landscapes and socio-political process, *Antiquity*, 73, 632–634, 1999.

Carmichael, D.L., Hubert, J., Reeves, B. and Schanche, A. Eds. *Sacred Sites, Sacred Places*, Routledge, One World Archaeology 23, London, 1994.

Cosgrove, D. and Daniels, S. Eds. *The Iconography of Landscape*, Cambridge University Press, Cambridge, 1988.

Curry, M.R., *Digital Places: Living with Geographic Information Technologies*, Routledge, London, 1998.

Daly, P. and Lock, G., Time, space and archaeological landscapes: establishing connections in the First Millennium BC, in *Spatially Integrated Social Science: Examples in Best Practice*, Goodchild, M.F. and Janelle, D.G., Eds., Oxford University Press, Oxford, 2004, pp. 349–365.

Ebert, J.I., The state of the art in "inductive" predictive modeling, in *Practical Applications of GIS for Archaeologists: A Predictive Modeling Toolkit*, Wescott, K.L. and Brandon, R.J., Eds., Taylor and Francis, London, 2000, pp. 129–134.

Edmonds, M., *Ancestral Geographies of the Neolithic: Landscape, Monuments and Memory*, Routledge, London, 1999.

Foley, R., Off-site archaeology: an alternative approach for the short-sited, in *Pattern of the Past: Studies in Honour of David Clarke*, Hodder, I., Isaac, G., and Hammond, N., Eds., Cambridge University Press, Cambridge, 1981, pp. 157–183.

Gaffney, V. and Tingle, M., The tyranny of the site: method and theory in field survey, *Scottish Archaeological Review*, 3, 134–140, 1984.

Gaffney, V. and van Leusen, P.M., GIS, environmental determinism and archaeology, in *Archaeology and Geographic Information Systems: A European Perspective*, Lock, G. and Stančič, Z., Eds., Taylor and Francis, London, 1995, pp. 367–382.

Gosden, C. and Lock, G., Prehistoric histories, *World Archaeology*, 30 (1), 2–12, 1998.

Gozdzik, G., Bergeron, S., and Rouse, J., A Predictive Model Integrating GIS and Archaeology for the Wyoming-Cloverdale 765-kV Transmission Line Project in McDowell, Mercer, Monroe, Raleigh, Summers, and Wyoming Counties West Virginia, Horizon Research Consultants, Morgantown, WV, 1997.

Harris, T.M., Weiner, D., Warner, T.A., and Levin, R., Pursuing social goals through participatory geographic information systems: redressing South Africa's historical political ecology, in *Ground Truth: The Social Implications of Geographic Information Systems*, Pickles, J., Ed., Guildford Press, New York, 1995, pp. 196–222.

Hasenstaab, R.J. and Resnick, B., GIS in historical predictive modelling: the Fort Drum project, in *Interpreting Space: GIS and Archaeology*, Allen, K.M.S., Green, S.W., and Zubrow, E.B.W., Eds., Taylor and Francis, New York, 1990.

Hirsch, E. and O'Hanlon, M., Eds., *The Anthropology of Landscape*, Oxford University Press, Oxford, 1995.

Ingold, T., The temporality of landscape, *World Archaeology*, 25: 152–74, 1993.

Judge, W.J. and Sebastian, L., Eds., Quantifying the Present and Predicting the Past: Theory, Method, and Application of Archaeological Predictive Modeling, U.S. Department of the Interior, Bureau of Land Management, Denver, 1988.

Kohler, T.A. and Parker, S.C., Predictive models for archaeological resource location, in *Advances in Archaeological Method and Theory*, Vol. 9, Schiffer, M.B., Ed., Academic Press, New York, 1986, pp. 397–452.

Kuiper, J.A. and Wescott, K.L., A GIS Approach for Predicting Prehistoric Site Locations, presented at 19th Annual ESRI User Conference, San Diego, CA, July 26–30, 1999; available on-line at http://www01.giscafe.com/technical_papers/Papers/paper057/, accessed 12 October 2002.

Kvamme, K.L., The fundamental principles and practice of predictive modelling, in *Mathematics and Information Science in Archaeology: A Flexible Framework*, Voorrips, A., Ed., Vol. 3 in *Studies in Modern Archaeology*, Holos-Verlag, Bonn, 1990, pp. 257–295.

Llobera, M., Exploring the topography of mind: GIS, social space and archaeology, *Antiquity*, 70, 612–622, 1996.

Llobera, M., Understanding movement: a pilot model towards the sociology of movement, in *Beyond the Map: Archaeology and Spatial Technologies*, Lock, G., Ed., IOS Press, Amsterdam, 2000, pp. 65–84.

Lock, G., Theorising the practice or practicing the theory: archaeology and GIS, *Archaeologia Polona*, 39, 153–164, 2001.

Lock, G. and Harris, T., Introduction: return to Ravello, in *Beyond the Map: Archaeology and Spatial Technologies*, Lock, G., Ed., IOS Press, Amsterdam, 2000, pp. xiii–xxv.

Mark, D.M., On the Ethics of Representation: or Whose World Is It Anyway? presented at Geographic Information and Society: A Workshop, National Center for Geographic Information and Analysis, Friday Harbor, WA, 11–14 November 1993.

Muir, R., *Approaches to Landscape*, Macmillan Press, London, 1999.

Rodaway, P., *Sensuous Geographies: Body, Sense and Place*, Routledge, London, 1994.

Rouse, L.J., Data Points or Cultural Entities: A GIS-Based Archaeological Predictive Model in a Post-Positivist Framework, unpublished M.A. thesis, Department of Geology and Geography, West Virginia University, 2000.

Taylor, P.J. and Overton, M., Further thoughts on geography and GIS, *Environment and Planning A*, 23, 1087–1094, 1991.

Thomas, J., Archaeologies of place and landscape, in *Archaeological Theory Today*, Hodder, I., Ed., Polity Press, Oxford, 2001, pp. 165–186.

Tilley, C., *A Phenomenology of Landscape: Places, Paths and Monuments*, Berg, Oxford, 1994.

Trifkovic, V., *The Construction of Space in Early Holocene Iron Gates*. Oxford: Unpublished D. Philosophy thesis, 2005, University of Oxford.

Wells, O.O. and Schmidtling, R.C., *Platanus occidentalis L.* Sycamore, 2001; available on-line at http://forestry.about.com/science.forestry/libarary/silvics/blsilsye.htm, accessed 12 October 2002.

Wescott, K.L. and Brandon, R.J., Eds., *Practical Applications of GIS for Archaeologists: A Predictive Modeling Toolkit*, Taylor and Francis, Philadelphia, 2000.

Wescott, K.L. and Kuiper, J.A., Using a GIS to model prehistoric site distributions in the Upper Chesapeake Bay, in *Practical Applications of GIS for Archaeologists: A Predictive Modeling Toolkit*, Wescott, K.L. and Brandon, R.J., Eds., Taylor and Francis, Philadelphia, 2000, pp. 59–72.

Wheatley, D.W., Going over old ground: GIS, archaeological theory and the act of perception, in *Computing the Past: Computer Applications and Quantitative Methods in Archaeology*, CAA 92 proceedings, Andresen, J., Madsen, T., and Scollar, I., Eds., Aarhus University Press, Aarhus, Denmark, 1993, pp. 133–138.

Wheatley, D.W., Spatial technology and archaeological theory revisited, in *Computer Applications and Quantitative Methods in Archaeology,* CAA 96 proceedings, Lockyear, K., Sly, T.J.T., and Mihilescu-Bîrliba, V., Eds., BAR International Series, 845, Archaeopress, Oxford, U.K., 2000, pp. 123–131.

Wheatley, D. and Gillings, M., Vision, perception and GIS: developing enriched approaches to the study of archaeological visibility, in *Beyond the Map: Archaeology and Spatial Technologies*, Lock, G., Ed., IOS Press, Amsterdam, 2000, pp. 1–27.

Wise, A.L., Building theory into GIS-based landscape analysis, in *Computer Applications and Quantitative Methods in Archaeology,* CAA 96 proceedings, Lockyear, K., Sly, T.J.T., and Mihilescu-Bîrliba, V., Eds., BAR International Series, 845, Archaeopress, Oxford, U.K., 2000, pp. 141–148.

Witcher, R.E., GIS and landscapes of perception, in *Geographical Information Systems and Landscape Archaeology*, Gillings, M., Mattingly, D., and van Dalen, J., Eds., Vol. 3, *The Archaeology of Mediterranean Landscapes*, Oxbow Books, Oxford, U.K., 1999, pp. 13–22.

3

One Step Beyond: Adaptive Sampling and
Analysis Techniques to Increase the Value
of Predictive Models*

Konnie L. Wescott

CONTENTS

ABSTRACT   A critical part of the planning process in developing an archaeological predictive model is defining the strategy for testing and refining
model results. Ideally, predictive modeling is a dynamic process that provides more than just a static map that is outdated as soon as new data become
available. The issue of uncertainty in the accuracy of available information
affects confidence in the results of any predictive model. Under certain
conditions, the resulting model may be too conservative to be useful for even
basic planning applications. An assessment of data uncertainty and a method
for developing a data-acquisition and -testing strategy to reduce that uncertainty can produce a more reliable and cost-effective model. Use of an iterative modeling and testing regime that is based on an adaptive sampling
approach is a possible solution. Bayesian techniques and adaptive sampling
methods have been used successfully to support the characterization and
remediation of hazardous-waste sites, and these have resulted in significant

* Work supported by U.S. Department of Energy under contract W-31-109-Eng-38.

cost savings in the cleanup of those sites. A similar approach may prove beneficial for determining the likelihood that archaeological sites are present in a given area. The ability to optimize sampling efforts to reduce the uncertainty will provide managers and planners with a higher degree of confidence in the model's accuracy, especially as new data are received and additional iterations of the model are run.

Preface

The conference held at Argonne National Laboratory (Argonne) was a unique opportunity to share some ideas about geographical information systems (GIS) and predictive modeling in a relaxed and collegial, but extremely focused, setting. For me, it was an opportunity to express some thoughts I had about assessing the varying levels of confidence in predictive model results and tie those ideas to some innovative work that was ongoing at Argonne using adaptive sampling methods to characterize hazardous-waste sites. I was not exactly sure how the two ideas could be compatible, but the underlying concepts seemed worth investigating and sharing, if for no other reason than to obtain feedback from the many experts attending the conference. In writing this chapter, I continued my struggle to link the concepts proven effective in the context of hazardous-waste characterization and remediation with the practice of archaeological predictive modeling, particularly in a cultural resources-management context. Although I have not had the resources available to me to test the concepts in the field, I believe they are worth adding to the discussion on archaeological modeling brought forth in these proceedings.

3.1   Introduction: Current Use and Value of Predictive Models

The value of using GIS-based predictive models for efficient land management and planning and cultural resources protection and stewardship is apparent in many contexts. Two specific examples follow based on U.S. regulatory requirements. The first example looks at some specific goals that a federal manager may have while embarking on an environmental assessment process for a proposed construction project. The second example follows the inventory and evaluation requirements of the National Historic Preservation Act (NHPA), which are better suited to long-term facility management approaches, rather than project management and administrative goals for a specific proposed project.

An environmental management goal of a federal project manager operating within the National Environmental Policy Act might be to minimize or prevent impacts to significant natural and cultural resources while achieving more specific project goals (e.g., construction of a scientific facility). An administrative goal of the same manager might be to keep project costs to a minimum. A specific cultural resources goal in this particular situation might then be to protect archaeological sites, so that they remain intact for future generations, while still meeting facility management and administrative goals. This cultural resources goal can be met by limiting (not necessarily eliminating) an intrusive survey and its associated costs (e.g., if a previously surveyed location meets project needs) and avoiding known and unknown/unrecorded archaeological sites to the extent feasible. The knowledge base to achieve this goal is typically derived from compliance-based survey projects with accelerated timelines and limited financial resources. Modeling is one logical approach for streamlining the planning and management process with regard to the presence or absence of archaeological sites and maximizing the value of the available data. This, in turn, provides a structure for maximizing the value of data collected in the future.

In addition to project-specific requirements for cultural resource compliance, as illustrated in the previous example, U.S. federal managers are also tasked under the NHPA to inventory and evaluate all cultural resources under federal jurisdiction. This task is acknowledged to be nearly impossible using traditional inventory methods, given the vast amount of federal acreage and the manpower and time required to survey that land and record and evaluate sites. Costs simply become too exorbitant to accomplish the task. However, with a modeling structure to help guide future inventory events, the task of expanding the knowledge base of the federal land unit can become less daunting.

The purpose of this chapter is not to discuss the nuts and bolts of predictive modeling, as modeling remains a popular topic in the literature and likely will continue to be a source of debate and practice. Rather, this chapter is reflective, intended to offer a possible approach for land managers to apply in order to prioritize their cultural resources management (CRM) activities, to be effective stewards of cultural resources under their jurisdiction, and to maximize their limited financial resources by enhancing both the data value of their surveys and the value of their predictive models that are either already in existence or are in various stages of development.

There are property types (federal facilities or land-management units) for which GIS-based predictive models may not be necessary: small facilities, facilities with extensive survey data already accumulated, and facilities with seasoned archaeological managers who know more about the area than can be communicated electronically (however, keep in mind that institutional knowledge has been known to leave). In these instances, priorities for additional sampling and data collection likely can be made easily and intuitively without a computer-generated model. However, for large land holdings, even skilled archaeologists may find GIS-based models useful and cost-effective for not just managing cultural resources under their jurisdiction, but for communicating their needs and management/compliance goals to nonarchaeologists. On the other hand, there are many instances in which the managers responsible for meeting stewardship and compliance requirements regarding cultural resources may not have a background or an education in archaeology. In the latter context, predictive models and a decision-support approach for maximizing the value of model output can be of great benefit, as long as qualified archaeological guidance is being provided.

3.2 Testing and Improving Predictive Models

One tangible product of a predictive model is a static map that illustrates the potential (typically some variation of high, medium, and low) of encountering sensitive cultural resources within a given area. These maps can be very useful for facility planning and management, as well as for assessment activities within a given planning horizon. However, these maps provide a "snapshot" that is based only on data available at a particular point in time. The predictive maps can become outdated very quickly, especially if the facility engages in an active survey or data-collection program. The models could require frequent updating or revision to incorporate new archaeological information, as well as results from other environmental updates to the facility GIS (e.g., updates from soil samples, geologic characterizations, data from new construction projects, etc.).

In current practice, archaeological predictive models are tested and improved through the use of traditional sampling methods (e.g., pedestrian surveys, shovel tests). A CRM manager is typically not able to dictate where surveys will be completed or how the data will be used beyond the immediate need, as survey activities are often project-driven for only compliance purposes. Recognizing that there may be administrative barriers, attempts should be made to actively use the newly acquired survey data to validate and test previous model results. In situations where staffing and financial resources are available, or can be made available, to more proactively manage and protect cultural resources, and thereby direct inventory efforts, alternative approaches may be employed for selectively sampling those areas with the greatest potential for refining an existing predictive model.

Other issues also may affect the need for model revision. Significant data gaps may have been present during model development, or it may be known or discovered that some of the data collected were suspect or biased in some fashion. Alternatively, the data used in the original model may have represented current environmental conditions rather than the paleoenvironmental conditions present when the site locations were first utilized, resulting in the possible need for a different model. New models could also be developed, if needed, to accommodate technological advances and new ways of looking at archaeological or environmental data.

Modeling is imprecise by nature, and the results are fraught with varying levels of uncertainty. Understanding the uncertainty and how it affects day-to-day operations is a key component to using a model successfully. Once a model is no longer fulfilling its intended purpose (i.e., it loses its predictive power and levels of uncertainty become too great for the model to be useful), it becomes necessary to be able to adapt to the changing conditions brought forth by new data and possibly employ new approaches. As stated above, selectively sampling those areas with the greatest potential for refining an existing predictive model (or for proving it is inadequate) is one way to reduce the uncertainty in the model. The remainder of this chapter will focus on this point.

Ideally, the modeling process should be dynamic and could even be automatic as new data become available. Each archaeological survey, each soil sample, or each monitoring event should trigger an update or refinement to the model. An adaptive system like this would assist the facility manager or the cultural resource manager in applying the most current information to projects by providing an opportunity for reinterpreting the model results, determining and quantifying uncertainty in the results, and evaluating the latest survey and testing strategies. However, to automate this process, a larger integrated data-management and decision-support system would be needed. The dynamic archaeological model could then be interfaced with data updates as sampling progressed. Ultimately, the decision-support system could assist the user in increasing the value of the predictive model by determining the optimal locations to survey on a priority basis, thereby reducing data uncertainty and increasing confidence in the model results.

3.3   Quantifying Uncertainty and Bayesian Statistics

Assuming a model has been developed for a particular facility that accommodates the available data, a three-dimensional cost surface can be produced to indicate levels of confidence (or certainty) in the predictions. For example, areas previously surveyed and known to contain sites are indicated as areas of high probability and high confidence (low uncertainty). Areas previously surveyed and known not to contain sites are indicated as low probability and high confidence. Areas not surveyed are modeled as high or low probability and low confidence (high uncertainty). Varying degrees of confidence are possible, depending on the methods employed. The confidence in surveyed areas with no sites may vary based on the possibility of buried remains not discovered during the initial survey. Confidence in unsurveyed areas may vary on the basis of other factors, such as suitability of the location based on environmental factors, expert knowledge of site distributions in the area, autocorrelation or proximity to high- (or low-) confidence areas, and physical clues from maps or remote sensing. Such "soft" types of data can be accommodated in a statistical framework following a Bayesian approach.

The Bayesian statistical approach is a potentially promising method for addressing uncertainty in a modeling context. As part of a dynamic modeling process, it allows combining data with existing knowledge and expertise (prior beliefs). It is the one statistical method that allows you to explicitly state what elements of the model are subjective, and it allows you to change your beliefs on the basis of new evidence. There are obviously pros and cons to this type of approach, and arguments can and will be made on either side of the debate. For example, expert knowledge will vary by expert, and who you choose to believe will affect the outcome. However, by making these subjective choices explicit in the model, there could be an advantage over classical statistics in how the output is interpreted by others (i.e., it may be less likely to be misinterpreted). To date, the most accepted use of Bayesian statistics in archaeology is in calibrating and interpreting scientific dating determinations; however, it appears to be getting some attention in other areas of archaeology, such as for predictive modeling (Millard 2003; Orton 2003; Van Leusen 2002). Further discussion of using Bayesian statistics for developing archaeological predictive models can also be found in Chapter 9 (Verhagen, this volume).

Given the model and the ability to apply expert knowledge through Bayesian statistics to quantify levels of uncertainty, such as for producing a cost surface as described above, the next step is to develop decision rules that will guide a priority-based sampling strategy for collecting data to maximize the value of the model's output (reduce the level of uncertainty).

Scientists working on the characterization and remediation of hazardous-waste sites have developed one possible approach using adaptive sampling methods and Bayesian statistics that may be of value in archaeological contexts (depending on the appropriateness and applicability of geostatistical models to specific project goals). More discussion on this possible application follows.

3.4   Adaptive Sampling and Analysis Programs

Although hazardous-waste characterization and archaeological research may vary greatly in their ultimate goals, the two activities share a need for obtaining accurate field data as efficiently as possible. Within the field of hazardous-waste characterization and remediation, a possible alternative approach to traditional field sampling has been developed for the U.S. Department of Energy (DOE 2001). The Adaptive Sampling and Analysis Programs (ASAPs) were developed to solve a particular problem related to the expense (in money and time) of site characterization. The traditional process was to prepare a detailed work plan for a sampling program, deploy technicians to collect a prespecified number of samples from specific, gridded locations, and send the samples off-site for analysis. On the basis of the off-site results, it would be determined if another work plan needed to be developed and field technicians redeployed to start the same process over again to resolve inevitable uncertainties in the data. To avoid some of the risk of needing to repeat the process, there has been a tendency to oversample an area to make sure it is adequately covered. However, this approach is still costly and does not always eliminate the need to resample if data anomalies are present. ASAPs have proven to be a good solution to this problem. They rely on a dynamic work plan that incorporates the new sample data with available GIS data and modeling techniques that allow the sampling team to evaluate the data in the field and provide a basis for field-level decision support for an "on the fly" sampling program. (Since the time of my conference presentation, the Environmental Protection Agency has introduced the "triad" approach, which applies similar concepts for streamlining data collection and addressing decision uncertainty; see Crumbling et al. 2001 and http://www.epa.gov/tio/triad.)

The ASAPs approach uses a GIS-based system with components for data integration and decision support. The data-integration software (e.g., Arc-View® GIS) is used to integrate, manage, and display real-time data for site characterization as the data are collected. A decision-support software tool, named Plume™, was developed at Argonne specifically for characterizing hazardous-waste sites by reducing uncertainty in the sampling collection process. Plume collectively accounts for "soft" site data (historical data, maps and photographs, past experience, model results) and "hard" in-field sample results to determine where the next samples should be taken. Plume incorporates analyst-developed prior probabilities of threshold concentration levels of the contaminant (generally on the basis of "soft" data) and advanced Bayesian and indicator geostatistical techniques to provide an image of the known contamination based on the samples taken, quantitative measures of the inherent uncertainty and the benefits of additional sampling, and the new sampling locations that will provide the most value in reducing uncertainty.

The advantages of the ASAPs over more traditional site-characterization methods are that fewer samples and sampling events are needed, which reduces overall project cost and the potential for worker exposure to contaminants. ASAPs result in better characterization because of the ability to rapidly visualize data as it is generated along with the ability to evaluate the value of additional data collection in the field (DOE 2001).

3.5   ASAPS Approach to Archaeological Predictive Modeling

Despite some obvious differences between hazardous-waste sites and archaeological sites, the concept of using Bayesian statistics and an adaptive sampling program is interesting and holds promise for application in archaeology. The following is an example of how an approach similar to ASAPs might work in an archaeological context for testing an existing predictive model.

Using a Bayesian approach, a new, more adaptive model can be developed that takes into account "soft" and "hard" data. Site presence is given an indicator value of 1, and site absence is given an indicator value of 0. These values are only achieved if the area has been adequately surveyed and either a site was recorded or no site was found. All unsurveyed areas would be given an initial indicator value of 0.5, as there would be equal opportunity for site presence or absence without the benefit of applying any additional information. However, because additional information is available for some locations, it can be used to change the 0.5 values toward either 1 or 0. The existing predictive model can be applied, although the values may only change slightly in the absence of rigorous model testing. For example, an area designated as having a high potential for containing a site might be assigned a value of 0.6 or 0.7 just based on the initial model. However, for that same area, on the basis of expert opinion or possibly an anecdotal recollection of artifacts being present in that vicinity, the assigned value of that area could be increased to 0.8 or 0.9, meaning there is much higher confidence that a site exists in that location. In locations where it is possible that deeply buried sites may be present but were not discovered during initial survey, the level of confidence in the result of site absence may not be as high, and the initial value of that area for site absence might change from 0 to possibly 0.15 because of the potential for deeply buried remains. As this cost-surface model is produced across the entire area, the goal of future surveys (whether they be directed or simply compliance-based) is to change as many of the intermediate values as possible (i.e., values within the range of 0.4 to 0.6) to values closer to 0 or 1. Areas for which there are higher levels of confidence (e.g., >0.8 for site presence and <0.2 for site absence) would not be prime candidates for directed sampling for planning purposes. However, these areas would be surveyed if selected for future ground-disturbing activities and the results of the survey fed back into the system.

As new data become available, the indicator values are allowed to change in response to that new data to improve the model results. As more 1s and 0s are entered and confidence builds in other areas, the initial model should be reevaluated, adjusted, and reapplied to change some of the underlying intermediate values, as appropriate. Future sampling efforts may provide sufficient validation of the initial model that increases confidence in the model output and allows additional adjustments in favor of the model output (e.g., areas previously designated 0.6 on the basis of the high-potential designations of the model could become 0.7 because there is increased confidence in the model results). Alternatively, the initial model may be determined ineffective, and indicator values based on that model may need to return to a value of 0.5.

As far as optimizing locations to sample next, the hazardous-waste site example uses Bayesian and geostatistical procedures to develop the decision rules about where to sample (e.g., rules to determine if the area is clean or contaminated). Kvamme (Chapter 1, this volume) makes reference to the potential for geostatistics and autocorrelation in archaeological predictive modeling, but the particulars on when and how to apply these approaches were not discussed. Particulars of the sampling program are likely to be model- and location-specific; therefore, there is little more for me to add to this discussion. Although I had hoped to actually demonstrate the utility of this approach with real data and a sampling program (possibly something similar to Plume) designed specifically for archaeology, support for these specific efforts has not yet been secured. However, I believe the conclusions, although reflective rather than substantive, are still relevant for the discussions on predictive modeling presented in this book. Ultimately, the goal of the sampling program is to reach an acceptable level of confidence to achieve a "statistical maximum" (Van Leusen 2002) for predicting and observing the same thing (either a site or no site) and to minimize occurrences of predicting and observing different things (predicting no site, but observing a site, and predicting a site, but observing no site). As stated by Van Leusen (2002), efforts for archaeological resource management seem to focus on instances where site absence is predicted but sites are actually present, rather than on the full set of model outcomes. This is mostly the case because it has costly consequences. However, to increase the accuracy and efficiency of decisions based on a particular model, one must have confidence that the entire system is operating correctly.

3.6 Conclusions

A clear statement of the specific context in which a model is developed and of the intended goals of a model is extremely important for evaluating model effectiveness. In a facility-planning and cultural resources management context, the goals might be to limit intrusive surveys and avoid sites to the extent feasible. The most efficient modeling for these purposes is a dynamic process that incorporates a sampling strategy to continually refine model results. Uncertainties in the accuracy of the data could be addressed within the sampling strategy. Bayesian statistics and an adaptive sampling approach have the potential to provide a minimally intrusive and cost-effective strategy for improving model results.

The general concept that expert knowledge can be used to generate decision rules to negotiate an adaptive sampling program is something that should be analyzed further. An adaptive sampling strategy and a decision-support system could streamline CRM activities to help land managers avoid planning development projects in areas of high potential with a high degree of confidence in the predictions, thereby avoiding many of the associated costs that accompany those areas in terms of required surveys and the possible excavations of sites. The most helpful information would be in continually reevaluating the level of confidence in the results of an existing predictive model that has been updated to reflect new information. The manager then could apply the logic that high confidence in high probability means potential high cost. Conversely, high confidence in a low-probability area means the likelihood that costs would not exceed those for a traditional survey, with the expectation that no significant sites would be encountered. Increases in uncertainty regarding site presence on either end of the scale would result in increases in uncertainty regarding the total cost of the project.

Using adaptive sampling and a dynamic modeling process increases management efficiency in planning and decision support by incorporating the most up-to-date information and the most accurate-to-date model results while reducing the amount of uncertainty and increasing the confidence level in the predictions.

References

Crumbling, D.M., Groenjes, C., Lesnik, B., Lynch, K., Shockley, J., Van Ee, J., Howe, R., Keith, L., and McKenna, J., Managing uncertainty in environmental decisions, *Environmental Science and Technology*, 35, 19, 404a–409a, October 2001.

DOE (U.S. Department of Energy), Adaptive Sampling and Analysis Programs (ASAPs), Innovative Technology Summary Report, DOE/EM-0592, U.S. Government Printing Office, Washington, D.C., August 2001.

Millard, A., What Can Bayesian Statistics Do for Archaeological Predictive Models? position paper for the Expert Meeting on Predictive Modeling Techniques, Amersfoort, Netherlands, May 22–23, 2003.

Orton, C., The Fourth Umpire: Risk in Archaeology (inaugural lecture), 2003; available on-line at http://www.ucl.ac.uk/archaeology/special/orton-inaugural-2003, accessed 17 September 2004.

Van Leusen, M., Patterns to Process: Methodological Investigations into the Formation and Interpretation of Spatial Patterns in Archaeological Landscapes, Ph.D. thesis, Rijksuniversiteit Groningen, Groningen, Netherlands, May 2002.

Section 3:

Issues of Scale

4

Modeling for Management in a Compliance World

Christopher D. Dore and LuAnn Wandsnider

CONTENTS

ABSTRACT In practice, compliance-driven cultural resource "management" and its requirements for resource location, evaluation, impact assessment, and mitigation manifests a fundamentally different use of geospatial predictive modeling than do research-oriented investigations. This difference primarily results from the lack of an iterative research design. In research-oriented modeling, iterations of model building and model testing gradually build a more robust model and lead to an increased understanding of the variables that condition human spatial behavior in the past. In a compliance environment, spatial models are rarely built and evaluated; rather, once built, they are applied in a single iteration. An assumption is made that the model being used will accurately predict behavior in space. Yet, in most settings, our knowledge of the factors that condition the spatial organization of activities — and under what conditions these factors are relevant — is just beginning to develop. Coupled with the methodological issues of sample size, changing environmental conditions, functional differences in resource types, the fact that most archaeological deposits represent depositional (as opposed to functional) sets that have accumulated over hundreds of years, spatial variability caused by nonenvironmental factors, etc., compliance modeling certainly does not represent best practice, even though it is legal under federal cultural resource law.

Rather than modeling the past, a more productive approach to modeling for cultural resource managers is to model the present. Instead of reacting to development and infrastructure projects that have taken the place of our stewardship responsibility, geospatial technologies can be used to design a proactive approach to resource management. With such an approach, present conditions, both natural and cultural, are modeled to predict site and feature visibility and to identify potential threats to surface sites and features. At a regional scale, the use of vegetation, slope, and sediment data can be used to develop erosion models for current and future conditions. Cultural resources can be compared with these models to categorize and prioritize the resources most at risk. At the scale of individual resources, aerial photography and new higher resolution satellite imagery can be used to establish the baseline condition of resources and, with follow-up visits, to establish and compare rates of change from erosion, all-terrain vehicles, and vandalism. At the intrasite scale, new processing techniques can be used with geophysical data to predict the nature of actual cultural features rather than identify data anomalies that then require excavation. These techniques will ultimately lead to absolute, rather than relative, signatures for properties of the archaeological record and provide a truly nondestructive archaeology. We illustrate this geospatial management framework with archaeological examples from western, southwestern, and midwestern North America.

4.1   Introduction

Although "cultural resource management" (CRM) is the term used to describe applied archaeology[1] within the United States, in fact, however, there is very little management of archaeological resources, at least in a stewardship context. Landholding federal agencies, while tasked with this responsibility under Section 110 of the National Historic Preservation Act (NHPA), Executive Order 11593, and others, are largely unable to meet this responsibility due to vast landholdings (especially in the western United States), numerous resources, small budgets, and the pragmatic priority of fulfilling compliance obligations such as those required by Section 106 of the NHPA. The Advisory Council on Historic Preservation (2001) recently stated,

"In spite of the important stewardship responsibility entrusted to Federal agencies for much of our Nation's heritage, other agency mission priorities often force historic preservation activities, programs, funding, and staffing to take a back seat."

Compliance with Section 106 requires that federal agencies take into account the effects of undertakings on historic properties and that the Advisory Council on Historic Preservation (Advisory Council) be given the opportunity to comment. In practice, four steps are usually taken to fulfill compliance responsibilities with Section 106 and other environmental laws, and thereby "manage" cultural resources: identification of resources, evaluation of resources, assessment of the effects of a project on significant resources, and an identification of ways to lessen effects that are deemed adverse.

Predictive modeling, done both within and outside of geographic information systems (GIS), has long been a part of federal cultural resources compliance (e.g., Ambler 1984). When modeling is implemented in the compliance process, it is almost exclusively used in the resource-identification phase. Driven by the high cost of systematic field surface surveys, federal agencies and nonagency project proponents have searched for ways to reduce the costs of resource identification. Sample surveys have been the cost-saving strategy of choice, and predictive modeling, sometimes less formally called sensitivity analysis, is the method most often utilized to spatially define sampling strata.

From a legal perspective, there is no mandate to comprehensively survey a project area, called the area of potential effects (APE). Likewise, when undertaking a compliance investigation, there exists no requirement necessitating that all resources be found within the APE. The legal burden is that a reasonable and good-faith effort be made to identify resources within the project area (36 CFR 800.4(b)(1)). Using predictive modeling to identify areas most likely to contain resources is not only allowable (U.S. Secretary of Interior 1983), it has been informally advocated by the Advisory Council (McCulloch 1999). As we will outline below, while legal for compliance purposes, we believe that the use of predictive modeling within a compliance framework is not best practice and actually perpetuates stagnation in our understanding of past human land use.

Most predictive modeling to identify resources is not best practice for a variety of reasons discussed below. Insofar as this is true, that aspect of "cultural resource management" encompassed by "identification" is similarly challenged. But, with the emergent perspective of landscape management coupled with widely available geospatial technologies (emphasized here), management in general and, especially, two other common compliance activities — assessment of project effects and ways to lessen adverse effects — become approachable. In what follows, we identify some of the problems with modern modeling applications in compliance-driven cultural resource management, concluding with examples of the application of geospatial modeling to stewardship-oriented management of cultural resources at a variety of scales. Through these efforts, we aim to put the "M" back in CRM.

4.2   Predictive Modeling and Compliance

Critiques and discussions of the methods involved in predictive modeling and sampling have permeated our professional literature over the last 30 years. Many of the issues we identify here have been outlined by others, including Kohler and Parker (1986), Kvamme (1989, 1990), contributors to Judge and Sebastian (1988), Church et al. (2000), and Ebert (2000). While, collectively, we are well-informed about the theoretical and methodological issues in modeling, this knowledge rarely seems to be considered in the design and application of models in the compliance community. With geographic information system software on the desktop of most agency cultural resource managers and cultural resource consultants, and with pressure to lessen the cost of compliance, the lure of technology has made predictive modeling vogue in the compliance world. Unfortunately, many of these modeling efforts have been flawed by methodological and application mistakes. Given these oversights, we feel that it is beneficial to briefly restate these issues with an emphasis on compliance applications, focusing especially on model building, model testing, and the theoretical issues that underlie each of these tasks.

4.2.1   Model Building

Archaeologists (Altschul 1988; Ebert and Kohler 1988) often distinguish between inductively and deductively derived models. No matter the mode of model building, decisions about data inclusion and data quality affect model performance. The appropriateness of the environmental base data used to build a model is rarely scrutinized sufficiently. Research projects may factor in data adequacy as a prerequisite to selection of a study area or incorporate building environmental data sets into the research program, but compliance investigations rarely have this luxury. The project area for a compliance project has been selected *a priori* by the nature of the undertaking, and investigators have little choice in the availability of environmental base data. Custom-designed data sets are virtually never created due to limitations in project schedules and budgets. Base data for model building in a compliance investigation almost always means using "off the shelf" data, usually from the United States Geological Survey (USGS), one of the private firms in the new value-added spatial data industry, or clients. Insufficient time is spent evaluating the metadata and asking if these data are appropriate at all for the scale of human landscape utilization of interest. Quite apart from issues of data scale, resolution, and algorithms used to create data sets (e.g., Hageman and Bennett 2000; Kvamme 1990), the actual accuracy, error, and precision of these data as expressed in the National Map Accuracy Standards is in fact too low to support high-resolution modeling efforts attempted in compliance exercises (Marozas and Zack 1990).

Another problem arises from the oversimplification of the natural environment as related to human land use (Church et al. 2000; Wescott and Kuiper 2000). For example, the distance to nearest water is frequently used as source data for models, but rarely do model builders consider the type of water. Is the modeled water snow, a stream, lake, spring, or ocean? If a body of water, is it brackish or fresh? If a stream, is it annual or perennial? Is it habitat for anadromous fish or other resources? These types of distinctions have very different ramifications for how people use the natural landscape.

Similar problems exist with the archaeological data used to establish the correlations. These data sets usually come from the records of the landholding federal agency. The geospatial controls for the spatial component of these data come from a wide range of sources, and accuracy metadata often do not even exist. In determining the accuracy of the Nebraska statewide archaeological database, for instance, we found that the archaeological resource database data error ranged dramatically over an order of magnitude in the hundreds of meters (Wandsnider and Dore 1995). To ensure at a 90% confidence interval that a site was actually located where records claimed it was, sites recorded with universal transverse Mercator (UTM) coordinates had to be buffered by 353 m, and sites recorded by legal description needed a 1000-m buffer for the same accuracy confidence — and this was after discarding sites with larger errors clearly originating from coding and data entry.

Additional problems, besides those of spatial accuracy, also exist with the archaeological component of modeling data. Many times, the number of available sites used to build an inductive model is insufficient to draw statistically meaningful correlations between resources and landscape features. This is particularly problematic in compliance investigations where project areas can be quite small and good spatial data sets in adjacent areas are lacking. Likewise, the functional class of archaeological sites is too often ignored. That is, sites are treated as unifunctional; the investigator fails to consider that habitation sites, processing sites, quarry sites, etc. are located on the landscape using different, and sometimes contradictory, criteria (but see Hasenstab and Resnick 1990; Savage 1990; Wescott and Kuiper 2000). Further, temporal distinctions are often slighted, especially beyond the simple historic/prehistoric division (but see Altschul 1990). These oversights exist even though, after doing archaeology for over 100 years, we have learned that human land use did change with time in response to social, economic, and environmental dynamics. Unfortunately, when a savvy model builder does in fact discriminate along temporal and functional dimensions, the sample size within each class can be reduced to meaningless levels, making a bad situation worse.

Finally, most archaeological sites that are known, and that exist in spatial databases for use by model builders, are sites discovered through surface survey. While this is less problematic in some portions of the desert west where 10,000 years of human land use is visible on the surface, in most places only a fraction of the resources have surface signatures. Thus, in most cases, we can only state where sites can be found, not where sites are not found, and models built upon these data best predict the *visibility* of a resource on the ground surface as opposed to the actual presence of a resource, whether surface or subsurface. (See discussion in Warren and Asch [2000: 27–28] and Cashmere and Wandsnider [1995] for explicit attempts to model surface visibility.)

   Combining environmental and archaeological data sets presents problems of its own. Do the two data sets even belong together? How representative is the environmental data of the landscape that existed when locations and landscapes were utilized (Church et al. 2000)? As previously mentioned, most data from compliance projects is off-the-shelf data, and almost all of these data are from the post-Landsat era (post-1972). These data may or may not be appropriate for modeling depending upon the degree of environmental change that has taken place. From a compliance perspective, attempting to draw correlations between the modern environment and the locations of archaeological sites is desirable. As archaeologists and scientists, however, what we really want to understand is how people interacted with past environments. Further, the correlations that may be established between the present environment and archaeological resources may be "false" correlations that may really be showing areas where past and present landscapes correspond (Duncan and Beckman 2000: 55). A second concern when combining environmental and archaeological data sets arises from stacking, or the vertical layering, of data sets. As Marozas and Zack (1990) have pointed out, the overall horizontal error is additive: the error of each layer is added together to produce composite error. Given the accuracy of individual data layers and their degree of heterogeneity, the error can quickly affect any possible associations produced by the model. This problem can be quite substantial if, for example, the accuracy figures we calculated for the Nebraska data set are representative of other archaeological data sets. This is unfortunately a likely scenario.

4.2.2   Model Testing

The U.S. Secretary of the Interior's Standards for Identification (1983) state that the accuracy of the model must be verified and that predictions should be confirmed through field testing. If necessary, the model must be redesigned and retested. Such actions, however, are virtually never taken within compliance investigations. The common scenario is that a model is built based upon resources in surveyed portions of the APE or upon surveyed areas in the general region. This model is then applied to unsurveyed portions of the APE to stratify the APE into areas likely to contain resources, as well as areas unlikely to contain resources. Field surveys are then conducted in these areas to find resources. In the worst cases, field surveys are only conducted in high-probability areas. In better-quality compliance investigations, sample surveys are conducted in all stratified areas to actually test the predictive power of the model. Even in compliance investigations that conduct field surveys in all stratified areas, a common methodological error is that areas of high site probability are surveyed more intensively than areas of lower site probability, and resource totals are not adjusted to reflect the search intensity. The result is that field surveys are self-fulfilling and almost always confirm the model; more sites are found in higher probability areas than are found in lower probability areas (but see Dalla Bona 2000).

One of the reasons that this methodological error is ignored is the disjuncture between the paradigm and units employed in a compliance investigation versus those in predictive modeling. In a compliance investigation, the tangible data unit, as defined by law, is a building, structure, object, or site.[2] In contrast, the meaningful unit in a predictive model is a region or land parcel: an area within which there exists a probability for finding a building, structure, object, or site (Kvamme 1988, 1989). The priority in a field survey of a modeled probability area for a compliance investigation is not to evaluate the probability; it is to find sites. When sites are found, further work is spent evaluating the resource for its significance and assessing the effects of the project on the resource rather than closing the iterative loop by reassessing the model.

Under this scenario for the application of a predictive model, there is a single iteration. A model is developed and applied in an attempt to limit the amount of field survey that must be done to identify archaeological resources. This kind of modeling is problematic for two reasons.

First, the model is not tested; it is applied. In doing this, an assumption is made that an adequate understanding of the factors that condition human land use exists for the APE. Although an argument can be made that the role of the compliance archaeologist is not to build theory but, rather, to apply theory constructed by research-oriented academic colleagues, it is clear that we are only beginning to understand the variables at play in conditioning human land use. Because a compliance investigation most often will result in the damage or destruction of archaeological resources from either archaeological excavation or the construction of the project, is it wise to use predictive modeling in this way? We believe not.

Second, one of the criteria for evaluating a resource for its eligibility to be listed in the National Register of Historic Places, Criterion D, is the resource's ability to have yielded, or its likely ability to yield, information important in prehistory or history. The degree to which a resource meets this criterion is inversely related to the resource's predictability in a predictive model. For example, if a resource is found in a location specified by a model, the factors conditioning the resource's placement on the landscape are understood. Therefore, it has less potential to provide data about the past, at least from a land-use context. Alternatively, a resource that is found where it is not predicted has great potential to provide information important in prehistory or history because of the fact that it was found where it was predicted not to be (Altschul 1990). This is one of the reasons why the methodological error of surveying less intensively, or even not at all, in low-probability areas is of concern.

4.2.3 Theoretical Issues

In addition to the problems we have pointed out in the areas of model building and model testing, there are some additional theoretical issues of predictive modeling that are worth mentioning briefly. First, most models assume that the selection and utilization of a place on a landscape is based upon environmental criteria. While environmental criteria are important for the location and performance of many activities, it is erroneous to build site-location models on these criteria alone, or at least for all activities. While cognitive and other perceptual criteria can and have been incorporated into models, working with nonenvironmental variables is not widely done in North America, although this has been explored extensively in Europe (contributors to Lock and Stančič 1995; Gaffney et al. 1996).

Second, the emphasis in archaeological predictive modeling is on sites normally assumed to be residential settlements and special-use locations (quarries, rock art, etc.). Two problems follow from this practice. Low-density archaeological remains are rarely considered. While it is not useful to revisit the site–nonsite debate here, suffice it to note that the nonsite approach has merit as a framework for understanding human land use even though this framework is not usually used in predictive models. The primary reason that nonsite data are not used is because of the paucity of available nonsite spatial data sets. Even if such data sets existed, within a compliance context, isolated or low-density evidence of human land use is routinely held to lack significance by the very nature of its being isolated or low density and is therefore slated for dismissal. Yet, the low-density archaeological record comprises substantially high numbers of discarded tools, usually taken to be great sources of information on past place use (Wandsnider 1988).

More critically, however, "sites as settlements" denies the temporal and taphonomic (Dunnell 1992; Kelly 1988) nature of site archaeological deposits. That is, when we find Nebraska-phase ceramics at a particular location, what settlement temporality can we infer for that location? A season? Many seasons? Extended or intermittent occupation over many years? Many decades? A constellation of other information — the presence/absence of structures, middens, and so forth — are commonly employed to "temporalize" settlement assessments. But this temporal information, beyond coarse chronology (i.e., "Central Plains Tradition settlement") is not commonly incorporated in settlement-modeling attempts. Yet, long-term Central Plains occupation and reoccupation is a very different kind of place use than brief, nonrecurring occupation. It may be that we must wait for the development of accessible temporal GIS (TGIS; Langran 1992) to fully deal with the temporal and taphonomic variation that our archaeological site deposits actually contain.

Third, correlation is not explanation. Correlating variables in a predictive model may establish relationships among data, but it does not, by itself, explain the dynamics of human land use (Church et al. 2000). What we really want to understand are the "whys" that led to the performance of different sets of activities at different places at different times. How are places on a landscape linked together through human organizational systems? Additional theoretical constructs and bridging arguments must be used to supplement the correlation of landscape features to provide explanation.

Fourth, in a compliance context, the current application and use of predictive modeling actually leads to a stagnation of our understanding about the past. This is due to the lack of model building, model testing, and model refinement iterations. When models are only created and applied, nothing new about the past is learned. The current state of knowledge about land use is quantified into a model, and then fieldwork, because of some of the application problems we have noted, usually confirms the model. Sites in low-probability areas, the ones that have the highest potential to be significant to our understanding about the past, but that are usually not found, are destroyed by the project that is undertaken. Thus, we rarely learn anything new and essentially continue to build the same model from project to project (Ebert and Kohler 1988; Ebert 2000).

4.2.4 Summary

In the preceding, we have criticized the use of predictive modeling in compliance investigations by pointing out many of the problems in model building, application, and theory. Nevertheless, we do not advocate discarding predictive modeling in archaeology. To the contrary, predictive models, both within and outside of a GIS environment, provide a very robust tool for understanding past human–land interactions. Within a research framework, when iterations of model building, testing, and refinement can be undertaken, this tool has been shown to advance our understanding of the past. In a compliance framework, however, where predictive modeling is characterized by a lack of iterations, we feel that predictive modeling serves neither the compliance process nor the advancement of knowledge about the past.

4.3 Managing with Geospatial Technologies

We believe that with a different orientation, predictive modeling can have a productive role in cultural resource management. As we noted at the beginning of this chapter, the management of archaeological resources has been forced to a low priority by many landholding federal agencies due to vast landholdings, numerous resources, small budgets, and the pragmatic priority of fulfilling compliance obligations such as those required by Section 106 of the NHPA. Although predictive modeling is largely unsuitable for the identification component of compliance, we believe that such models can be used to better purpose to put the "M" back in CRM.

To borrow from Judge and Sebastian (1988), who titled their publication *Quantifying the Present and Predicting the Past*, rather than using contemporary data to model the past, we propose a framework that consists of modeling the present and predicting the future. Using this framework avoids most of the methodological problems mentioned earlier and can easily and economically be implemented by federal cultural resource managers even with large land areas, small budgets, and little time. To illustrate this framework, we will present examples at the regional, site, and feature scales. All of these examples have in common the use of contemporary data about the archaeological record and natural environment to characterize the present and predict future conditions.

4.3.1   Regional Scale

Our first example comes from northwest Nebraska, on a portion of the Nebraska National Forest, and illustrates how the threat of natural erosion on archaeological resources can be assessed, predicted, and managed. In this example we have identified two of the major variables contributing to sediment erosion: steep slope and lack of vegetation cover. The principal variable, precipitation, can be assumed to be even over this region that covers 142 km². Another major variable, soil type, was not factored in even though these soil data were available. Lacking this data layer does not negate the results of our analysis, but using it would certainly have enhanced and refined the results. We did use off-the-shelf data for this analysis: a 7.5-min digital elevation model (DEM) from the USGS and a multispectral Landsat thematic mapper (TM) image (Figure 4.1).

To calculate the quantity of vegetation, we used the transformed vegetation index (TVI) on TM bands 3 (0.63–0.69 µm, red) and 4 (0.76–0.90 µm, near infrared). The TVI is one of several vegetation indices that can provide a rough, relative indication of the amount of vegetation. In this image (Figure 4.2), the quantity of vegetation is shown grading from none (white) to dense (black). Note that the northern portion of this area consists of agricultural fields crosscut by riparian corridors, while the southern portions are predominantly covered in pine forest. The DEM was used to compute the degree of slope (Figure 4.3). White indicates low slope; black indicates high slope. Then the inverse of the vegetation values was computed so that high values represent low vegetation. The slope and TVI values were then rescaled into the same 8-bit data space (256 distinctions). These two data sets were then added together to produce a numerical index representing the relative threat of erosion. As seen in Figure 4.4, the threat values grade from low (white) to high (black). Known archaeological sites were then added to the analysis and can now be ranked according to their potential for erosion.

A federal cultural resource manager, with little time to monitor sites and a small budget to spend on preservation, can use these results to predict which sites are at the greatest risk and where, perhaps, cattle grazing might

0   1   2 kilometers

Sparce Vegetation   Dense Vegetation

N

FIGURE 4.1
Transformed vegetation index for the 7.5-min study area calculated from Landsat TM data.

be reduced. Similar models can be constructed for looting, recreational damage, military training, etc. Scarce resources can then be spent most effectively
on the sites that really need the attention. This erosion model that we have
presented is, admittedly, simplistic and could certainly be refined by better
data assessment, more careful model building, ground truthing, and iterative
refinements. Our point, however, is that even these simplistic models — this
one completed in less than two hours — can offer the cultural resource
manager effective tools for proactively managing archaeological resources.

4.3.2   Site Scale

Similar techniques can be applied at the site scale to help the cultural resource
manager monitor the condition of sites. In the western United States, erosion,
vandalism, and recreational activities such as the use of all-terrain vehicles
(ATVs) can irreparably damage archaeological sites. At Vandenberg Air Force
Base in California, a systematic aerial monitoring program is being used to
maximize limited CRM resources. Cultural resource managers responsible
for large federal land parcels, although short on funds, often have access to
aircraft. Even "casual" aerial photography done out of the side of a plane

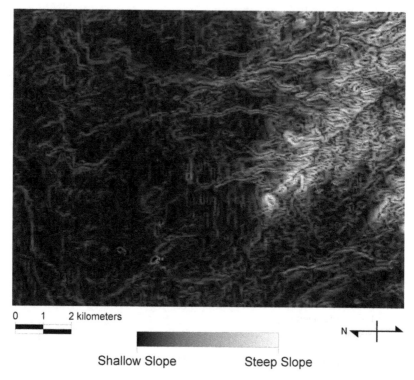

0   1   2 kilometers

Shallow Slope   Steep Slope   N

FIGURE 4.2
Slope model calculated from a USGS 30-m digital elevation model.

with a 35-mm or video camera can provide extremely valuable management results.

This example shows two images. The first was taken in 1997 (Figure 4.5), and the second was taken in 1998 (Figure 4.6). Note that the oblique angle, scale, and camera position are different in each image. Using image-analysis techniques, the two images can be placed in the same geometry. In this case, the 1997 image was transformed into the geometry of the 1998 image. This analysis was done relatively, but with ground-control points and absolute geographic coordinates obtained from the global positioning system, both images can be placed into geographic space (Figure 4.7).

Following the transformation, the limit of the bank erosion was marked for each year. With the limits of erosion identified, the lines are simply subtracted from each other, leaving polygons that represent the amount of the site lost to erosion (Figure 4.8). Because the time that elapsed between the two photographs is known, the rate of erosion can be determined. As in the previous example, this rate can then be compared with other sites in the area to determine the resources that are most at risk (Figure 4.9). With knowledge of rates of change, cultural resource managers are then in a position to predict future site damage and can direct resources appropriately.

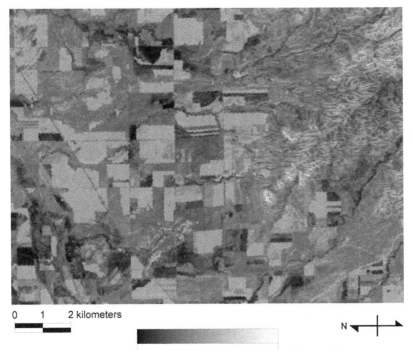

0 1 2 kilometers

Low Erosion Potential   High Erosion Potential

N

FIGURE 4.3
Model of erosion potential.

4.3.3 Feature Scale

Our last example is an intrasite example and is at the scale of the individual feature. This prototype study was completed for the City of Albuquerque and illustrates how a predictive model can work in the present. The city has purchased a prehistoric archaeological site to protect it from development. While the initial goal was to create an active archaeological park with ongoing excavations, Native American objections caused the city to reconsider their plans. Subsurface remote sensing was then proposed as a nondestructive option to map the architectural remains of the pueblo. However, because excavation could not be used to verify and identify geophysical anomalies, an alternative geophysical methodology needed to be developed. An additional problem that needed to be overcome on this project was that the architectural features of interest were unburned adobe. Adobe that is unburned does not usually have properties that make it readily distinguishable from the surrounding sediment matrix, at least in terms of most geophysical properties.

The key to developing our approach was the fact that the city's archaeologist had noticed that, under the right conditions, several wall segments

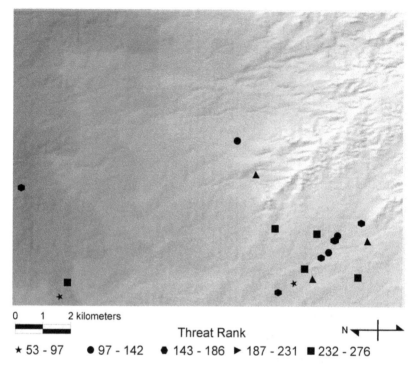

Threat Rank

★ 53 - 97   ● 97 - 142   ● 143 - 186   ► 187 - 231   ■ 232 - 276

FIGURE 4.4
Archaeological site-erosion threat assessment and ranking.

could occasionally be seen faintly exposed in the ground surface. Over a
period of several years, a number of wall segments were mapped to the
extent that both walls and room fill could be spatially defined over a small
area. With known features identifiable, we designed an approach based upon
multispectral satellite remote sensing using supervised classification. A similar approach using unsupervised classification had been used by Ladefoged
et al. (1995) in New Zealand. To cope with the unburned-adobe problem,
we decided to use three geophysical techniques to raise the discriminatory
potential above what any single method can achieve. We used magnetics
(gradiometer), resistance, and time-sliced radar data as the "spectral bands"
(Figure 4.10). Given the thickness of the known wall segments, about 20 cm,
particular attention was given to both the spatial resolution of data and the
spatial control of data. It was essential that any error in correspondence
between all data layers be less than half the wall thickness, about 10 cm.

The supervised classification method is essentially a model-building and
prediction technique. In the computer, classes of phenomena are identified
and marked on top of a stack of data layers. In this case, walls and room fill
were the two classes of interest (Figure 4.11). There are a variety of classification
algorithms that can be used to differentiate features. For this study, we used
the Mahalanobis classification algorithm, which is based upon neural-network
classification principals. Regardless of the particular algorithm, however, the

FIGURE 4.5
Oblique aerial photography used for monitoring and the 1998 aerial image. (Courtesy of Applied EarthWorks with support of Vandenberg Air Force Base. With permission.)

FIGURE 4.6
The 1997 aerial image placed in the geometry of the 1998 aerial image. (Courtesy of Applied EarthWorks with support of Vandenberg Air Force Base. With permission.)

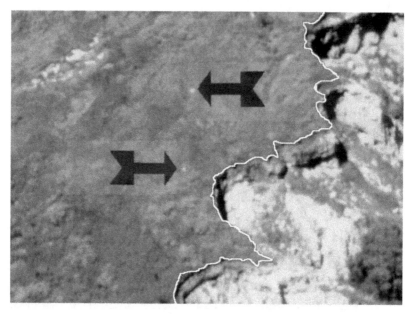

FIGURE 4.7
After georeferencing, the edge of the bank was defined in each image. (Courtesy of Applied EarthWorks with support of Vandenberg Air Force Base. With permission.)

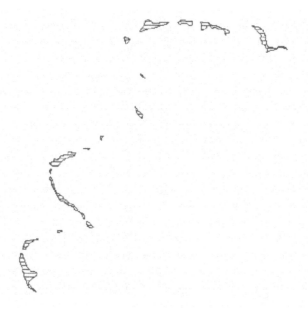

FIGURE 4.8
The lines defining the edge of the bank are subtracted from each other, leaving polygons that define the bank erosion that took place between the 1997 and 1998 photographs.

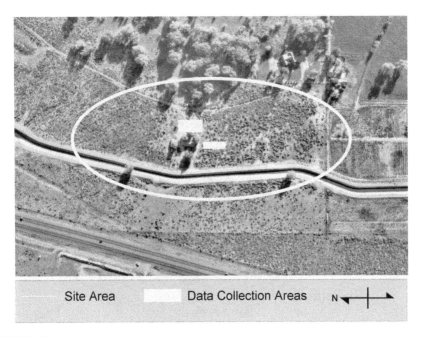

Site Area   Data Collection Areas   N

FIGURE 4.9
Aerial photograph showing the approximate site area and the areas of geophysical data collection.

strategy of each is identical: to examine the variability in the data for the known features, referred to as the training set, and develop mathematical criteria for distinguishing each feature from the others. These criteria form the predictive model. In a second phase of analysis, the model is applied to unknown areas of the data set and predicts, or classifies, data into the typology that was defined. In our example, this would be either adobe wall or room fill. In an ideal situation, of course, there would be iterations of prediction, testing through excavation, and model refinement, but in this case there is no immediate means of obtaining additional verification. The final step is to evaluate the classification results against the original training data (Figure 4.12). In this study, a 69.6% success rate was obtained, quite good given the nature of unburned adobe, a small sample size, and some problems with the radar data.

This technique illustrates one way in which predictive models can be used at the intrasite scale to manage resources in a nondestructive way. Additionally, it takes the important step of realizing the nondestructive potential of geophysics by beginning to develop absolute signatures for particular materials and feature types. Archaeological geophysics, at least as it is most commonly practiced, involves identifying an unknown anomaly that is then excavated to determine what it is. The geophysics technique may be nondestructive, but the application of the technique is no less destructive than traditional excavation without using remote sensing. We would hope that, in the future, a library of absolute signatures would exist

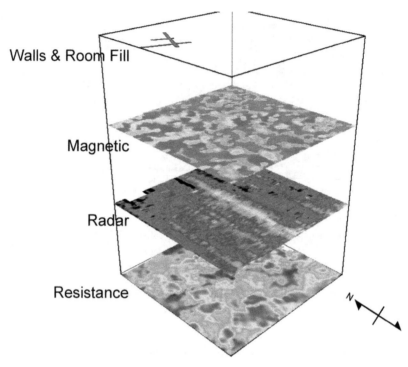

FIGURE 4.10
Different types of geophysical data are treated as if they were different bands of multispectral data. Using the known sample of walls and room fill as the training set, these geophysical data are then classified using a supervised classification technique.

for subsurface archaeological phenomena similar to those available for many plants, minerals, and sediment types on the ground surface (e.g., ASTER Spectral Library, Johns Hopkins University Spectral Library, NASA Jet Propulsion Laboratory Spectral Library, USGS Spectral Library [Clark et al. 1993]).

4.4 Summary

In this chapter, we have attempted two things. First, we have argued that, for many reasons, the use of predictive modeling in cultural resource compliance, at least as it is most frequently applied, is not best practice. As with any other method, we encourage our colleagues to critically evaluate the appropriateness of predictive modeling for each particular application and not to use the method when it is not warranted. We understand the desire to reduce field time and labor costs in the resource-identification

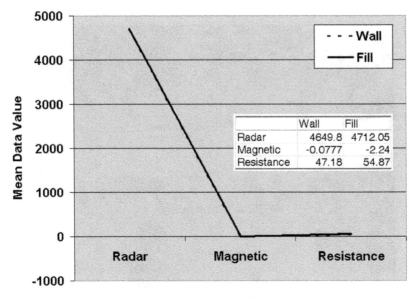

FIGURE 4.11
Graph showing the lack of difference between the mean data value for walls and room fill across three geophysical techniques. Thus, the potential for any one individual technique to discriminate between these two material classes is low.

phase, but we hope that our integrity as scientists and as stewards of the archaeological record supersedes the pragmatic realities of the business of compliance archaeology.

Second, we have tried to provide, through the examples presented here, a different perspective on predictive modeling in archaeology. In this framework, the present is not characterized to retrodict the past, but to predict the future and contingent state of extant resources. We believe that this framework can productively be used by cultural resource managers, even within their current constraints and compliance responsibilities, to regain their stewardship responsibilities by intelligently assessing, prioritizing, and responding to the needs of the resources they manage.

Notes

1. As well as a number of other applied disciplines, including architectural history, ethnology, history, etc.
2. While landscapes and districts do exist in the compliance world, these are actual entities as opposed to areas of probability.

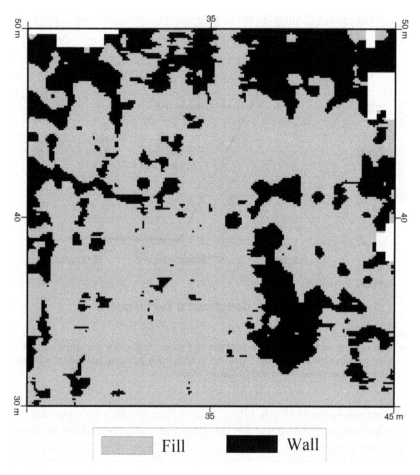

FIGURE 4.12
Results of supervised classification using the Mahalanobis classification algorithm. Evaluating the classification model against the original data places the accuracy of the model at 69.6%. Rectilinear room blocks can be seen in the left portion of the image; the right portion is an area of architectural collapse.

References

Advisory Council on Historic Preservation, Caring for the Past, Managing for the Future: Federal Stewardship and America's Historic Legacy, Advisory Council on Historic Preservation, Washington, DC, 2001.

Altschul, J.A., Models and the modeling process, in *Quantifying the Present and Predicting the Past: Theory, Method, and Application of Archaeological Predictive Modeling*, Judge, W.J. and Sebastian, L., Eds., U.S. Department of the Interior, Bureau of Land Management, Denver, 1988, pp. 61–96.

Altschul, J.A., Red flag models: the use of modelling in management contexts, in *Interpreting Space: GIS and Archaeology*, Allen, K.M.S., Green, S.W., and Zubrow, E.B.W., Eds., Taylor and Francis, London, 1990, pp. 226–238.

Ambler, J.R., The use and abuse of predictive modeling in cultural resource management, *American Archeology*, 14 (2), 140–146, 1984.

Cashmere, C. and Wandsnider, L., GIS modeling of surface visibility at agate fossil bed, in *Agate Fossil Beds: Prehistoric Archaeological Landscapes, 1994–1995*, Wandsnider, L. and MacDonell, G.H., Eds., prepared under Cooperative Agreement 6115-4-8026 for the Midwest Archeological Center, National Park Service, Lincoln, NE, 1995, pp. 13–25.

Church, T., Brandon, R.J., and Burgett, G.R., GIS applications in archaeology: method in search of theory, in *Practical Applications of GIS for Archaeologists*, Wescott, K.L. and Brandon, R.J., Eds., Taylor and Francis, Philadelphia, 2000, pp. 135–155.

Clark, R.N., Swayze, G.A., Gallagher, A.J., King, T.V.V., and Calvin, W.M., The U.S. Geological Survey, Digital Spectral Library, Version 1: 0.2 to 3.0 microns, U.S. Geological Survey Open File Report 93-592, 1993. http://speclab.cr.usgs.gov/spectral-lib.html

Dalla Bona, L., Protecting cultural resources through forest management planning in Ontario using archaeological predictive modeling, in *Practical Applications of GIS for Archaeologists*, Wescott, K.L. and Brandon, R.J., Eds., Taylor and Francis, Philadelphia, 2000, pp. 73–99.

Dunnell, R.C., The notion site, in *Space, Time, and Archaeological Landscape*, Rossignol, J. and Wandsnider, L., Eds., Plenum Press, New York, 1992, pp. 21–41.

Duncan, R.B. and Beckman, K.A., The application of GIS predictive site location models within Pennsylvania and West Virginia, in *Practical Applications of GIS for Archaeologists*, Wescott, K.L. and Brandon, R.J., Eds., Taylor and Francis, Philadelphia, 2000, pp. 33–58.

Ebert, J.I., The state of the art in "inductive" predictive modeling: seven big mistakes (and lots of smaller ones), in *Practical Applications of GIS for Archaeologists*, Wescott, K.L. and Brandon, R.J., Eds., Taylor and Francis, Philadelphia, 2000, pp. 129–134.

Ebert, J.I. and Kohler, T.A., The theoretical basis of archaeological predictive modeling and a consideration of appropriate data-collection methods, in *Quantifying the Present and Predicting the Past: Theory, Method, and Application of Archaeological Predictive Modeling*, Judge, W.J. and Sebastian, L., Eds., U.S. Department of the Interior, Bureau of Land Management, Denver, 1988, pp. 97–172.

Gaffney, V., Stančič, Z., and Watson, H., Moving from catchments to cognition: tentative steps toward a larger archaeological context for GIS, in *Anthropology, Space, and Geographic Information Systems*, Aldenderfer, M. and Maschner, H.D.G., Eds., Oxford University Press, New York, 1996, pp. 132–154.

Hageman, J.B. and Bennett, D.A., Construction of digital elevation models for archaeological applications, in *Practical Applications of GIS for Archaeologists*, Wescott, K.L. and Brandon, R.J., Eds., Taylor and Francis, Philadelphia, 2000, pp. 113–128.

Hasenstab, R.J. and Resnick, B., GIS in historical predictive modelling: the Fort Drum project, in *Interpreting Space: GIS and Archaeology*, Allen, K.M.S., Green, S.W., and Zubrow, E.B.W., Eds., Taylor and Francis, London, 1990, pp. 284–306.

Judge, W.J. and Sebastian, L., Eds., *Quantifying the Present and Predicting the Past: Theory, Method, and Application of Archaeological Predictive Modeling*, U.S. Government Printing Office, Washington, DC, 1988.

Kelly, R.L., Hunter-gatherer land use and regional geomorphology: implications for archeological survey, in Issues in archeological surface survey: meshing method and theory, Wandsnider, L. and Ebert, J.I., Eds., *American Archeology*, 7 (1), 49–57, 1988.

Kohler, T.A. and Parker, S.C., Predictive models for archaeological resource location, *Advances in Archaeological Method and Theory*, 9, 397–452, 1986.

Kvamme, K.L., Development and testing of quantitative models, in *Quantifying the Present and Predicting the Past: Theory, Method, and Application of Archaeological Predictive Modeling*, Judge, W.J. and Sebastian, L., Eds., U.S. Department of the Interior, Bureau of Land Management, Denver, 1988, pp. 325–428.

Kvamme, K.L., Geographic information systems in regional archaeological research and data management, in *Archaeological Method and Theory*, Vol. 1, Schiffer, M.B., Ed., University of Arizona Press, Tucson, 1989, pp. 139–203.

Kvamme, K.L., GIS algorithms and their effects on regional archaeological analysis, in *Interpreting Space: GIS and Archaeology*, Allen, K.M.S., Green, S.W., and Zubrow, E.B.W., Eds., Taylor and Francis, London, 1990, pp. 112–126.

Ladefoged, T.N., McLachlan, S.M., Ross, S.C.L., Sheppard, P.J., and Sutton, D.G., GIS-based image enhancement of conductivity and magnetic susceptibility data from Ureturituri Pa and Fort Resolution, New Zealand, *American Antiquity*, 60, 471–481, 1995.

Langran, G., *Time in Geographic Information Systems*, Taylor and Francis, New York, 1992.

Lock, G. and Stančič, Z., *Archaeology and Geographical Information Systems: A European Perspective*, Taylor and Francis, London, 1995.

McCulloch, T., Section 106 and Archaeology: Council Archaeologists Discuss How Select Issues Will Be Treated under Revised Regulations, forum presented at the 64th annual meeting of the Society for American Archaeology, Chicago, 1999.

Marozas, B.A. and Zack, J.A., GIS and archaeological site location, in *Interpreting Space: GIS and Archaeology*, Allen, K.M.S., Green, S.W., and Zubrow, E.B.W., Eds., Taylor and Francis, London, 1990, pp. 165–172.

Savage, S.H., Modelling the Late Archaic social landscape, in *Interpreting Space: GIS and Archaeology*, Allen, K.M.S., Green, S.W., and Zubrow, E.B.W., Eds., Taylor and Francis, London, 1990, pp. 330–355.

U.S. Secretary of Interior, Secretary of the Interior's Standards for Identification, 1983.

Wandsnider, L., Cultural resources "Catch 22" and empirical justification for discovering and documenting low-density archaeological surfaces, in *Tools to Manage the Past: Research Priorities for Cultural Resources Management in the Southwest*, Tainter, J. and Hamre, R.H., Eds., U.S. Department of Agriculture, Forest Service, Rocky Mountain Forest and Range Experiment Station, Fort Collins, CO, 1988.

Wandsnider, L. and Dore, C., Creating cultural resource data layers: experiences from the Nebraska Cultural Resources GIS Project, Part II: combining GPS and GIS data, *Bulletin of the Society of American Archaeology*, 13 (5), 16–18, 1995.

Warren, R.E. and Asch, D.L., A predictive model of archaeological site location in the Eastern Prairie Peninsula, in *Practical Applications of GIS for Archaeologists*, Wescott, K.L. and Brandon, R.J., Eds., Taylor and Francis, Philadelphia, 2000, pp. 5–32.

Wescott, K.L. and Kuiper, J.A., Using GIS to model prehistoric site distributions in the upper Chesapeake Bay, in *Practical Applications of GIS for Archaeologists*, Wescott, K.L. and Brandon, R.J., Eds., Taylor and Francis, Philadelphia, 2000, pp. 59–72.

5

Problems in Paleolithic Land Evaluation: A Cautionary Tale

Hans Kamermans

CONTENTS

ABSTRACT   Land evaluation is a technique developed by soil scientists to generate different models for land use on the basis of ecological and social economic data. In archaeology, land evaluation can be used as a deductive form of predictive modeling. In this chapter, land evaluation is applied to Pleistocene data from the Agro Pontino (Lazio, Italy), a coastal plain ca. 80 km southwest of Rome. The data were collected by the Agro Pontino survey project between 1979 and 1989. After an initial inventory of the paleoenvironment, socioeconomic models are constructed using ethnographic, historic, and archaeological data. Land units are ranked according to their suitability for a certain type of land use, and finally an expected form of land use is compared with the archaeological record. Land evaluation seems to be a very suitable technique for models of simple types of agriculture (Kamermans 2000). For hunter-gatherer societies, the application is more problematic. Some methodological problems are presented here. In the end, it is shown to be impossible to detect (assumed) differences in land use between the Middle and Upper Paleolithic in the area.

5.1   Introduction

There are two different reasons for applying predictive modeling in archaeology. The first is to predict archaeological site location as a guide to future developments in the modern landscape — an archaeological heritage management application. The second is to gain insight into former human behavior in the landscape — an academic research application.

Modern developments are changing the European landscape rapidly. In many areas the archaeological record is under threat. In 1992 a number of European countries signed in Valletta (Malta) a treaty to protect the European archaeological heritage. This treaty is now known as the Convention of Malta or the Valletta Convention. The aim of the convention is "to protect the archaeological heritage as a source of the European collective memory and as an instrument for history and scientific studies" (Council of Europe 1992). To reach this goal, governments want archaeologists to identify areas with a high density of archaeological find spots in order to protect the archaeology that is left, and to record and study what is under threat of being demolished. To do this, archaeologists will have to reconstruct the original spatial patterning of the material culture of the past. One way of doing this is by means of predictive modeling. However, this practice has not been without criticism (cf. Ebert 2000).

The interpretation of behavior and material culture over space has always been one of the fundamental aspects of archaeology (Green 1990: 3). Nowadays the concept of landscape is very popular, as seen in the work of scholars like Barbara Bender (1993), Julian Thomas (1993), Christopher Tilley (1994), Richard Bradley (1997), and Gabriel Cooney (2000). But especially during the late 1960s and the early 1970s — the time of New Archaeology with its emphasis on explanation, quantitative thinking, and a scientific perspective of the past — archaeologists turned to other fields, notably geography, for tools and ideas for spatial analysis (Aldenderfer 1996). Examples are von Thunen's model of agricultural land use, Weber's model of industrial location, Christaller's central place model, and Hägerstrand's model of innovation. For geographers, these models were described in Haggett's book *Locational Analysis in Human Geography* (1965). This book was "translated" for archaeologists by Ian Hodder and Clive Orton (1976) as *Spatial Analysis in Archaeology*. Predictive modeling can also play a role in this more academic part of archaeological research.

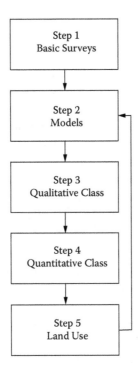

FIGURE 5.1
Different steps in the application of land evaluation in archaeology.

The following application of land evaluation in the Italian Agro Pontino hopes to contribute to the debate on differences in land use during the Middle and Upper Paleolithic. Especially the subsistence strategies of the Neanderthals (the Ancients as Stringer and Gamble [1993] call all premodern humans) are still far from clear.

5.2   Land Evaluation

The technique of land evaluation was originally developed by soil scientists (Brinkman and Smyth 1973) and generates different models for land use on the basis of ecological and social economic data. In an archaeological application of land evaluation, there are five steps to be considered (Kamermans et al. 1985, 1990; Kamermans 1993, 1996, 2000) (Figure 5.1):

Step 1 is an inventory of the natural environment collected by field surveys and reviews. These data form the basis for a reconstruction of the natural environment at different times in the past.

Step 2 is the construction of socioeconomic models for early forms of
land use with ethnographic, historic, and archaeological data.

Step 3 is the classification of the area into different land-mapping units
on the basis of physical factors. These units are described in terms
of their properties to provide a qualitative land classification.

Step 4 is a semiquantitative land classification: the measurement of the
suitability of an area for a certain type of land use on the basis of
the requirements for that type of land use.

Step 5 is an expected form of land use for every chosen socioeconomic
model based on results from steps 2 to 5. These models are then
confronted with the archaeological database. The comparison of the
expected form of land use with the archaeologically recorded land
use provides a basis for modifying the model and repeating steps 2
to 5.

The application of land evaluation in archaeology requires some (unfortunately arguable) assumptions:

• Humans in the past exploited the environment according to the
principle of least effort.

• The combination of environment and human behavior creates a
specific spatial pattern in particular types of areas.

• There is a relationship between prehistoric land use and artifact or
find-spot density.

• The economic system during each archaeologically distinct period
was, broadly speaking, constant.

Archaeological land evaluation is a form of deductive predictive modeling.
Predictive modeling is a technique used to predict archaeological site locations in a region on the basis of observed patterns or on assumptions about
human behavior (Kohler and Parker 1986; Kohler 1988; Kvamme 1985, 1988,
1990). There are two different approaches to predictive modeling, an inductive and a deductive one.

With the inductive approach, a model is constructed based on the correlation between known archaeological find spots and attributes (mostly) from
the current physical landscape. On the basis of correlation, causality is
assumed, and the model is then used to predict site location.

With the deductive approach, a model is constructed on the basis of *a priori*
knowledge (social, mainly anthropological, historical, and archaeological
knowledge), and the known find spots are then used to evaluate the model.
Archaeological land evaluation is an example of this last approach. Kamermans (2000) demonstrates the differences between the two approaches of
predictive modeling.

5.3   The Agro Pontino Survey Project

During the 1980s a team of Dutch, American, and Italian scholars and students studied the archaeology and past environment of the Agro Pontino
(Lazio, Italy) (Voorrips et al. 1983, 1991). The two main research themes of
the project were the transition from the Middle to the Upper Paleolithic
(Loving et al. 1990/91, 1992; Loving 1996a, 1996b) and the application of
land evaluation in archaeology (Kamermans et al. 1985, 1990; Kamermans
1993, 1996, 2000).

The Agro Pontino is a coastal plain along the Tyrrhenian Sea ca. 80 km
southeast of Rome (Figure 5.2). Half of its surface consists of a low-lying
graben filled with peat; the other half is formed by a complex of stable marine
terraces (Segre 1957; Sevink et al. 1982, 1984, 1991; Kamermans 1991). Due
to this special combination, the Agro Pontino is a perfect area for the study
of the relationship between humans and their natural environment in the
past: palynological data from the graben was used to reconstruct the past

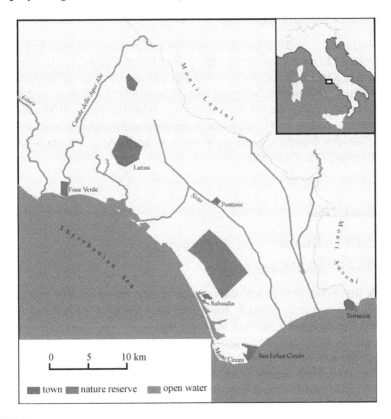

FIGURE 5.2
The Agro Pontino.

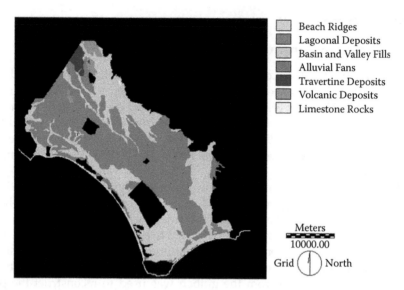

Beach Ridges
Lagoonal Deposits
Basin and Valley Fills
Alluvial Fans
Travertine Deposits
Volcanic Deposits
Limestone Rocks

Meters
10000.00
Grid North

FIGURE 5.3
The Agro Pontino geomorphological map.

environment (Eisner et al. 1984, 1986; Hunt and Eisner 1991; Eisner and Kamermans 2004), and the stable surfaces yield archaeological evidence from the Paleolithic onwards (Zampetti and Mussi 1988; Loving et al. 1990/91). The area is famous for its Neanderthal finds from the caves of Monte Circeo (Blanc 1957; Ascenzi 1990/91). Data from past and ongoing research in the caves of nearby Monte Circeo were used as additional information (Caloi and Palombo 1988, 1990/91).

5.4 Land Evaluation and Archaeology

5.4.1 Step 1: Basic Surveys

Step 1 is the reconstruction of the natural environment at different times in the past. The reconstruction is based on an inventory of natural features identified through field surveys and reviews.

Geologically speaking, the Agro Pontino consists of two parts, a low-lying graben at the foot of the mountains, mainly filled with peat, and a dune area along the coast, both dating from the Quaternary (Figure 5.3). A soil survey by Sevink distinguished four marine terraces along the coast (Figure 5.4): the Latina level at ca. 25 m a.s. (above see level); the Minturno level at ca. 16 m a.s.; the Borgo Ermada level at ca. 6 m a.s.; and the youngest, still active marine complex, the Terracina level. Estimated dates are: Latina level 560,000 BP

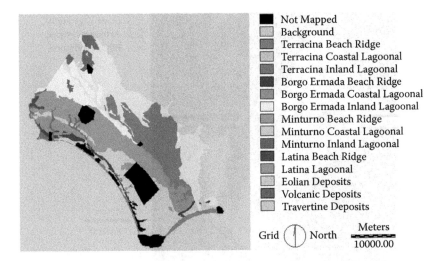

Not Mapped
Background
Terracina Beach Ridge
Terracina Coastal Lagoonal
Terracina Inland Lagoonal
Borgo Ermada Beach Ridge
Borgo Ermada Coastal Lagoonal
Borgo Ermada Inland Lagoonal
Minturno Beach Ridge
Minturno Coastal Lagoonal
Minturno Inland Lagoonal
Latina Beach Ridge
Latina Lagoonal
Eolian Deposits
Volcanic Deposits
Travertine Deposits

Grid ⊕ North   Meters
10000.00

FIGURE 5.4
The Agro Pontino land units.

(Tyrrhenian I), Minturno level 125,000 BP (Tyrrhenian II), Borgo Ermada level 90,000 BP (Tyrrhenian III), and Terracina level postglacial (pre-Neolithic). Each terrace consists of a sandy beach ridge and a clayey lagoon. The graben is partly covered with Holocene alluvial and slope deposits (Sevink et al. 1982, 1984, 1991). The calcaric mountains along the northern part of the Agro Pontino, the Monti Lepini and the Monti Ausoni, and Monte Circeo in the extreme south, an isolated part of the Apennines, were formed during the Mesozoic. In the northern part of the area are volcanic rocks, a result of volcanic activity 700,000 to 10,000 years ago (Segre 1957; Kamermans 1991).

During the last glacial period the youngest terrace was not yet formed, and the sea level was 100 m below its present level. This means that the Agro Pontino was about twice as large as today. The graben has a very young surface, leaving only a small sample of the original surface available for research into the Middle and Upper Paleolithic.

Palynological research of a 9-m sediment core from the graben in an area called Mezzaluna, close to the fault along the Monti Lepini, provided a well-dated record of environmental change from the full glacial to recent times (Eisner et al. 1984, 1986; Hunt and Eisner 1991; Eisner and Kamermans 2004).

During the part of the full glacial before 15,800 BP, a dry herb steppe was the dominant regional vegetation (Figure 5.5). A number of woodland species sporadically occupied moister ecological niches. The vegetation mosaic of steppe-forest and mesic woodland suggests that although the climate was drier and cooler than today, there was more moisture than at the glacial maximum. A freshwater lake occupied at least part of the Agro Pontino graben during the full and late Pleistocene.

Dune Vegetation
Artemisia Steppe
Birch/Pine wood

Meters
10000.00

Grid North

FIGURE 5.5
Reconstructed ecological zones of the Agro Pontino during the early glacial.

An episode can be postulated that drained the lagoon between 35,000 and 16,000 BP. A sand layer 9 m below the present surface may be all that remains of a massive erosional event, and upper layers of the mid-Pleistocene clays may have been scoured by the covering sands.

The lake that developed during the last part of the full glacial (15,800 to 13,000 BP) on top of the pollen-poor sand layer was not the same as the earlier lake. Lagoonal deposits were replaced by peaty clays. The species found in the pollen core are typical of shallow open water. The regional vegetation consists of predominantly steppe assemblages, suggesting that this full glacial landscape was more severe in terms of aridity, with warm summers and dry, cool winters.

The late glacial period (13,000 to 10,000 BP) represents the most severe conditions of the Late Quaternary. Open water continued to shrink, and as the land dried it was occupied by wet and mesic herb assemblages. The driest and best-drained soils were covered with steppe vegetation, as well as drought-adapted species of Compositae. *Quercus* (oak) and *Pinus* (pine) managed to survive during this period but were unable to spread to the expanding available land. Increasing *Artemisia* and Chenopodiaceae (goose-foot) indicates a climate of cold winters and low precipitation too severe to permit drought-adapted woodland species to flourish.

The early Holocene is characterized by a rapid expansion of the forest vegetation and the disappearance of typical steppe taxa (Figure 5.6).

Because virtually no faunal remains are known from the plain, the fauna had to be reconstructed from animal remains excavated in the caves of Monte Circeo and the Lepini and Ausoni mountains (see Caloi and Palombo 1988, 1990/91; Stiner 1994). During the last interglacial, the following animals were

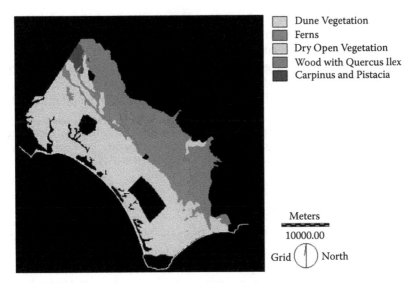

Dune Vegetation
Ferns
Dry Open Vegetation
Wood with Quercus Ilex
Carpinus and Pistacia

Meters
10000.00
Grid ( ) North

FIGURE 5.6
Vegetation reconstruction of the Agro Pontino during the Holocene.

among others living in Latium: hare, wolf, fox, hyena, elephant, rhinoceros, horse, hippopotamus, fallow deer, red deer, roe deer, and aurochs.

At the beginning of the last glacial, hippopotamus disappears and ibex appears. During the pleniglacial, elephant and rhinoceros disappear, and wild boar, red deer, and roe deer are dominating the fauna. Other large mammals during that period are horse, wild ass, fallow deer, aurochs, and ibex. In the Agro Pontino, chamois and red deer are characteristic for the late glacial.

5.4.2 Step 2: Models

Step 2 is the construction of socioeconomic models for early forms of land use with ethnographic, historic, and archaeological data.

With respect to their food management strategies, human hunter-gatherers can, analogous to other organisms, be divided into generalists and specialists. A generalist eats a greater range of food types or a greater variance of types or a greater "breadth" of types; it has a great repertoire of feeding behavior (Schoener 1971: 384). Winterhalder (1981a: 27; 1981b: 69) studied the spatial implications of this distinction and concluded (1981b: 69):

> This model predicts that organisms living in a relatively small-scaled environmental mosaic will develop a broad use of habitats; they will be patch type generalists. Conversely, those exposed to a large-scaled environment will specialize and forage within relatively few habitats. As scale increases the optimum patch choice shifts towards specialization.

For the purpose of this study two models are used, defined in terms of mobility and food management strategies: the generalist practicing residential mobility and the specialist practicing logistic mobility.

The characteristics of a generalist are: foraging in an area with a great variability in land units and a high residential mobility. The archaeological correlate will be a low visibility of camp sites, resulting in small find scatters dispersed over many small land units.

The characteristics of a specialist are: high logistic mobility, foraging in large land units. The archaeological correlate will be bigger find complexes dispersed over one or a few land units.

5.4.3   Step 3: Qualitative Classification

Step 3 consists of the classification of the area into different land-mapping units on the basis of physical factors. These units are described in terms of their properties to provide a qualitative land classification.

It is not easy to construct a qualitative land classification for prehistoric hunter-gatherer societies. To a certain extent Brinkman and Young's land qualities related to domestic animal productivity can be used (Brinkman and Young 1976: 16).

This produces, in an adapted form, the following list:

- The productivity of an area in terms of edible vegetation for animals
- Climatic hardship affecting animals
- Endemic pests and diseases
- Toxicity of the vegetation

It is difficult to go any further because, for most factors Brinkman and Young used in their original work, there is a lack of data for a prehistoric situation. Furthermore, in the Middle and Upper Paleolithic, animals were not forced to stay in a certain area as is the case with domestic animals. Wild animals will simply move if the circumstances are not right. So the qualitative land classification for prehistoric hunter-gatherer societies has to be replaced by a reconstruction of appropriate ecological zones for the various time periods (Figure 5.5 and Figure 5.6).

5.4.4   Step 4: Quantitative Classification

The next step is a semiquantitative land classification: the measurement of the suitability of an area for a certain type of land use on the basis of the requirements for that type of land use.

During step 2, two socioeconomic models were constructed: the generalist practicing residential mobility and the specialist practicing logistic mobility. A semiquantitative land classification was formulated for these two models.

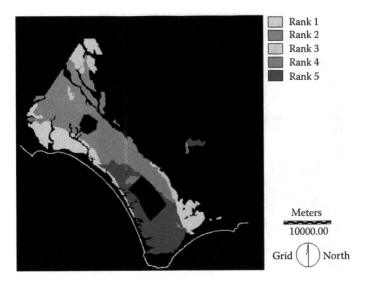

FIGURE 5.7
Semiquantitative land classification of the Agro Pontino for the generalist hunter-gatherer for the Paleolithic (see also Table 5.1).

TABLE 5.1

Semiquantitative Land Classification for the
Generalist Hunter-Gatherer for the Paleolithic[a]

Land Unit	Predicted Rank
Coastal terraces	1
Small lagoonal	2
Volcanic and travertine	3
Latina lagoonal	4
Eolian	5

[a] See Figure 5.7.

The characteristics of a generalist are: hunting various species of animals in an area with a great variability in land units and a high residential mobility. In an effort to identify the generalist, land units are grouped together to construct units with a great variability. The smaller marine terraces along the coast are grouped together, as are the younger inland lagoonal deposits and the volcanic and travertine deposits. The Terracina level and the more recent alluvial/colluvial deposits are left out of the analysis because they did not exist during the Paleolithic and the Epipaleolithic (Mesolithic).

Figure 5.7 and Table 5.1 give the semiquantitative land classification for the generalist hunter-gatherer for the Paleolithic, and Figure 5.8 and Table 5.2 for the Epipaleolithic. The most suitable area would seem to be a combination of the younger marine terraces characterized by a diverse environment, that is, sandy ridges alternated by clayey plains. Also, the more inland lagoonal areas would be suitable for the general hunter-gatherer, followed

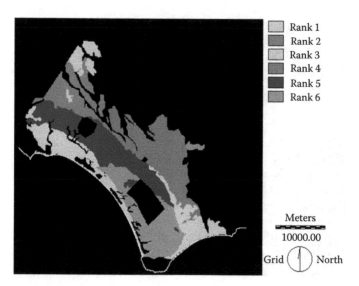

FIGURE 5.8

Semiquantitative land classification of the Agro Pontino for the generalist hunter-gatherer for the Epipaleolithic (see also Table 5.2).

TABLE 5.2

Semiquantitative Land Classification for the
Generalist Hunter-Gatherer for the Epipaleolithic[a]

Land Unit	Predicted Rank
Coastal terraces	1
Small lagoonal	2
Volcanic and travertine	3
Terracina lagoonal	4
Latina lagoonal	5
Eolian	6

[a] See Figure 5.8.

by the volcanic and travertine deposits and the big lagoonal and eolian deposits.

The characteristics of a specialist hunter-gatherer are: high logistic mobility and foraging in large land units. The environment in the land units should be less diverse than for the generalist hunter-gatherer. In this case the smaller marine terraces along the coast, the younger inland lagoonal deposits, and the volcanic and travertine deposits are not grouped together. Table 5.3 and Table 5.4 show that the large Latina lagoonal deposit would be the most suitable land unit for hunter-gatherers during the Paleolithic and the Epipaleolithic, respectively (Figure 5.9 and Table 5.3 for the Paleolithic, and Figure 5.10 and Table 5.4 for the Epipaleolithic).

TABLE 5.3

Semiquantitative Land Classification for the
Specialist Hunter-Gatherer for the Paleolithic[a]

Land Unit	Predicted Rank
Latina lagoonal	1
Borgo Ermada inland lagoonal	3
Minturno beach ridge	3
Eolian	3
Borgo Ermada lagoonal	6
Volcanic	6
Travertine	6
Borgo Ermada beach ridge	9
Minturno lagoonal	9
Minturno inland lagoonal	9

[a] See Figure 5.9.

TABLE 5.4

Semiquantitative Land Classification for the
Specialist Hunter-Gatherer for the Epipaleolithic[a]

Land Unit	Predicted Rank
Latina lagoonal	1
Borgo Ermada inland lagoonal	3
Minturno beach ridge	3
Eolian	3
Terracina inland lagoonal	5.5
Borgo Ermada lagoonal	5.5
Volcanic	5.5
Travertine	5.5
Terracina beach ridge	10
Terracina lagoonal	10
Borgo Ermada beach ridge	10
Minturno lagoonal	10
Minturno inland lagoonal	10

[a] See Figure 5.10.

5.4.5   Step 5: Land Use

The product of the last step is an expected form of land use for every chosen socioeconomic model based on results from steps 2 to 4. A comparison of the expected form of land use with the archaeologically recorded land use provides a basis for modifying the model and repeating steps 2 to 5.

The archaeological data used for the land evaluation research was collected during seven surveys over ten years; three small ones with two to four people and four larger ones with a crew of up to twenty scholars and students. The total area of the Agro Pontino is huge, approximately 750 km², so it was decided to sample (Loving et al. 1991). In the then-prevailing tradition of processual archaeology, a multistage approach (Redman 1973) was used

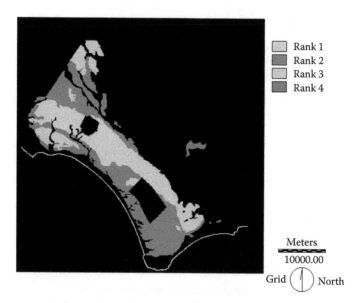

FIGURE 5.9
Semiquantitative land classification of the Agro Pontino for the specialist hunter-gatherer for
the Paleolithic (see also Table 5.3).

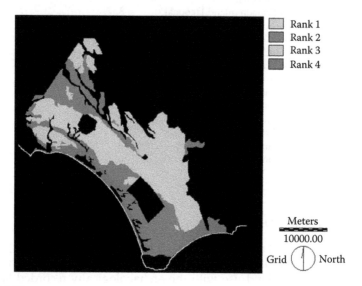

FIGURE 5.10
Semiquantitative land classification of the Agro Pontino for the specialist hunter-gatherer for the
Epipaleolithic (see also Table 5.4).

consisting of three stages: an exploratory phase, a probabilistic phase, and
a problem-oriented phase, with the results of one phase being used for
making decisions about the next phase. During the exploratory phase, fields
were surveyed to get an impression of the presence of archaeological material

TABLE 5.5

Comparison of Predicted and Observed Preferences for General Hunter-Gatherers during the Middle and Upper Paleolithic

Land Unit	Predicted Rank	Observed Middle Paleolithic	Observed Upper Paleolithic
Coastal terraces	1	4	4
Small lagoonal	2	2	1
Volcanic and travertine	3	3	3
Latina lagoonal	4	1	2
Eolian	5	5	5

and the distribution over the different soil units as defined by the soil survey. The information collected from these fields was used to develop methods for assessing factors affecting visibility (Verhoeven 1991) and to estimate the size of a randomly selected sample, which required making statements about the population of fields in the entire Agro Pontino. During this same phase, field techniques were developed (Loving and Kamermans 1991a). For the probability sampling phase, a systematic nonaligned transect design was selected to ensure a sufficient sample size for making probability statements about the archaeological populations in the Agro Pontino as a whole. During the problem-oriented phase, additional materials were collected to help accomplish specific research goals.

During all the surveys, the same field and analysis procedures were used. The result is a very well-controlled sample from the open-air find spots from the Agro Pontino. All phases of the survey together yielded in total 360 find spots; most of these, 289, were multiperiod. The material from the Paleolithic and Mesolithic find spots consisted exclusively of stone implements, mostly made from small-size flint and quartzite pebbles (Loving and Kamermans 1991b).

For the application of land evaluation in the Agro Pontino, find-spot density per land unit was used instead of artifact density, although Foley (1981) convincingly demonstrated that artifact density should be given preference. It was impossible to use artifact density because it would require dating the individual artifacts, and because most find spots are a palimpsest, this was not only impracticable but even impossible. The rank order of the different land units, based on data on find-spot density collected during the survey, was compared with the expected rank order of the land units for the different models. Apart from that, a test was carried out to see if there was a significant difference in find-spot density between the different land units for every separate time period.

Table 5.5 gives the expected and observed rank order for the generalist hunter-gatherer during the Middle and Upper Paleolithic. Both the Spearman test and Kendall's test were used to test the rank order (Table 5.6). With an α of 0.1, none of the rankings was significant, which means that none of

TABLE 5.6

Spearman's and Kendall's Test for the Data in Table 5.5

Period	Spearman Test			Kendall's Test		
	r	t(3)	signif.	Tau-c	ASE1	t
Middle Paleolithic	.10000	0.17408	.87289	.00000	.48990	.00000
Upper Paleolithic	.30000	0.54470	.62384	.20000	.45607	.43853

TABLE 5.7

Comparison of Predicted and Observed Preferences for Specialized Hunter-Gatherers during the Middle and Upper Paleolithic

Land Unit	Predicted Rank	Observed Middle Paleolithic	Observed Upper Paleolithic
Latina lagoonal	1	2	4
Borgo Ermada inland lagoonal	3	3	3
Minturno beach ridge	3	6	5
Eolian	3	10	8
Borgo Ermada lagoonal	6	8	7
Volcanic	6	4	9
Travertine	6	9	2
Borgo Ermada beach ridge	9	5	6
Minturno lagoonal	9	7	10
Minturno inland lagoonal	9	1	1

TABLE 5.8

Spearman's and Kendall's Test for the Data in Table 5.7

Period	Spearman Test			Kendall's Test		
	r	t(8)	signif.	Tau-c	ASE1	t
Middle Paleolithic	−.00629	−0.01780	.98623	.00000	.33092	0.00000
Upper Paleolithic	.13217	0.37714	.71588	.10667	.30591	0.34868

the observed rankings correspond to the predicted ranking for general hunter-gatherers.

Table 5.7 gives the expected and observed rank order for the specialist hunter-gatherer during the Middle and Upper Paleolithic. Table 5.8 gives the results for both the Spearman test and Kendall's test, which were used to test the rank order. Again, with an α of 0.1, none of the rankings was significant, which means that none of the observed rankings corresponds to the predicted ranking for specialized hunter-gatherers.

For the Middle and Upper Paleolithic, none of the expected rank orders for either the generalist or the specialist fits with the observed rank order. For the Epipaleolithic, the results are more promising. The rank order based on find-spot density of the land units does not correlate with the order for the generalist hunter-gatherers (Table 5.9 and Table 5.10), but it does for the specialist hunter-gatherers ($t > 2$ and significance $< .1$, see Table 5.11 and Table 5.12).

TABLE 5.9

Comparison of Predicted and Observed Preferences of Generalist Hunter-Gatherers for the Epipaleolithic

Land Unit	Predicted Rank	Observed Epipaleolithic
Coastal terraces	1	5
Small lagoonal	2	2
Volcanic and travertine	3	4
Terracina lagoonal	4	6
Latina lagoonal	5	3
Eolian	6	1

TABLE 5.10

Spearman's and Kendall's Test for the Data in Table 5.9

Period	Spearman Test			Kendall's Test		
	r	t(4)	signif.	Tau-c	ASE1	t
Epipaleolithic	−.42857	−0.94868	.39650	−.33333	.36107	0.92319

TABLE 5.11

Comparison of Predicted and Observed Preferences of Specialized Hunter-Gatherers for the Epipaleolithic

Land Unit	Predicted Rank	Observed Epipaleolithic
Latina lagoonal	1	5
Borgo Ermada inland lagoonal	3	4
Minturno beach ridge	3	3
Eolian	3	2
Terracina inland lagoonal	5.5	10
Borgo Ermada lagoonal	5.5	9
Volcanic	5.5	8
Travertine	5.5	6
Terracina beach ridge	10	12
Terracina lagoonal	10	12
Borgo Ermada beach ridge	10	7
Minturno lagoonal	10	12
Minturno inland lagoonal	10	1

TABLE 5.12

Spearman's and Kendall's Test for the Data in Table 5.11

Period	Spearman Test			Kendall's Test		
	r	t(11)	signif.	Tau-c	ASE1	t
Epipaleolithic	.54254	2.14208	.05540	.48915	.22991	2.12755

Land evaluation as a form of deductive predictive modeling fails to identify a dominant form of land use for the Middle and Upper Paleolithic, but it succeeds for the Epipaleolithic. Elsewhere it was shown that land evaluation works in the Agro Pontino for the Neolithic period (Kamermans 1993, 2000). Would the results for Paleolithic hunter-gatherer societies be more conclusive if an inductive way of predictive modeling was used?

5.5   Inductive Predictive Modeling

With the inductive approach to predictive modeling, a model is constructed on the basis of correlations between known archaeological find spots and attributes that are predominantly taken from the current physical landscape. It is only a predictive model when the observed correlations are extrapolated. These extrapolation models are most commonly used in archaeological heritage management archaeology, but may also have their use in scientific research, for example to analyze anomalies in an observed spatial pattern. In order to explain the failure of land evaluation to detect differences in land use in the Agro Pontino between the Middle and the Upper Paleolithic, an inductive approach is used to see if there is a correlation between find-spot density and land units.

The archaeological hypothesis for the Middle Paleolithic is that human hunter-gatherers had no preference for any of the constructed land units. The null hypothesis is that there is no difference in find-spot density between the defined land units.

The Attwell-Fletcher test was used to test this hypothesis. This test (Attwell and Fletcher 1985, 1987) is designed to test the existence of a significant association between a point pattern distribution and categories of an environmental variable. It compares an observed pattern with a simulated random pattern. Two sets of hypotheses are tested. The null hypothesis for the first set is "no association," the alternative hypothesis is that at least one category is favored. In the other case, the null hypothesis is of course the same, but the alternative hypothesis is that at least one category is avoided. The main advantages of this test over, for instance, the χ^2 test are that it can be applied to small samples, that it indicates the strength of the association, and that it is directional (i.e., is a category favored or avoided?). The often-used and misused χ^2 test can do nothing more than demonstrate the existence of a relationship. A weakness of the Attwell-Fletcher test is that the simulation does not take into account the problem of autocorrelation, that is, the inherent relationship between different aspects of the natural environment. For both the Attwell-Fletcher test and the χ^2 test, one should be aware that the existence of an association does not necessarily imply a causal relationship.

Table 5.13 shows that one land unit, eolian, has a value below the fifth percentile such that the null hypothesis of no association is rejected because there is an association. This means that there are fewer than expected find spots in the eolian area. This is, however, very easily explained by geological

TABLE 5.13

Attwell-Fletcher Test to Compare the Find-Spot Density and Geomorphological Land Units for Hunter-Gatherers during the Middle Paleolithic in the Agro Pontino

Land Unit	Number of Find Spots	Expected Proportion	Observed Proportion	Category Weight
Coastal terraces	13	0.2904	0.23	0.16
Small lagoonal	12	0.1483	0.21	0.30
Latina lagoonal	23	0.2769	0.40	0.31
Eolian	2	0.1410	0.04	0.05
Volcanic and travertine	7	0.1435	0.12	0.18

Note: Number of find spots = 57; number of categories = 5; number of simulations = 1000; 95th percentile = 0.34 ± 0.005; 5th percentile = 0.07 ± 0.013.

TABLE 5.14

Attwell-Fletcher Test to Compare the Find-Spot Density and Geomorphological Land Units for Hunter-Gatherers during the Upper Paleolithic in the Agro Pontino

Land Unit	Number of Find Spots	Expected Proportion	Observed Proportion	Category Weight
Coastal terraces	7	0.2904	0.22	0.15
Small lagoonal	7	0.1483	0.22	0.30
Latina lagoonal	12	0.2769	0.38	0.28
Eolian	2	0.1410	0.06	0.09
Volcanic and travertine	4	0.1435	0.13	0.18

Note: Number of find spots = 32; number of categories = 5; number of simulations = 1000; 95th percentile = 0.38 ± 0.008; 5th percentile = 0.03 ± 0.005.

phenomena such as the old surface being no longer available, or find spots having disappeared under eolian deposits.

For the Upper Paleolithic the hypotheses are the same. The archaeological hypothesis is that human hunter-gatherers had no preference for any of the constructed land units. The null hypothesis is that there is no difference in find-spot density between the land units. Table 5.14 shows that, for the Upper Paleolithic, the null hypothesis is not rejected and that there is no difference in find-spot density between the land units.

5.6 Discussion

For the Middle and Upper Paleolithic none of the expected rank orders for either the generalist or the specialist fits with the observed rank order. Furthermore, no statistically significant difference in find-spot density for both time periods could be found between the defined land units, with the exception of the eolian land unit for the Middle Paleolithic. The absence of find spots in this land unit, however, has nothing to do with land use but can be explained on the basis of the geology. So is there a problem to be solved or not? No pattern whatsoever could be detected. Does that mean that the wrong research tool (land evaluation) was used, or does it mean that the way in which Middle and Upper Paleolithic people used the Agro Pontino did not create a distinct patterning of the material record?

Was the wrong research tool used? How do we know that the land evaluation technique works? For the Epipaleolithic, the rank order of the land units correlates with the order for the specialist hunter-gatherers. This is what was to be expected for the Mesolithic. The land evaluation technique also works for societies with a subsistence based on simple types of agriculture (Kamermans 2000).

So assuming that the land evaluation technique works, a different explanation for the fact that none of the models work has to be found. Is there something wrong with the assumptions? The assumptions are:

- Humans in the past exploited the environment according to the principle of least effort.
- The combination of environment and human behavior creates a specific spatial pattern in particular types of areas.
- There is a relationship between prehistoric land use and artifact or find-spot density.
- The economic system during each archaeologically distinct period was, broadly speaking, constant.

There could be a problem with the second assumption. Perhaps the definition of the land units is too detailed for hunter-gatherers. It is possible that during the Middle and Upper Paleolithic hunter-gatherers considered the Agro Pontino as one land unit. There is also a possibility that there was a difference in exploitation between the coastal zone (Agro Pontino) and the hinterland. Earlier research points in this direction (Voorrips et al. 1985).

The even spread of the material culture throughout the landscape makes the model of generalists practicing residential mobility in the Agro Pontino for both periods most likely, although the number of artifacts per find spot for these periods is very low. The archaeological correlate should be larger find complexes dispersed over one or a few land units.

Do these results contribute to the discussion on Middle and Upper Paleolithic subsistence practices in the Agro Pontino? There is a debate between American and Italian scholars about the hypothesis that a major shift in land use took place in the region during the Middle Paleolithic (Mussi 1999). In the late 1980s two American scholars studied the faunal and the lithic material from the cave sites of Monte Circeo. The results were sensational. Both Stiner and Kuhn (Kuhn 1991, 1995; Stiner 1991, 1994; Stiner and Kuhn 1992) see a major change in subsistence during the Middle Paleolithic in Latium. Before 55,000 BP, scavenging was the main activity for subsistence, while after 55,000 BP, hunting was. They base their conclusions mainly on the fact that head parts of medium-sized ungulates dominate the pre-55,000 BP collections. The range of formal tool types in the Mousterian sample stays the same across the 55,000-year boundary, but the reduction technique changes. Mussi (1999) expressed surprise that scavenging continued until that late a date in the Agro Pontino and ascribes the differences in, notably, the faunal material to differences in excavation techniques. Indeed, all the sites dated before 55,000 BP were largely excavated before the Second World War, the later sites after the war.

When using land evaluation, it is necessary to assume that the economic system during each archaeologically distinct period (defined on the basis of flint typology) was constant. Unfortunately, this makes it impossible to test the Kuhn and Stiner hypothesis. But the lack of difference between both the density and the distribution of find spots with typologically Middle and Upper Paleolithic material does not support the current view of a major cultural break during the Middle or between the Middle and Upper Paleolithic. The cautious conclusion from the land-evaluation exploration is that both the Middle Paleolithic Ancients and the Upper Paleolithic Moderns used the same way of exploiting the Agro Pontino: as generalist hunter-gatherers practicing residential mobility.

5.7 Conclusions

The tentative conclusion we can draw is that the land evaluation technique as a form of deductive predictive modeling does work. It works for societies with a subsistence based on simple types of agriculture (Kamermans 2000), and it works for hunter-gatherer societies, as demonstrated by the fact that, for the Epipaleolithic in the Agro Pontino, the rank order of the land units correlates with the order for specialist hunter-gatherers.

There are, however, problems with the Middle and Upper Paleolithic. None of the subsistence models for those periods fit with the distribution of the archaeological material in the plain, and it looks as if there is no significant difference in find-spot density between the land units for Middle and Upper Paleolithic material. This can be attributed to a number of causes:

- The definition of the land units was not in accordance with Paleolithic land use. The generalist hunter-gatherers operated on a different scale and exploited the whole of the coastal zone as one land unit.
- The area dating from the Pleistocene that is available for research is unrepresentative. The sea level was 100 m lower than today, and consequently the coastal plain was twice as large. Furthermore, the surface of the graben consists of Holocene deposits.

- The archaeological sample was unrepresentative. No use was made of the archaeological data from the caves of Monte Circeo or the Lepini and Ausoni mountains. Only data from the sample collected by the surveys was used.

Unfortunately, due to the above-mentioned methodological flaws, this research cannot clarify the problem of a possible occurrence of a major cultural break during the Middle or between the Middle and Upper Paleolithic. But the tentative conclusion, that both the Middle Paleolithic Ancients and the Upper Paleolithic Moderns exploited the Agro Pontino as generalist hunter-gatherers practicing residential mobility, makes a major change improbable.

Acknowledgments

The Agro Pontino survey project was a joint venture by the Instituut voor Pre- en Protohistorische Archeologie Albert Egges van Giffen (IPP), of the University of Amsterdam, and the Instituut voor Prehistorie of the State University of Leiden (both The Netherlands). It was under the auspices of the Istituto Olandese in Rome and supported by research grants from the Netherlands Organisation for Scientific Research (NWO). The following institutes participated in the project: Soprintendenza Archeologica del Lazio, the Department of Anthropology of the Universit· di Roma "La Sapienza," the Laboratory for Physical Geography and Soil Science of the University of Amsterdam, the Museo Nazionale Preistorico Etnografico "Luigi Pigorini" in Rome, the Istituto Italiano di Paleontologia Umana in Rome, DIGITER geo-informatica in Rome and the Hugo de Vries Laboratory of the University of Amsterdam.

I would like to thank my codirectors of the Agro Pontino survey project Susan Loving and Bert Voorrips and all the other participants in the surveys for their help in and out of the field, and Prof. Jan Sevink of the Laboratory for Physical Geography and Soil Science of the University of Amsterdam for his inspiration to use land evaluation in archaeology. Finally I would like to thank Kelly Fennema for correcting the English text.

References

Aldenderfer, M., introduction to *Anthropology, Space, and Geographic Information Systems*, Aldenderfer, M. and Maschner, H.D.G., Eds., Oxford University Press, Oxford, 1996, pp. 3–18.

Ascenzi, A., A short account of the discovery of the Mount Circeo Neandertal cranium, *Quaternaria Nova*, 1, 69–80, 1990/91.

Attwell, M. and Fletcher, M., A new technique for investigating spatial relationships: significance testing, in *To Pattern the Past, PACT 11*, Voorrips, A. and Loving, S.H., Eds., Conseil de l'Europe, Strasburg, 1985, pp. 181–189.

Attwell, M.R. and Fletcher, M., An analytical technique for investigating spatial relationships, *Journal of Archaeological Science*, 14, 1–11, 1987.

Bender, B., Ed., *Landscape: Politics and Perspectives*, Berg, Oxford, 1993.

Blanc, A.C., On the Pleistocene sequence of Rome, paleoecologic and archeologic correlations, *Quaternaria*, 4, 95–109, 1957.

Bradley, R., *Rock Art and the Prehistory of Atlantic Europe: Signing the Land*, Routledge, London, 1997.

Brinkman, R. and Smyth, A.J., Eds., *Land Evaluation for Rural Purposes*, publication 17, ILRI, Wageningen, Netherlands, 1973.

Brinkman, R. and Young, A., Eds., *A Framework for Land Evaluation*, publication 22, ILRI, Wageningen, Netherlands, 1976.

Caloi, L. and Palombo, M.R., Large Paleolithic mammals of Latium (central Italy): palaeoecological and biostratigraphic implications, in *L'Homme de Neandertal*, Vol. 2, *L'Environnement*, Laville, H., Ed., Université de Liége, Liége, 1988, pp. 21–44.

Caloi, L. and Palombo, M.R., Les grands mammifères du Pléistocène supérieur de Grotta Barbara (Monte Circeo, Latium Meridional): encadrement biostratigraphique et implications paleoecologiques, *Quaternaria Nova*, 1, 267–276, 1990/91.

Cooney, G., *Landscapes of Neolithic Ireland*, Routledge, London, 2000.

Council of Europe, European convention on the protection of the Archaeological Hertitage (Revised) European Treaty Series no. 143, Valeta 16.1, 1992.

Ebert, J.I., The state of the art in "inductive" predictive modeling: seven big mistakes (and lots of smaller ones), in *Practical Applications of GIS for Archaeologists: A Predictive Modeling Toolkit*, Wescott, K.L. and Brandon, R.J., Eds., Taylor and Francis, London, 2000, pp. 129–134.

Eisner, W., Kamermans, H., and Loving, S.H., Resultati preliminari di una ricerca palinologica nell'Agro Pontino, Atti della XXIV riunione scientifica dell'Istituto Italiano di preistoria e protostoria nel Lazio, Firenze, 1984, pp. 207–211.

Eisner, W., Kamermans, H., and Wymstra, T.J., The Agro Pontino survey: results from a first pollen core, *Dialoghi di Archeologia*, 2, 145–153, 1986.

Eisner, W.R. and Kamermans, H., Late Quaternary Vegetation History of Latina, Italy: A Final Report on the Mezzaluna Core, in *The Agro Pontino Archaeological Survey*, Holstrom, S., Voorips, A., Kamermans, H., Eds., Archaeological Studies Leiden University (ASLU) 11, Leiden University, Leiden, 2004. CD-ROM..

Foley, R., *Off-site Archaeology and Human Adaptation in Eastern Africa*, BAR International Series 97, 1981.

Green, S.W., Approaching archaeological space: an introduction to the volume, in *Interpreting Space: GIS and Archaeology*, Allen, K.M.S., Green, S.W., and Zubrow, E.B.W., Eds., Taylor and Francis, London, 1990, pp. 3–8.

Haggett, P., *Locational Analysis in Human Geography*, Edward Arnold, London, 1965.

Hodder, I. and Orton, C., *Spatial Analysis in Archaeology*, Cambridge University Press, Cambridge, 1976.

Hunt, C.O. and Eisner, W.R., Palynology of the Mezzaluna core, in *The Agro Pontino Survey Project: Methods and Preliminary Results*, Voorrips, A., Loving, S.H., and Kamermans, H., Eds., Vol. 6 of Studies in Prae- en Protohistorie, Amsterdam, 1991, pp. 49–59.

Kamermans, H., Faulted land: the geology of the Agro Pontino, in *The Agro Pontino Survey Project: Methods and Preliminary Results*, Voorrips, A., Loving, S.H., and Kamermans, H., Eds., Vol. 6 of Studies in Prae- en Protohistorie, Amsterdam, 1991, pp. 21–30.

Kamermans, H., Archeologie en Landevaluatie in de Agro Pontino (Lazio, Italië), Ph.D. dissertation, University of Amsterdam, Amsterdam, 1993.

Kamermans, H., The Agro Pontino survey project: a summary of the application of land evaluation, in *Archaeology, Methodology and the Organisation of Research*, Milliken, S. and Peretto, C., Eds., A.B.A.C.O. s.r.l., Forli, Italy, 1996, pp. 51–60.

Kamermans, H., Land evaluation as predictive modelling: a deductive approach, in *Beyond the Map: Archaeology and Spatial Technologies*, Lock, G., Ed., NATO Science Series, Series A: Life Sciences, Vol. 321, IOS Press, Amsterdam, 2000, pp. 124–146.

Kamermans, H., Loving, S.H., and Voorrips, A., Changing patterns of prehistoric land use in the Agro Pontino, in *Papers in Italian Archaeology*, Vol. 4, Part 1, The Human Landscape, Malone, C. and Stoddart, S., Eds., BAR International Series 243, 1985, pp. 53–68.

Kamermans, H., Loving, S.H., and Voorrips, A., Archaeology and land evaluation in the Agro Pontino, in *La valle pontina nell'antichità: Atti del convegno*, Cherchi, M., Ed., Casa Editrice Quasar s.r.l., Roma, 1990, pp. 9–11.

Kohler, T.A., Predictive locational modelling: history and current practice, in *Quantifying the Present and Predicting the Past: Theory, Method, and Application of Archaeological Predictive Modeling*, Judge, W.L. and Sebastian, L., Eds., U.S. Bureau of Land Management, Denver, 1988, pp. 19–59.

Kohler, T.A. and Parker, S.C., Predictive models for archaeological resource locations, in *Advances in Archaeological Method and Theory*, Vol. 9, Schiffer, M.B., Ed., Academic Press, New York, 1986, pp. 397–452.

Kvamme, K.L., Determining empirical relationships between the natural environment and prehistoric site locations: a hunter-gatherer example, in *For Concordance in Archaeological Analysis: Bridging Data Structure, Quantitative Technique, and Theory*, Carr, C., Ed., Westport Publishers, Kansas City, MO, 1985, pp. 208–238.

Kvamme, K.L., Development and testing of quantitative models, in *Quantifying the Present and Predicting the Past: Theory, Method, and Application of Archaeological Predictive Modeling*, Judge, W.L. and Sebastian, L., Eds., U.S. Bureau of Land Management, Denver, 1988, pp. 324–428.

Kvamme, K.L., The fundamental principles and practice of predictive archaeological modeling, in *Mathematics and Information Science in Archaeology: A Flexible Framework*, Vol. 3, Voorrips, A., Ed., Studies in Modern Archaeology, Verlag-Holos, Bonn, 1990, pp. 257–295.

Kuhn, S.L., "Unpacking" reduction: lithic raw material economy in the Mousterian of West-Central Italy, *Journal of Anthropological Archaeology*, 10, 76–106, 1991.

Kuhn, S.L., *Mousterian Lithic Technology: An Ecological Perspective*, Princeton University Press, Princeton, NJ, 1995.

Loving, S.H., Estimating the age of stone artifacts using probabilities, in *Interfacing the Past: Computer Applications and Quantitative Methods in Archaeology*, CAA 95 proceedings, Kamermans, H. and Fennema, K., Eds., *Analecta Praehistorica Leidensia*, 28, 251–261, 1996a.

Loving, S.H., Discard of cores assigned to the Middle Paleolithic in the Pontine region of West Central Italy, *Quaternaria Nova*, 6, 505–532, 1996b.

Loving, S.H. and Kamermans, H., Field trials and errors: field methods used in the Agro Pontino survey, in *The Agro Pontino Survey Project: Methods and Preliminary Results*, Voorrips, A., Loving, S.H., and Kamermans, H., Eds., Studies in Prae-en Protohistorie 6, Amsterdam, 1991a, pp. 79–86.

Loving, S.H. and Kamermans, H., Figures from flint: first analysis of lithic artifacts collected by the Agro Pontino survey, in *The Agro Pontino Survey Project: Methods and Preliminary Results*, Voorrips, A., Loving, S.H., and Kamermans, H., Eds., Studies in Prae- en Protohistorie 6, Amsterdam, 1991b, pp. 99–116.

Loving, S.H., Kamermans, H., and Voorrips, A., Randomizing our walks: the Agro Pontino survey sampling design, in *The Agro Pontino Survey Project: Methods and Preliminary Results*, Voorrips, A., Loving, S.H., and Kamermans, H., Eds., Studies in Prae- en Protohistorie 6, Amsterdam, 1991, pp. 61–78.

Loving, S.H., Voorrips, A., and Kamermans, H., Old finds in new fields: first results of the Agro Pontino archaeological survey, *Bullettino di Paletnologia Italiana*, 83, 361–390, 1992.

Loving, S.H., Voorrips, A., Koot, C.W., and Kamermans, H., The Pontinian on the plain: some results from the Agro Pontino survey, *Quaternaria Nova*, 1, 453–477, 1990/91.

Mussi, M., The Neanderthals in Italy: a tale of many caves, in *The Middle Palaeolithic Occupation of Europe*, Roebroeks, W. and Gamble, C., Eds., University of Leiden, Leiden, Netherlands, 1999, pp. 49–80.

Redman, C.L., Multistage fieldwork and analytical techniques, *American Antiquity*, 38, 61–79, 1973.

Schoener, T.W., Theory of feeding strategies, *Ann. Rev. Ecol. Syst.*, 2, 369–404, 1971.

Segre, A.G., Nota sui rilevamenti esguiti nel Foglio 158 Latina della carta geologica d'Italia, *Bolletino del servizio geologico d'Italia*, 73, 569–584, 1957.

Sevink, J., Vos, P., Westerhoff, W.E., Stierman, A., and Kamermans, H., Sequence of marine terraces near Latina (Agro Pontino, Central Italy), *Catena*, 9, 361–378, 1982.

Sevink, J., Remmelzwaal, A., and Spaargaren, O.C., *The Soils of Southern Lazio and Adjacent Campania*, University of Amsterdam, Amsterdam, 1984.

Sevink, J., Duivenvoorden, J., and Kamermans, H., The soils of the Agro Pontino, in *The Agro Pontino Survey Project: Methods and Preliminary Results*, Voorrips, A., Loving, S.H., and Kamermans, H., Eds., Studies in Prae- en Protohistorie 6, Amsterdam, 1991, pp. 31–47.

Stiner, M.C., The faunal remains at Grotta Guattari: a taphonomic perspective, *Current Anthropology*, 32 (2), 103–117, 1991.

Stiner, M.C., *Honor among Thieves: A Zooarchaeological Study of Neandertal Ecology*, Princeton University Press, Princeton, NJ, 1994.

Stiner, M.C. and Kuhn, S.L., Subsistence, Technology, and Adaptive Variation in Middle Paleolithic Italy, *American Anthropologist*, 94, 306–339, 1992.

Stringer, C. and Gamble, C., *In Search of the Neanderthals*, Thames and Hudson, London, 1993.

Thomas, J., The politics of vision and the archaeologies of landscape, in *Landscape: Politics and Perspectives*, Bender, B., Ed., Berg, Oxford, 1993, pp. 19–48.

Tilley, C., *A Phenomenology of Landscape: Places, Paths and Monuments*, Berg, Oxford, 1994.

Verhoeven, A.A.A., Visibility factors affecting artifact recovery in the Agro Pontino survey, in *The Agro Pontino Survey Project: Methods and Preliminary Results*, Voorrips, A., Loving, S.H., and Kamermans, H., Eds., Studies in Prae- en Protohistorie 6, Amsterdam, 1991, pp. 87–97.

Voorrips, A., Loving, S.H., and Kamermans, H., An archaeological survey of the Agro Pontino (prov. of Latina, Italy), in *Archaeological Survey in the Mediterranean Area*, Keller, D.R. and Rupp, D.W., Eds., BAR International Series 155, 1983, pp. 179–181.

Voorrips, A., Loving, S.H., and Kamermans, H., Eds., *The Agro Pontino Survey Project: Methods and Preliminary Results*, Studies in Prae- en Protohistorie 6, Amsterdam, 1991.

Voorrips, A., Loving, S.H., and Strackee, J., The gamma mix density: a new statistical model for some aspects of system organization, in *To Pattern the Past*, PACT 11 proceedings, Voorrips, A. and Loving, S.H., Eds., Conseil de l'Europe, Strasburg, 1985, pp. 323–346.

Winterhalder, B., Optimal foraging strategies and hunter-gatherer research in anthropology: theory and models, in *Hunter-Gatherer Foraging Strategies: Ethnographic and Archaeological Analyses*, Winterhalder, B. and Smith, E.A., Eds., University of Chicago Press, Chicago, 1981a, pp. 13–35.

Winterhalder, B., Foraging strategies in the boreal forest: an analysis of Cree hunting and gathering, in *Hunter-Gatherer Foraging Strategies: Ethnographic and Archaeological Analyses*, Winterhalder, B. and Smith, E.A., Eds., University of Chicago Press, Chicago, 1981b, pp. 66–98.

Zampetti, D. and Mussi, M., Du Paléolithique moyen au Paléolithique supérieur dans le Latium (Italie centrale), in *L'Homme de Neandertal*, Vol. 8, Kozlowski, J.K., Ed., La Mutation, Université of Liége, Liége, 1988, pp. 273–288.

6

Regional Dynamics of Hunting and Gathering: An Australian Case Study Using Archaeological Predictive Modeling

Malcolm Ridges

CONTENTS

6.1   Introduction

Archaeological predictive modeling (APM) comprises a diverse set of approaches that have found application, primarily, in cultural resource management (CRM). The general aim of these approaches is to estimate the occurrence of archaeological material throughout a landscape given what is known about an existing archaeological sample (Sebastian and Judge 1988). Such knowledge has proved useful for improving the management of cultural heritage by permitting more effective regional conservation strategies (Hall and Lomax 1996; Kincaid 1988). However, it is also possible to outline some wider applications for the set of approaches that APM comprises. For the study outlined in this chapter, APM was used as an exploratory data tool to examine the factors determining the location of archaeological finds within a region in northwest central Queensland, Australia.

The high degree of correlation between patterns in the natural environment and the distribution of hunter-gatherer archaeological features is now well-established (Ebert and Kohler 1988; Jochim 1981; Veth et al. 2000). Such findings highlight the importance that environmental context has on the location of hunter-gatherer activities. However, an emerging issue is that environmental variables, on their own, are not appropriate for explaining all the variation in the location (or composition) of archaeological features (Gaffney and van Leusen 1995). Although environmental context is readily incorporated into GIS systems and spatial models of the distribution of archaeological features (Kvamme 1985), other factors are also important if the full range of variation in the location and form of archaeological features is to be understood (Whitley 2000).

Ebert (2000: 133) recently flagged this issue by suggesting that the frequency with which archaeological predictive models return an accuracy on the order of 60 to 70% may be indicative of an important component being omitted from models. If this is the case, then it also suggests that there is a need to expand the range of variables commonly employed in archaeological predictive modeling (Gaffney and van Leusen 1995). Addressing this issue requires a continuing focus upon the relationship between hunter-gatherer behavior and its archaeological expression (Binford 1982). However, for archaeological predictive modeling, it also means identifying spatial variables that are relevant to the location of hunter-gatherer activity, but that are not necessarily drawn directly from ecological patterns in the environment.

In part, the historical emphasis upon the use of variables describing patterns in the natural environment can be traced to the types of logic used to develop predictive models (Salmon 1976), which has led to a gap between models that predict versus those that explain (Altschul 1988). However, the position taken here is that, in reality, it is common for archaeologists to use both forms of logic in a feedback loop during research (VanPool and VanPool 1999), and that neither approach, on its own, provides a complete mechanism for understanding the factors driving archaeological patterns. From such a perspective, APM might be better thought of as a suite of tools that attempts to elucidate generalizations about the location and distribution of archaeological features, which can also be used to predict the spatial context in which they occur.

A related issue is the persistence of applying APM to "sites," in particular, in the case of hunter-gatherer studies, open lithic scatters. The types of hunter-gatherer behavior associated with the activities preserved in open lithic scatters generally demonstrate a high degree of correlation with patterns in the natural environment due to the requirements of subsisting through hunting and gathering. But this does not necessarily mean that all hunter-gatherer archaeological features will also correlate with patterns in the environment to the same degree or in the same way. It therefore remains possible that the importance of environmental context may be overestimated because the contribution of other factors has not been thoroughly explored separately for different kinds of archaeological evidence. When different types of archaeological evidence are modeled independently, significant differences have been observed in the models produced and in the importance of individual variables (Ridges 2003).

In this study, a large number of open lithic scatters were considered, along with a significant corpus of rock art that existed in the region. To some extent, an understanding of the roles that rock art and stone-tool production played in regional behavior could be derived by studying them independently. However, such an approach would overlook the fact that both were components of a regional system of behavior, and that some dependent relationship probably existed between them, despite their distinctiveness. The approach adopted in this study was to develop their respective spatial contexts, since this potentially provided an avenue to understanding the commonality between them. It was here that APM demonstrated its utility as an important research tool.

In many instances, generalizations about where archaeological features occur can be used to predict the location of undiscovered archaeological features. But equally important, they can also be used to refine theories about past behavior. With this in mind, APM was utilized in this study as a tool to explore the spatial trends in archaeological data. In turn, these patterns were important in understanding the kinds of behavior that produced the regional pattern of archaeological feature distribution.

6.2 Background

The work reported in this chapter formed part of a project conducted in northwest central Queensland, Australia (see Figure 6.1). This is a remote part of Australia, on the fringe of the arid zone. Aridity in the region results from a low and highly unpredictable rainfall pattern. Rain mostly occurs during the summer months in the form of thunderstorms. Evaporation far exceeds precipitation, so that standing water is rare, and there are no continuously flowing rivers in the region. Topographically, the region consists of low rugged uplands and broad alluvial plains. The region demonstrates little relief, varying by only about 200 m in elevation. Nonetheless, where relief is appreciable, it is generally abrupt, and although not sufficient to completely hinder the movement of people in any direction, it is enough to have strongly influenced the routes people traveled. Ethnographic reports (Roth 1897) indicate that the movement of people closely followed the drainage lines, which also provided the most reliable and plentiful sources of water when it was available.

In the upland zones, the vegetation is generally low, open woodland interspersed with hummock grasses. On the plains, this gives way to extensive areas of tussock grassland. The geology consists of Cambrian and Precambrian sedimentary units that have been intruded by granite. The

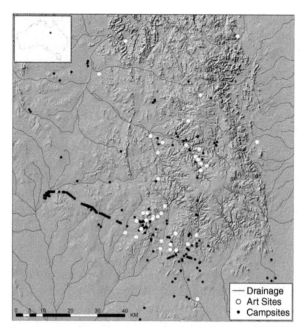

FIGURE 6.1
Location of the study region.

upwarping of the sedimentary units produces a low tableland forming abrupt mesas at its margins. It is in rock shelters along these mesas that many of the rock-art sites encountered in the region are located. A wide variety of rock types occur in the surrounding region, enabling stone artifacts to be manufactured from quartz, quartzite, chalcedony, chert, silcrete, and metabasalt.

Aboriginal occupation in the region extends back at least 15,000 years, although it appears to have been punctuated by periods of abandonment (Davidson et al. 1993). These people adopted a hunting and gathering lifestyle that was typical for the arid zone of Australia (e.g., Gould 1980). Their population density was very low, and they were highly mobile. Importantly, Roth, the most prominent ethnographic source for the region, suggests that the movement of people related as much to social and economic reasons as it did to subsistence (Roth 1897: 132).

The sparseness of the vegetation means that archaeological visibility in the region is very good. The most common archaeological feature is open lithic scatters. In an investigation of sampling methods, Davidson observed stone artifacts in every 500-m quadrant he examined (Davidson 1993b), demonstrating the abundance of stone artifacts occurring in the region. Stone artifacts were also flagged by Hiscock as one of the significant research interests of the region (Hiscock 1988), and stone technology has been the focus of several theses (Drury 1996; James 1993; Kippen 1992). The other most commonly encountered archaeological features include rock-art sites (Davidson et al. in press; Ross 1997), ceremonial stone arrangements, and stone and ocher quarries.

The region, which provides a rich and well-preserved archaeological record, also played host to two kinds of behavioral processes that are important for understanding the prehistory of the region: trade and rock art. Each of these is outlined briefly below.

Roth described in some detail the movement of items into and out of northwest central Queensland. The most important of these items were stone axes manufactured within the region (Davidson et al. in press); ocher, which was also available within the region (Jones 1984); and the narcotic plant pituri (*Duboisi hopwoodii*) (Watson 1980), which occurs just to the south of the region. Along with naturally available items, rights to hold ceremonial events were also important in the network of regional exchange (Roth 1897: 120–125). The harvesting of pituri each year facilitated several months of ceremonies and markets at various places throughout the region (Roth 1897: 133). The region has been implicated in the vast network of exchange that extended from the Gulf of Carpentaria to the Flinders Ranges in Australia's south (McBryde 1987; Mulvaney 1976). In turn, these extended links formed part of the trunk routes crisscrossing the Australian continent (McCarthy 1939) that facilitated connections between Aboriginal people throughout mainland Australia.

Northwest central Queensland also contains a diverse and underreported assemblage of rock art about which there is little oral tradition. About 180 art sites containing paintings or engravings have been recorded in the region. Morwood (1985) outlined the distinctiveness of a recurring anthropomorphic figure that set the region apart from the other major centers of rock art in Australia. Detailed analysis of these figures by Ross (1997) showed that they were spatially restricted to the region. Ross also demonstrated that the anthropomorphic figures are depicted with stylistic elements that, while adhering to a regional style, potentially encoded additional information about group affiliation. Ross concluded that the method of portrayal of anthropomorphic figures indicated a social system that sought to distinguish itself from its neighbors. For whatever reason, it was important to mark places throughout the region with a prolific array of iconic pictures.

A social system that sought to distinguish itself from its neighbors contrasts with the freedom of movement associated with trade. Similarly, if anthropomorphic figures were part of a boundary maintenance system, the process was not straightforward given that they occur throughout the region rather than at its periphery. Thus, some relatively complex spatial processes are evident in the behavior of Aboriginal people in the region. Importantly, these spatial processes involve social and economic interactions, so that the location of archaeological features cannot be understood purely from the viewpoint of subsistence. Understanding these spatial processes necessarily involves invoking procedures capable of characterizing economic and social spatial relationships along with environmental context.

6.3 Approach

The approach adopted in this study aimed to explore some of the concepts that have been developed by Zubrow (1994) and van Leusen (in Gaffney and van Leusen 1995). Doing so involved three steps:

1. Constructing a model of the location of open campsites using conventional archaeological predictive-modeling procedures and variables
2. Examining the spatial arrangement of residuals in this model to identify where and how the model was deficient
3. Deriving a new variable that addressed these deficiencies and that produced a model with better predictive power and improved explanatory power

The first step is covered only briefly because it is not the focus of this paper, and the methods employed have been well described in other places (Judge and Sebastian 1988; Kvamme 1989; Warren 1990; Wescott and Brandon 2000). In summary, the first step involved forming an inductive predictive model using logistic regression, where the dependent variable was the presence or absence of open campsites. The input data was derived from archaeological surveys conducted over a ten-year period (Davidson 1993a) and comprised 840 100-m squares that have been surveyed, 322 of which contained campsites and associated artifact assemblages. In addition, 46 art sites have been recorded in the study area.

The independent variables included geology (Blake et al. 1983), vegetation (Neldner 1991), elevation,[1] slope, aspect, wetness[2] (Wolock and McCabe 1995), topographic position (Jenson and Domingue 1988), and proximity to streamline. The last variable was defined especially for this study to describe the particular importance of water in semiarid landscapes (see Figure 6.2).

Streamline proximity was measured using the accumulated cost distance of slope away from drainage lines. In addition, the cost distance was measured separately for each of the stream orders resulting from a Strahler (1952) classification of the drainage pattern. Each of these cost-distance surfaces was then rescaled to have values ranging from 0 to 1, and inverted so that areas close to streams had a value of 1. The layers were then combined arithmetically using a weight corresponding to stream order. Thus:

Streamline proximity = (cost distance from stream order 1)
+ (2 × cost distance from stream order 2)
+ (3 × cost distance from stream order 3) ... etc.

The advantage of measuring streamline proximity in this way was that it captured the greater likelihood of finding water in downstream areas.

FIGURE 6.2
Streamline proximity variable.

Because none of the streams in the region flow regularly, and because water holes are intermittent in the sandy bottoms of the drainage channels, measuring streamline proximity in this way better reflected the nature of water availability. A second advantage of the approach was that it identified the confluence of several middle-order streams. In cost-benefit terms, these areas offered the greatest potential for finding multiple sources of water within close proximity to one another.

A presence-only approach was used to develop the model, so that a set of 500 randomly distributed points comprised the comparison, or nonsite, data. It is acknowledged that such an approach incurs the potential for misrepresenting a "site" as a "nonsite" (Kvamme 1988). However, because the important element of this study was to examine the trends in the occurrence of archaeological features, and in the absence of surveyed "nonsites," this was deemed the best approach possible. After performing appropriate statistical tests to confirm the appropriateness of each variable for distinguishing campsites versus noncampsite locations (following Warren and Asch 2000: 15), a model was formed using all the input variables. This model was then used to predict the probability of a campsite occurring in any of the 1-Ha grid cells used to describe the study area.

The most important aspect of the approach used in this study concerns the steps taken once the model was built. Although the approach adopted in step one routinely results in a model with a useful degree of predictive accuracy, in this study it was known *a priori* that the variables outlined above were unlikely to be the only factors determining the location of archaeological features. The difficulty, in terms of developing a methodology for addressing the social factors that were known to be playing a role in the region, is how such factors could be identified, and how variables describing them could be incorporated into the model.

It is here that the ideas developed by van Leusen (in Gaffney and van Leusen 1995) and Zubrow (1994) become important. Van Leusen suggested that human settlement patterns can be understood in terms of two components: environment and culture. Through using environmental variables to describe the location of activities, van Leusen argued that any remaining variation should reflect cultural factors unrelated to, in this case, subsistence through hunting and gathering. Thus he states (Gaffney and van Leusen 1995: 370): "By applying an ED [environmental determinism] model to a dataset, one can eliminate environmental patterning in the data, leaving a clearer view of whatever cultural factors may have influenced the data."

Van Leusen's idea has some appeal in that it captures the notion that although there may be strongly influential factors in determining where people choose to locate their activities in a given landscape, there can remain other significant factors determining more subtle levels of variation. In other words, despite the patterns in the environment to which hunter-gatherers must adapt, there still remains an influential overarching sociocultural system (Gamble 1986: 31). Pickering (1994) has argued that despite ample anthropological descriptions of the way social landscapes operated in Aboriginal societies, very little of hunter-gatherer social landscapes produces detectable evidence in the archaeological record. Thus, even with knowledge that an overarching sociocultural system is present, we may find that, given the nature of the archaeological signature of hunter-gatherers, the locational variation produced by many sociocultural processes is masked by the dominance of subsistence-related behavior. However, as Gamble (1986: 299) observed for the Palaeolithic in Europe, such a proposition remains largely untested archaeologically.

It is unlikely, therefore, that van Leusen's idea can be applied in such simplistic terms. For one thing, models produced using ecological context are subject to how well the subsistence system is captured in such an approach. For example, the remaining variation might be produced through subsistence behavior not described by the environmental variables employed. Similarly, as Gaffney (Gaffney and van Leusen 1995: 375) argues, it is difficult to separate behavior induced by patterns in the environment from a cultural decision to adapt to an environment in a particular way, thus precluding any simple, two-part separation.

The other important work in this discussion is Zubrow's (1994) approach to mapping cognitive space. Using a modeled optimal settlement pattern and that observed archaeologically for the Iroquois cultural area, Zubrow hypothesized that the differences between them reflected variation within an otherwise economically driven pattern. In this case, the variation was driven by the operation of a cognitive landscape. Zubrow attributed the degree of difference from the predicted location for a site to a host of local and regional cultural factors, unrelated to the economic system. Zubrow argued that by examining the differences between a predicted and observed settlement pattern, the effect of one aspect of a cultural system upon another aspect could be observed when the predicted pattern was based on explicit theoretical statements. In other words, Zubrow effectively describes an approach whereby the idea of van Leusen might be investigated.

An approach to applying Zubrow's methodology is to look at the residuals of the model produced in step one. Traditionally, the analysis of residuals is used to assess the quality of the fit of a regression model to the data, since the mathematics of regression seeks to minimize the size of the residuals. However, it is important to realize that in modeling spatial data, the residuals have a spatial component also, and systematic variation in the spatial arrangement of residuals can be informative about spatial variables that could be missing from a model. Or, as in this case, they can be used to identify nonenvironmental variables that have a systematic spatial pattern. Unfortunately, very few archaeological predictive modeling projects examine the spatial patterning of residuals of models to see whether any identifiable systematic variation remains.

Whereas in Zubrow's study the objective was to propose an approach to mapping a cognitive landscape, in this study the analysis of residuals was used as a guide in identifying where systematic spatial variation remained in the model. Equipped with such knowledge, variables demonstrating a high degree of correlation with the pattern of residuals could then be investigated. As will be illustrated below, a relationship was found between the size of the residual and the proximity of art sites. With additional variables identified, their inclusion into the model may then improve the quality of the resulting model in terms of both predictive and explanatory power.

6.4 Results

Figure 6.3 shows the results of applying the methodology outlined for step one. The calculated accuracy of the model (i.e., the number of campsites predicted by the model to have a probability greater than 0.5) was 71%. Important in the model is the emphasis it places on the highest probability zones being on the flatter terrain in the bottom and right of Figure 6.2. The areas in the center of Figure 6.3 are the uplands that comprise a series of broad sandstone mesas. From the mesa areas, the drainage largely follows a radial pattern, illustrating how these upland areas are at the headwaters of most creeks and have very low capacity for containing water. To a large

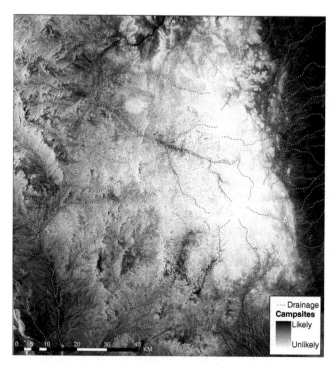

FIGURE 6.3
Conventional inductive model using environmental variables.

extent, then, the pattern produced in this model is consistent with what would be expected from a model primarily describing subsistence behavior.

Figure 6.3 might also be considered as a reflection the environmental component of archaeological location variation if we were to apply van Leusen's idea. However, it is worth exploring other variables because, with a predicted accuracy of 71%, the model is still within one standard deviation of what would be achieved by randomly setting locations as campsite or noncampsite. It would therefore appear that either a significant amount of variation remains unexplained by the model, or that a significant amount of the variation in the location of campsites is random.

The next step was to examine the residuals in the model, explore what other variables might be correlated with them, and provide an avenue for improving the model. As was explained earlier, the other important type of archaeological feature occurring in the study region is rock art. These rock-art sites are known to have played an important role in the maintenance of a social identity for the people in the region (Ross 1997), and as such were likely to have played an important role in structuring behavior at the regional level.

Figure 6.4 shows the distribution of art sites in the study area and, in conjunction, also shows the size of the residual at each of the recorded campsites in the study area by varying the size of the circle representing them. A visual examination of Figure 6.4 reveals that there is a tendency for

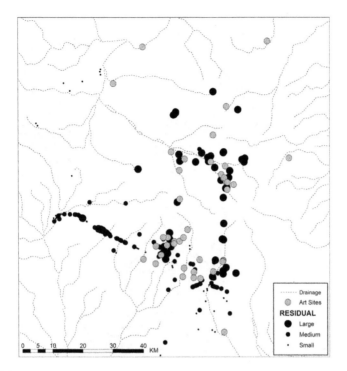

FIGURE 6.4
Comparison of art sites and residuals from the model based on environmental variables only.

the largest residuals, i.e., the largest white circles, to be located close to art sites, suggesting that proximity to an art site may be an important element in unexplained variation in the model. This was supported by performing a spatial autocorrelation analysis on the residuals, using the Greary measure (Cliff and Ord 1973). The result was a value of 0.68, which indicates there is a strong tendency for an archaeological location with a large residual to occur close to locations with similarly high residuals, and vice versa for locations with small residuals. It therefore became clear that there was a significant degree of systematic spatial variation in the residuals of the model.

To explore this idea further, a proximity-to-art-sites layer was derived using a cost-distance function and slope as a cost layer. The result was similar to a catchment analysis (Bailey and Davidson 1983), where high values indicate areas that are proximate to several art sites. Thus, if we were to think about access to art sites in a similar way as we might model access to resources (Winterhalder 1981), then areas in this layer with high values might be preferentially chosen for placing campsites if being close to art sites was important. The resulting layer is presented in Figure 6.5. It can be seen in this figure how areas along drainage lines that are close to several art sites produce values that are higher. Hence, in some cases at least, it would appear that sticking to drainage lines could serve a dual purpose, providing access to water and several art sites.

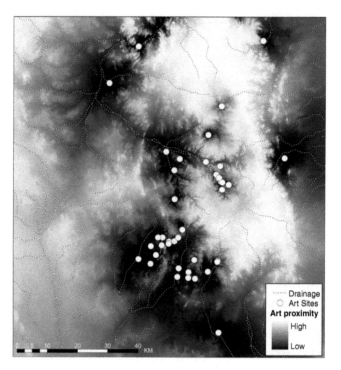

FIGURE 6.5
Proximity-to-art-sites layer.

Incorporating the distinctiveness of each site into the proximity layer extends this idea. The recording of art sites in the study region has revealed that some may contain just one or two paintings, whereas others can have hundreds, the implication being that some art sites may have been more important, or perhaps visited more often, than others. Consequently, it could also be suggested that the sphere of influence of each art site is likely to be different.

With this in mind, a procedure similar to that used to derive streamline proximity was employed. In this instance, each of the art sites was given a score for its diversity, measured as the number of motif types occurring at that site. The diversity values were then classed into five categories, and the cost distance derived for each class separately. The five resulting layers were then rescaled to have values ranging from 0 to 1 and inverted so that values of 1 occurred at art sites. The layers were then summed arithmetically in the same way as for stream orders, by applying a weight for the degree of diversity occurring at the art sites in each class. The result was the layer presented in Figure 6.6, which also shows the diversity of each art site as a function of the size of the dot representing it.

A final examination of the art-sites-proximity variable was to compare the proximity values occurring for each campsite with the size of the residual occurring from the initial model. This is produced as a scatter plot in Figure 6.7. In this figure, it can be seen that as the size of the residual becomes

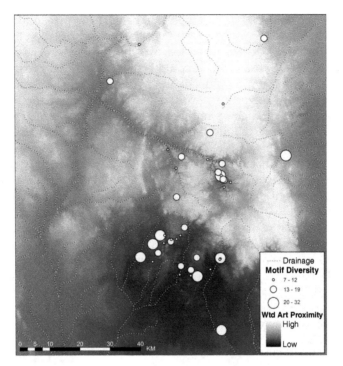

FIGURE 6.6
Art-site-proximity variable, weighted by site diversity.

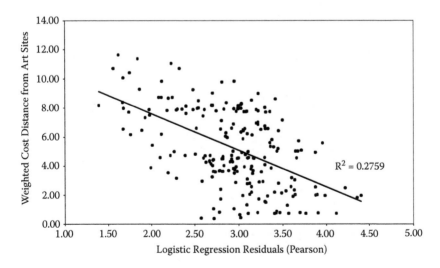

FIGURE 6.7
Plot of weighted art-site proximity against regression residuals.

larger, there is a tendency for a campsite to be in closer proximity to art sites that have a broad diversity. The result was a strong correlation between the size of the residual, and the weighted proximity to art site, as is indicated by an r^2 value of 0.3. From this it could be concluded that proximity to art sites was an important component in influencing where campsites were located, and that proximity to art sites was more important for art sites containing a diverse range of motifs.

The final step was to rebuild the model, this time including the weighted proximity-to-art-sites variable. This model is presented in Figure 6.8, where it can now be seen that higher probabilities occur for much of the upland areas where art sites are located. Significantly, this means that the model gives greater weight to areas along second- and third-order streams, which is more consistent with observations in the region whereby many campsites are encountered at the base of mesas. Similarly, there is less distinction along the course of drainage lines, as campsites in the new model are predicted to occur in more or less equal probability along their full course. This is consistent with the ethnographic observations of Roth (1897), who observed that the drainage lines were the main routes of travel. Correspondingly, the calculated accuracy of the model has improved to 75% of campsites now being predicted to have a probability greater than 0.5.

FIGURE 6.8
New model incorporating weighted proximity to art sites.

Interestingly, the eastern margin of the mesas in both models maintained very similar predictions. As may have been gathered from the distribution of recorded campsites and art sites in Figure 6.4, there is considerable bias in survey coverage in the region, with most emphasis being on the western and southern portion of the region. Consequently, an indirect benefit of approaching model development in the way it was for this study has been its utility for designing future work. Although survey in this region has been intensive, it has not been comprehensive, in the sense that there are several areas that could now be targeted for future survey using testable assumptions arising from the models (in particular, whether art sites that are likely to occur on the eastern margin of the uplands demonstrate a similar influence on the location of campsites). Such results have implications for cultural resource management in that not only is the ecological context of a model testable through revised research design, but also the explanatory utility of the model, as behavioral principles, can also be investigated.

6.5   Discussion

The position adopted in this study was that all archaeological evidence reflects a multitude of influences. Many of these influences are difficult to observe at the gross level because they can be swamped by one or two dominating factors. Nonetheless, this study proceeded with the intention of attempting to identify at least one example of how nonsubsistence-related behavior might be (a) important in determining the location of campsites and (b) incorporated into archaeological predictive modeling. The gain in predictive accuracy from 71% to 75% may not have been a substantial improvement in predictive terms, but the real benefit is in a model that improves in explanatory power as well. Thus the approach adopted in this study offers some scope for getting out of the mold of employing either deductive versus inductive logic for model formation (Ebert 2000; Salmon 1976).

Crucial to moving forward in this way is viewing the location of hunter-gatherer activities as being flexible, rather than rigidly adhering to ecological context. Although there undoubtedly were limitations in the way hunter-gatherers located their activities in semiarid environments such as western Queensland, it would appear that, within these limitations, there was considerable scope for other factors to come into play. It would therefore appear that van Leusen's notion of separating environmentally driven variation from other aspects of culturally determined variation has some merit. But equally, there is still quite a vast array of factors that could be important in this regard, and separating them is unlikely to be straightforward. Thus, what appears to be taking place in this case was that hunter-gatherers were primarily locating their activity within broad zones that are all relatively efficient in terms of providing access to resources. However, within those zones, there remains a degree of flexibility about the specifics driving the location of activities in individual cases.

This is consistent with anthropological observations of hunter-gatherer behavior. Woodburn (1972: 205) has noted that the movement of people between camps was primarily motivated by social forces. In particular, changing camp involved changing the associations between people, and often this was a means of relieving tensions in the group. As Lee (1972: 181–182) also observes, although scarcity of water may draw people together to camp at water holes, eventually it was tensions within the group that drove them apart again. Hence, although social factors were involved in where hunter-gatherers located their activities, we do not understand very well how they manifest themselves archaeologically. At issue for hunters and gatherers is that some degree of planning would clearly have been required to recover resources efficiently (Winterhalder and Smith 1981), but of equal importance was the flexibility needed to cope with changing environmental and social conditions (Smith 1988). Some of these issues have been borne out in the approach adopted in this study.

Through exploring the residuals of a conventional inductive model, it was possible to demonstrate some of the limitations of using only environmental variables to derive predictive models. This is not to make the case that cultural variables are more important, but to illustrate the advantages of using both in conjunction, which really should come as no surprise. The need to incorporate such variables has always been present, but this approach has been hampered by methodological difficulties. The utility of exploring the remaining spatial variation in residuals is that it can potentially provide important clues to where other variables might be sought.

In this study, it was known that the rock art was playing an important role in structuring social behavior, but how that influence filtered down to other forms of archaeological evidence and regional patterns was not clear. However, by approaching the problem as it was in this study, it can now be seen that the influence of social behavior can extend to forms of archaeological evidence where, traditionally, social behavior has been difficult to observe archaeologically (Pickering 1994). The consequence is that, even in the presence of a dominating correlation with ecological context, it may be that variation from such trends proves to be more informative about regional behavior than the existence of the trend itself.

This position bears some resemblance to Binford's (1987) opinion on what he referred to as the ambiguity of the archaeological record. Binford suggested that ambiguity in the archaeological record develops from alternative behavioral processes producing similar material patterns. Because he saw ambiguity as the main hindrance to archaeological interpretation, Binford argued that it was necessary to resolve the causes of ambiguity to improve our understanding of archaeological pattern using empirical generalizations. In contrast, the view taken here was that the ambiguity in the archaeological record represents a window into the complexity of human behavior, which itself is inherently ambiguous (Hodder 1982).

Developing a methodology for this approach meant reassessing what it is that archaeologists are trying to describe about the archaeological record. Rather than focusing solely on generalizations, as would be the case in a strictly positivist framework, the generalizations were used in this study to form the baseline from which variation could be measured. Hence, this study adopted the premise that the environment is determinant in the sense that it sets out, in broad terms, the constraints in adopting a particular mode of existence in a given landscape (Gamble 1986: 30). However, within that set of constraints lies another level where the social relations are ultimately dominant in specifying the particular behavioral response to the constraints imposed by the environment (Ingold 1981).

With so much variation present (at the end of this study, 25% still remains unexplained), it may be tempting to lean toward simple linear relationships between archaeological find spots and ecological context, which are easier to understand and predict. However, such behavioral complexity should be the challenge, not the problem.

6.6   Conclusions

Archaeological predictive modeling (APM) is well-placed to contribute to the broader study of regional hunter-gatherer behavior. As a method for describing the spatial pattern within regional data sets, APM provides a useful means of identifying generalizations about the distribution of archaeological features. At the moment, the application of APM in hunter-gatherer archaeology is somewhat dominated by inferences derived from subsistence strategies and the needs of cultural resource management. However, there is no reason why the suite of tools APM represents cannot be applied to a wider spectrum of behavioral issues. Thus, other social and economic processes can also be investigated with APM if issues with appropriate variables can be resolved (Gaffney and van Leusen 1995: 374). However, making the most of the potential contribution of APM requires viewing the approach from a slightly different perspective. Rather than seeing it as a theoretical statement about the nature of hunter-gatherer subsistence patterns (Church et al. 2000), APM might be better viewed as an approach that consists of a suite of tools designed to explore spatial trends in archaeological datasets. With this in mind, APM can potentially offer a great deal to a range of theoretical perspectives of hunter-gatherer behavior by providing a means for characterizing spatial trends in holistic and sophisticated ways.

Notes

1. A digital elevation model (DEM) was derived using the 20-m contours, spot heights, and 1:100,000 scale drainage available for the study area. These were used as input for the TOPOGRID function available in the grid module of Arc/Info, which produces a hydrologically correct DEM using the ANU DEM algorithm (Hutchinson 1989).
2. Wetness is an index derived from a DEM based upon the capacity of a single grid cell to retain surface flows of water based upon the slope at that point and the size of the water catchment leading into that cell.

References

Altschul, J.A., Models and the modeling process, in *Quantifying the Present and Predicting the Past: Theory, Method, and Application of Archaeological Predictive Modeling*, Judge, W.J. and Sebastian, L., Eds., U.S. Department of the Interior, Bureau of Land Management, Denver, 1988, pp. 61–96.

Bailey, G. and Davidson, I., Site exploitation territories and topography: two case studies from Palaeolithic Spain, *Journal of Archaeological Science*, 10, 87–115, 1983.

Binford, L.R., The archaeology of place, *Journal of Anthropological Archaeology*, 1, 5–31, 1982.

Binford, L.R., Researching ambiguity: frames of reference and site structure, in *Method and Theory for Activity Area Research*, Kent, S., Ed., Columbia University Press, New York, 1987, pp. 449–512.

Blake, D.H., Jacques, A.L., and Donchak, P.J.T., 1:100 000 Geological Map Commentary: Selwyn Range, Queensland, Australian Government Printing Service, Canberra, 1983.

Church, T., Brandon, R.J., and Burgett, G.R., GIS applications in archaeology: method in search of theory, in *Practical Applications of GIS for Archaeologists: A Predictive Modeling Toolkit*, Wescott, K.L. and Brandon, R.J., Eds., Taylor and Francis, Philadelphia, 2000, pp. 135–155.

Cliff, A.D., and Ord, J.K., *Spatial autocorrelation*, Pion, London, 1973.

Davidson, I., Archaeology of the Selwyn Ranges, in *People of the Stone Age: Hunter-Gatherers and Early Farmers*, Burenhult, G., Ed., Harper Collins, New York, 1993a, pp. 210–211.

Davidson. I., Sampling in Selwyn. Departmental Seminar, Department of Archaeology and Palaeoanthropology, University of New England, Armidale, NSW, Australia. 1993b.

Davidson, I., N.D.J. Cook, M. Fischer, M. Ridges, J. Ross, and S.A. Sutton, Archaeology in another country: exchange and symbols in North West Central Queensland, in *Many Exchanges: Archaeology, History, Community and the Work of Isabel McBryde*, I. Macfarlane, M.-J. Mountain, and R. Paton, Eds., Aboriginal History, in press.

Davidson, I., Sutton, S., and Gale, S.J., The human occupation of Cuckadoo 1 Rock-shelter, northwest Central Queensland, in *Sahul in Review: Pleistocene Archaeology in Australia, New Guinea and Island Melanesia*, Smith, M.A., Spriggs, M., and Fankhauser, B., Eds., Australian National University Press, Canberra, 1993, pp. 164–172.

Drury, T., An investigation into the Influence of Geological Landscapes and Land Systems on Stone Artefact Variability in the Selwyn Range, Northwest Queensland, Bachelor of Arts (Hons.) thesis, University of New England, Armidale, NSW Australia, 1996.

Ebert, J.I., The state of the art in "inductive" predictive modeling: seven big mistakes (and lots of smaller ones), in *Practical Applications of GIS for Archaeologists: A Predictive Modeling Toolkit*, Wescott, K.L. and Brandon, R.J., Eds., Taylor and Francis, Philadelphia, 2000, pp. 129–134.

Ebert, J.I. and Kohler, T.A., The theoretical basis of archaeological predictive modeling and a consideration of appropriate data-collection methods, in *Quantifying the Present and Predicting the Past: Theory, Method, and Application of Archaeological Predictive Modeling*, Judge, W.J. and Sebastian, L., Eds., U.S. Department of the Interior, Bureau of Land Management, Denver, 1988, pp. 97–171.

Gaffney, V. and van Leusen, M., GIS, environmental determinism and archaeology: a parallel text, postscript in *Archaeology and Geographic Information Systems: A European Perspective*, Lock, G. and Stančič, Z., Eds., Taylor and Francis, London, 1995, pp. 367–382.

Gamble, C., *The Palaeolithic Settlement of Europe*, Cambridge University Press, Cambridge, 1986.

Gould, R.A., *Living Archaeology*, Cambridge University Press, Cambridge, 1980.

Hall, R. and Lomax, K., A regional landscape approach to the management of stone artefact sites in forested uplands in Eastern Australia, *Australian Archaeology*, 42, 35–38, 1996.

Hiscock, P., Archaeological Investigations in the Boulia District, Southwestern Queensland, report to the Australian Heritage Commission, 1988.

Hodder, I., *Symbols in Action*, Cambridge University Press, Cambridge, 1982.

Hutchinson, M.F., A new procedure for gridding elevation and stream line data with automatic removal of spurious pits, *Journal of Hydrology*, 106, 211–232, 1989.

Ingold, I., The hunter and his spear: notes on the cultural mediation of social and ecological systems, in *Economic Archaeology*, Sheridan, A. and Bailey, A., Eds., BAR International Series, 96, Oxford, 1981, pp. 119–130.

James, R.E., Stones, Samples and the Stories We Tell, unpublished BA (Hons.) thesis, University of New England, Armidale, NSW Australia, 1993.

Jenson, S.K. and Domingue, J.O., Extracting topographic structure from digital elevation data for geographic information system analysis, *Photogrammetric Engineering and Remote Sensing*, 54, 1593–1600, 1988.

Jochim, M.A, *Strategies for Survival: Cultural Behaviour in an Ecological Context*, Academic Press, New York, 1981.

Jones, P., Red ochre expeditions: an ethnographic and historical analysis of Aboriginal trade in the Lake Eyre Basin, *Journal of the Anthropological Society of South Australia*, 22, 3–10, 1984.

Judge, W.J., and Sebastian, L., Eds., *Quantifying the Present and Predicting the Past: Theory, Method, and Application of Archaeological Predictive Modeling*, U.S. Department of the Interior, Bureau of Land Management, Denver, 1988.

Kincaid, C., Predictive modeling and its relationship to cultural resource management applications, in *Quantifying the Present and Predicting the Past: Theory, Method, and Application of Archaeological Predictive Modeling*, Judge, W.J. and Sebastian, L., Eds., Department of the Interior, Bureau of Land Management, Denver, 1988, pp. 549–569.

Kippen, K., An Analysis of Stone Artefacts from Twelve Open Scatters in the Selwyn Range, Northwest Central Queensland, unpublished B.A. (Hons.) thesis, University of New England, Armidale, NSW Australia, 1992.

Kvamme, K.L., Determining empirical relationships between the natural environment and prehistoric site locations: a hunter-gatherer example, in *For Concordance in Archaeological Analysis*, Carr, C., Ed., Westport Publishers, Kansas City, MO, 1985, pp. 208–238.

Kvamme, K.L., Development and testing of quantitative models, in *Quantifying the Present and Predicting the Past: Theory, Method, and Application of Archaeological Predictive Modeling*, Judge, W.J. and Sebastian, L., Eds., Department of the Interior, Bureau of Land Management, Denver, 1988, pp. 325–428.

Kvamme, K.L., Geographic information systems in regional archaeological research and data management, in *Archaeological Method and Theory*, Vol. 1, Schiffer, M.B., Ed., University of Arizona Press, Tucson, 1989, pp. 139–203.

Lee, R., Work effort, group structure and land-use in contemporary hunter-gatherers, in *Man, Settlement and Urbanism*, Ucko, P.J., Tringham, R., and Dimbleby, G.W., Eds., Duckworth, London, 1972, pp. 180–195.

McBryde, I., Goods from another country: exchange networks and the people of the Lake Eyre Basin, in *Australians to 1788*, Mulvaney, D.J. and White, J.P., Eds., Fairfax, Syme and Weldon Associates, Sydney, 1987, pp. 252–273.

McCarthy, F.D., "Trade" in Aboriginal Australia and "trade" relations with Torres Strait, New Guinea and Malaya, *Oceania*, 9, 405–438, 1939; 10, 80–104, 171–195, 1939.

Morwood, M.J., Facts and figures: notes on rock art in the Mt. Isa area, Northwest Queensland, *Rock Art Research*, 2, 140–145, 1985.

Mulvaney, D.J., The chain of connection: the material evidence, in *Tribes and Boundaries in Australia*, Peterson, N., Ed., Australian Institute of Aboriginal Studies, Canberra, 1976, pp. 72–94.

Neldner, V.J., Vegetation Survey of Queensland, Queensland Botany Bulletin No. 9, Queensland Department of Primary Industries, Brisbane, 1991.

Pickering, M., The physical landscape as a social landscape: a Garawa example, *Archaeology in Oceania*, 29, 149–161, 1994.

Ridges, M., Numerous Indications: The Archaeology of Regional Hunter-Gatherer Behaviour in Northwest Central Queensland, Australia, Ph.D. thesis, University of New England, Armidale, NSW Australia, 2003.

Ross, J., Painted Relationships: An Archaeological Analysis of a Distinctive Anthropomorphic Rock Art Motif in Northwest Central Queensland, unpublished B.A. (Hons.) thesis, University of New England, Armidale, NSW Australia, 1997.

Roth, W.E., *Ethnological Studies among the North-West Central Queensland Aborigines*, Government Printer, Brisbane, 1897.

Salmon, M.H., "Deductive" versus "inductive" archaeology, *American Antiquity*, 41, 376–381, 1976.

Sebastian, L. and Judge, W.J., Predicting the past: correlation, explanation, and the use of archaeological models, in *Quantifying the Present and Predicting the Past: Theory, Method, and Application of Archaeological Predictive Modeling*, Judge, W.J. and Sebastian, L., Eds., U.S. Department of the Interior, Bureau of Land Management, Denver, 1988, pp. 1–18.

Smith, E.A., Risk and uncertainty in the "original affluent society": evolutionary ecology or resource sharing and land tenure, in *Hunters and Gatherers: History, Evolution, and Social Change*, Vol. 1, Ingold, T., Riches, D., and Woodburn, J., Eds., Berg, New York, 1988, pp. 222–251.

Strahler, A.N., Quantitative analysis of watershed geomorphology, *American Geophysical Union Transactions*, 38, 913–920, 1952.

VanPool, C.S. and VanPool, T.L., The scientific nature of postprocessualism, *American Antiquity*, 64, 33–53, 1999.

Veth, P., O'Connor, S., and Wallis, L.A., Perspectives on ecological approaches in Australian archaeology, *Australian Archaeology*, 50, 54–66, 2000.

Warren, R.E., Predictive modeling in archaeology: a primer, in *Interpreting Space: GIS and Archaeology*, Allen, K.M.S., Green, S.W., and Zubrow, E.B.W., Eds., Taylor and Francis, London, 1990.

Warren, R.E. and Asch, D.L., A predictive model of archaeological site location in the Eastern Prairie Peninsula, in *Practical Applications of GIS for Archaeologists: A Predictive Modeling Toolkit*, Wescott, K.L. and Brandon, R.J., Eds., Taylor and Francis, Philadelphia, 2000, pp. 5–32.

Watson, P., The use of Mulligan River pituri, *Occasional Papers in Anthropology*, 10, 25–42, 1980.

Wescott, K.L. and Brandon, R.J., Eds., *Practical Applications of GIS for Archaeologists: A Predictive Modeling Toolkit*, Taylor and Francis, Philadelphia, 2000.

Whitley, T.G., Dynamical Systems Modeling in Archaeology: A GIS Evaluation of Site Selection Processes in the Greater Yellowstone Region, Ph.D. thesis, University of Pittsburgh, Pittsburgh, 2000.

Winterhalder, B., Optimal foraging strategies and hunter-gatherer research in anthropology: theory and models, in *Hunter-Gatherer Foraging Strategies: Ethnographic and Archaeological Analyses*, Winterhalder, B. and Smith, E.A., Eds., The University of Chicago Press, Chicago, 1981, pp. 13–35.

Winterhalder, B. and Smith, E.A., Eds., *Hunter-Gatherer Foraging Strategies: Ethnographic and Archaeological Analyses*, University of Chicago Press, Chicago, 1981.

Wolock, D.M. and McCabe, S.G., Comparison of single and multiple flow direction algorithms for computing topographic parameters in TOPMODEL, *Water Resources Research*, 31, 1315–1324, 1995.

Woodburn, J., Ecology, nomadic movement and the composition of the local group among hunter-gatherers: an East African example and its implications, in *Man, Settlement and Urbanism*, Ucko, P.J., Tringham, R., and Dimbleby, G.W., Eds., Duckworth, London, 1972, pp. 195–226.

Zubrow, E.B.W., Knowledge representation and archaeology: a cognitive example using GIS, in *The Ancient Mind: Elements of Cognitive Archaeology*, Renfrew, C. and Zubrow, E.B.W., Eds., Cambridge University Press, Cambridge, 1994, pp. 107–118.

Section 4:

Quantitative and Methodological Issues

7

Making Use of Distances: Estimating Parameters of Spatial Processes

Christian Mayer

CONTENTS

ABSTRACT   A general model of settlement structures by means of spatial processes and a specification of a special model for an archeological landscape is presented. The study is mainly concerned with the estimation of systematic effects, their proper formulation according to the general model applied, and methods for their estimation and their validation. As an example, the recent and historical settlement structure from a region northeast of the Austrian capital of Vienna has been analyzed.

7.1 Introduction

Using distances between sites as a tool to describe an archaeological landscape has some tradition in archaeology. Usually, distances between sites were supposed to be a measure of the size of the more or less circular shaped territories of sites. Consequently, the number of sites was estimated by the distances measured between them, thus being a kind of site prediction. From a current perspective, these approaches lacked a methodological background.

Although the development of the statistical theory for working with distances was fully developed only in the middle of the 1980s, its potential as a research tool in archaeology was recognized earlier (Hodder and Orton 1979; Orton 1982). Possibly due to the fact that the kind of statistics suitable for spatial data is different from commonly used statistics, working with distances between sites has not become a standard method for describing archaeological landscapes.

This chapter presents a model of an archaeological landscape in terms of distances by making explicit use of the concept of spatial processes. Of course, a spatial process may have a large number of parameters, but this chapter is mainly concerned with the estimation of systematic effects which play a key role in understanding an archaeological landscape as well as in the estimation of N, the number of sites in a region.

7.2. A General Model: Spatial Process

A spatial process is a set of data consisting of locations and measures of some property at these locations. A crucial point is that the location where the measurements are taken plays an important role for the phenomenon modeled. In this case, data is considered as spatially dependent and this dependency is the main difference from the more familiar statistics since that form of statistics does not regard the location where measurements are taken to be of importance. The traditional analysis of graveyards is an example of the two different approaches: one can analyze the graves as independent data by comparing their content and size without reference to their position in the graveyard and infer their relative age. By analyzing the graves' positions in the graveyard and inferring their relative dating from their location, one views the graveyard as a spatial process. This means that the graves are not independent from each other and the usual statistical techniques which are designed for independent data do not apply here. An example of the consequences of applying techniques for independent data to dependent data is given by Cressie (1993: 13).

7.2.1 Some Definitions

Cressie (1993: 8) gives a formal account of a spatial process

$$\{Z(s){:}s \in D\}$$

where s is a location and $Z(s)$ is any quantity of a property at that location; D is an index to identify the location that may be taken from a multidimensional space including the time dimension; the $Z(s)$ may take any form either being binary, discrete, or continuous.

Naturally, a process cannot be observed directly since it is a construction, only data generated by the process can be directly observed. This data is called a realization of a process. When the $Z(s)$ in the formal description of a given process is a collection of events, the process is called a spatial point process, a realization of a spatial point process is a spatial point pattern (Cressie, 1993: 577).

If it is equally likely for a point to occur at any location within a region A, the pattern is said to be random or denoted as a Poisson point process. Consequently, if one analyzes a point pattern, one tries to compare a realization of a Poisson point process to the data at hand.

A point process is characterized by the distribution of distances between points and the number of points within the pattern. The ratio between the size of an area A of a pattern and the number n of events is denoted as intensity λ. Distances and intensity are interchangeable since the intensity defines the average area around an event and, from the intensity, the average distance to nearest neighbor can be estimated for a pattern. If the distances between the events in a pattern and its counterpart, the intensity, vary systematically, then the process has a Local Intensity Function.

Of course, empirical data is much more complex than formal descriptions; therefore, it may happen that the data, seen from the generating process, is somewhat distorted. In this case the data is said to be inhomogeneous or nonstationary or simply infected by trend.

The object of further considerations can therefore be written as:

$$Process(observed) = (point)\,process \otimes trend \tag{1.1}$$

or

$$Process(observed) = point\,pattern \otimes trend \tag{1.2}$$

Both trend and Local Intensity Function are systematic effects and are not discernable in the first place. The difference between them is that the Local Intensity Function (LDF) is a property of the pattern while trend is considered to be an external impact on the pattern. Therefore LDF is not explicitly included in model 1.2.

7.2.2 Characterizing a Point Process

Parameters of a point process are derived from the distribution of the distances between points. To characterize these distributions, two techniques are possible. The first one is the computation of the K-function or its derivate the L-function. The other possibility is to study the distribution to the neighbor of order n (first, second .... n-th neighbor) using Thompson statistics. The difference between these two techniques is that Thompson statistics views a process by the distances ordered by neighbor of order n, while K-, L- and Paircorrelation functions view distances by lag. Naturally, fusing the distributions of distances ordered by neighbor and investigating them by lag, we obtain the K-, L- and Paircorrelation functions of the same data set.

If the data are infected by trend, trend removal is of high importance for the recognition of the process under investigation. To remove trend once again has two possibilities. The first one is to estimate the K- function for each point, replace the empirical function by a mathematical function with a - hopefully - low number of parameters, and remove the trend from these parameters. The second possibility is to compute the distances of order n and apply a detrending method to their spatial distributions. Replacing the individual K-functions by theoretical functions is complicated and seems less reliable, since the approximation doubtlessly will enhance estimation error. Therefore, the trend will be removed from the distances of order n. Model 1.2 therefore develops to

$$Process(observed) = U\,(distance(neighbor(order\,n)) \\ + trend(distance(neighbor(order\,n)))) \tag{1.2.1}$$

7.2.3 Edge Correction

When computing distances between points, each point should have an equal chance to be considered as a neighbor. If the point is near the boundary of the working area, this equality is not given, because there are no points outside the boundary. Consequently, points in the vicinity of that special point have higher chances to be considered as neighbors than the others, resulting in a distortion in the distribution of distances. This phenomenon is denoted as the edge effect and has to be corrected. From the various possibilities to perform edge correction, torus correction has been chosen: the data under consideration are shifted so that they are embedded in replications of the original data. From this, distances are computed.

7.2.4 Trend Removal

To analyze model 1.2.1, trend has to be removed first. From the various techniques available, median polish has been chosen, since the median is not as exposed to outliers as the use of the mean would be. By the decomposition given by Cressie (1993: 186), model 1.2.1 develops to

$$\text{Process(observed)} = U \text{ distance(neighbor(order(n)))} \qquad (1.2.2)$$

where

$$\text{distance (neighbor(order(n)))} = \text{median (distance(neighbor(order(n))))}$$

$$+ \text{ row (distance(neighbor(order(n))))}$$

$$+ \text{ column (distance(neighbor(order(n))))}$$

$$+ \text{ residual (distance(neighbor(order(n))))}$$

7.2.5 K-Functions, L-Functions, Thompson Statistics

From model 1.2.2 distances were recomputed by summing median (distance (neighbor (order (n)))) and residual (distance (neighbor (order (n)))), - the row and column effects stand for the trend in the x- and y- directions - and the sums are used as a new data set to compute K-, L- and Paircorrelation functions and Thompson statistics. Formulas applied are:

- K- and L- functions: Cressie 1993: 616, 617; Stoyan- Stoyan 1992: 270
- Paircorrelation function: Cressie 1993; Stoyan- Stoyan 1992: 275
- Thompson statistic: Thompson 1956; Cressie 1993: 611

To test for significance, 99 realizations of a random process are simulated and plotted against the empirical data. To estimate the Local Intensity Function, median polish can be applied.

7.2.6 Properties of a Pattern, Local Intensity Function

Usually, as a first stage of investigation, patterns are classified into regular, clustered, and random patterns by using one in more tests like the ones in section 7.2.5. Of course, patterns may be more complex, for instance, by being regular at short distances or by neighbors of low order, and these groups of regularly spaced points are located randomly in a study area (Stoyan- Stoyan 1992).

Consider a pattern that is detected as regular by means in section 7.2.5. If the pattern is perfectly regular, the points are arranged in equilateral triangles. But there could be another source of regularity. Imagine a simulation routine of a point pattern like this: start with a random point and generate an additional point with random coordinates. If the distance between the first point and the next point is below some fixed value, discard it, or retain it otherwise. Generate an additional point and check whether it is at least the fixed distance away from all existing points. Repeat this procedure until a certain number of points is placed. The result is a regular pattern generated by a Simple Inhibition Process (Diggle et al.: 1976), which shares no property with a triangular pattern except for regularity. Definitely, the goal of a complete account of a point pattern would be to know the sources of regularity, randomness, and clustering. To discern between different possibilities one can compare realizations of point process of known properties by adjusting their parameters to empirical data thus formulating a model of the generating process.

But there is a problem to be solved prior to the choice of the model of the point pattern: consider the placement of groups of regularly spaced points in an area. Such a pattern displays regularity as well as clustering. If the points in the pattern are settlements it would be of major interest where the empty space between the groups of points originates. There could be two assumptions: The first could involve a loss of sites, or the landscape in which the pattern that is realized cannot be completely settled in. The second assumption could be that there is a regulation so that members of the different groups are forced to settle in a certain distance to members of any other group. In both cases a systematic effect is involved but in the first case the systematic effect is an external impact on the pattern, in the second case the systematic effect is a property of the pattern. In the first case the systematic effect is denoted as trend, in the second case, as Local Intensity Function. Trend and Local Intensity Function cannot be discerned at first sight. Therefore, the parameters of a model applied to a data set have to be specified, so that the presence or lack of a Local Intensity Function can be determined.

7.2.7 Validation of the Model

Naturally, the validity of the model depends on the degree of reproduction of the original data by the model applied. Of course, the validation of a model is a twofold problem. Firstly, a model has to have a pertinent correspondence to the data under investigation. This problem has to be solved in advance of any computation and in connection with the actual problem. Secondly, there is a technical aspect of model validation. This aspect will be discussed in respect to model 1.2.2.

$$Process(observed) = U \ distance(neighbor(order(n))$$

where

$$distance(neighbor(order(n))) = median \ (distance \ (neighbor \\ (order(n))) + row \ (distance$$

$$(neighbor(order(n)))) + column \ (distance \ (neighbor(order(n))) + residual \\ (distance(neighbor(order(n))))$$

When this model is applied, one estimates not only the distances between points but the order of the distances, the median, and the systematic effects simultaneously. Assuming a correct estimation of the medians and the trend component, one can interpret an actual distance by estimated medians and trends by seeking iteratively for the minimum difference between estimated and observed distance. The distribution of the differences between estimated and actual order and the difference between estimated and actual distance have to equal the distribution of the residuals. If so, the model is valid.

In fact, this restriction is not realistic. Consider a regular pattern of a triangular prototype. The distances of the 1st to the 6th neighbor are equal and their order is not distinguishable. If some normally distributed variation - that is the distribution of the residuals - is added, one can order the distances, but the actual rank of a value would be random. Consequently, the recomputation of distances possibly will be exact, but - a least in a regular pattern - the estimation of the order of the distances will be inexact by virtue of regularity. Therefore, the estimated distances will be affected by a measurement error that is a property of the pattern under study. The measurement error of the estimation will certainly be different from the one stemming from the residuals. To ensure the nonrandomness of the results, a randomly distributed effect for the x- and y- axes can be generated and used for the recomputation of the distances.

7.3 A Model for an Archaeological Landscape

If an archaeological landscape is described by a collection of coordinates representing sites, the archaeological landscape is modeled by a point process.

Certainly, to achieve a meaningful result, the elements of the model must have a counterpart in the phenomenon studied. Therefore, the process (observed) will be denoted as settlement structure and the point pattern as settlement pattern. Of course, by interpreting an empirical phenomenon by a model, one is crossing the line between formal description and assumption, which is quite visible from the interpretation of model 1.2 in this paragraph: by applying model 1.2 to a settlement structure, one implicitly assumes a decomposition of the settlement structure into a settlement pattern and a trend element where the settlement pattern is regarded as a basic habit of the inhabitants of the sites which are influenced by some additional spatial phenomena. Since all elements of this model, settlement pattern and some additional spatial phenomena, are regarded as nonrandom, model 1.2 assumes that the process (observed) is in fact a combination of at least two processes:

$$\text{Process (observed)} = \text{process } 1 \otimes \text{process } 2 \otimes \text{.......} \otimes \text{process N} \qquad (2.0)$$

Traditionally, the settlement pattern is regarded as a social phenomenon that adapts to the natural conditions. Therefore, model 2.0 can be written as

$$\text{Process (observed)} = \text{process (soc)} \otimes \text{process (nat)} \qquad (2.1)$$

or

$$\text{settlement structure} = \text{settlement pattern} \otimes \text{process (nat)}$$

A more realistic model of an archaeological landscape would include a process that stands for the loss of sites for one reason or the another, therefore,

$$\text{settlement structure} = \text{settlement pattern} \otimes \text{process (loss)} \otimes \text{process (nat)} \quad (2.1.1)$$

Before actual computations will be undertaken, some implicit assumptions have to be discussed. Consider the process (soc). Firstly, a point has to represent the properties of a site. Of course, since a site is not a point but an area, the distances measured between sites are exact only to a certain degree and a measurement error will be present in the results of the computations. Technically, this can be easily compensated for by blurring a pattern thus producing a certain number of simulated realizations of the original data to get an idea of variation. Secondly, at a nontechnical level of consideration, the key assumption of the process (soc) as a point process is that the distances between sites are of importance. Usually, the reason to investigate a pattern of sites is the conception that a settlement structure is nonrandom, that is, distances are meaningful. In this case, the distances between sites measure the territories of sites. Then the coordinates used to localize the sites localize their territories as well, replacing the more common conception of site by a concept of space (Neustupný 1998: 9). Here the problem of dependent data comes in: if the territory size is of importance, then the location of a site and its territory is chosen with reference to all other sites and territories thus establishing a direct relation to all the other sites (Mayer 2001: 272). Under this assumption, the application of statistics for independent data is not appropriate.

Furthermore, if one uses 2.1 or 2.1.1 as a model for an archaeological landscape, one investigates the reaction of the process (soc) on the process (nat). This means that only one aspect of the process (soc) is considered.

Finally, to validate the model, one has to estimate the parameters of two processes. This complicates the situation, since process (nat) is considered as a systematic effect influencing the realization of the process (soc). Process (nat) may not be the only systematic effect in the process observed, since a Local Density Function may be inherent in process (soc). Consequently, the analysis has to be conducted in a way that systematic effects in the process (soc) and the process (nat) are estimated separately. If a Local Intensity Function found in the data can be replaced by the process (nat) and this hypothesis can be verified by the validation approach 1.7, then process (soc) has no Local Intensity Function.

7.4. A Case Study: Medieval Settlement Structure

7.4.1 The Model

To carry out the analysis of these data, model 2.1 has been chosen. Process (loss) of model 2.1.1 will not be discussed here. Therefore process (soc) and the process (nat) have to be estimated. Process (soc) is a point process since the medieval settlements enter the analysis as points.

The process (nat) is not a point process since the attributes of a landscape are spatial varying continuously. Therefore, process (nat) has to be turned into a point process implying a set of assumptions.

7.4.2 Basic Assumptions, Causative Elements, Direction of Analysis

Viewing an archaeological landscape as a point process only works if the points considered are a sample for the phenomenon under study. Points represent - in the statistical sense - villages, as long as only their geographical position is concerned.

To model a landscape, the choice of data is crucial. Of course, soils, climate, geology, and so forth are obvious candidates but the model of process (nat) would require a specification of the interaction of all these components, which, if successful, would certainly be a major contribution to the environmental sciences. Because the chances of obtaining such a solution are poor, data on landuse and crop yield have been chosen for modeling process (nat), since plants provide not only a good summarization of a landscape but illustrate a farming civilization's view of a landscape as well. In fact, that assumption infers a causative element into the model: the settlement pattern is a result of natural conditions.

Of course, properties of a landscape vary spatially continuously. To compare these data to a point process, it has to be turned into a point process by sampling the landscape's properties of interest by a set of points that are associated with a distribution of the properties investigated in a defined vicinity. To make the expression "defined vicinity" meaningful as a social element, it is defined as the area in which the requirements of an agricultural civilization are met. Of course, the defined vicinity in this context is the territory of a site and is estimated by the cadastral units as recorded in the land register.

According to this layout the analysis has to take the following steps:

- Estimate the distance distributions of the settlement pattern.
- Determine the influence of systematically distorting effects on the pattern.
- Estimate the Local Intensity Function.
- Estimate the distribution of distances after the removal of systematically distorting effects and characterize the pattern.
- Determine the trends of the variables collected to model process (nat).
- Determine a set of variables that gives the best account of the landscape under investigation.
- Compare the Local Intensity Function to the trends estimated from the variables that characterize the landscape.
- Compute a reproduction of the distance distributions of the pattern by substituting the Local Intensity Function by the trends of characteristic variables of process (nat).

7.4.3 Data

Data consist of 74 existent villages in a region of 1247 km² northeast of the Austrian capital of Vienna (Figures 7.1 and 7.2). Seventy-two of them lay on Austrian territory. Two Slovakian villages were added to provide a rectangular working area.

The region in which data has been collected changes from a hilly landscape in the north to a plain in the south. Elevations reach from 270 m in the north to 145 m in the south. In the middle of the region there is an area with a relative elevation of between 15 and 7 m stretching from east to west. Two rivers dominate the landscape. The River March in the east and the Danube River in the south both have deep impacts on the landscape. There are only

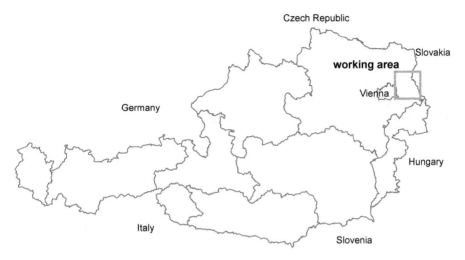

FIGURE 7.1
Map of Austria.

three more watercourses of some importance which cross the working area from northwest to southeast. During summertime, these three usually do not have much water but in spring and fall they bring water from the hilly regions outside the study area.

This part of Austria is extremely fertile except for a region in the middle of the working area, visible in Figure 7.2, by its relative elevation of about 15 m over the flood plain in the south. While the other parts of the working area are covered by black soils, this one is covered by soils originating from gravel and is extremely dry. Altogether, there are no mineral resources except for some oil in the north.

Today's appearance of the working area is extremely uniform. Originally, the landscape was structured by boggy areas, ponds, and flood areas of the watercourses. The transformation of the original landscape into the present situation started with massive deforestation in the Middle Ages resulting even in the uncovering of areas of drifting sand, a problem lasting until the end of the 18th century when trees of a special kind imported from Italy were planted in the endangered areas. By draining wet places, diverting watercourses since the middle of the 18th century, and damming up the Danube, the March, and their contributors at the beginning of the 20th century and the intensive use of heavy machinery, the landscape has created the present situation.

Due to the lack of mineral resources and the fertility of soils in the region, agriculture soon flourished resulting in the intensive cultivation of grain and its delivery to the markets in Vienna. Some dynamics entered the population development by the building of railways at the end of the 19th century and the following erection of some industrial plants, which are more or less

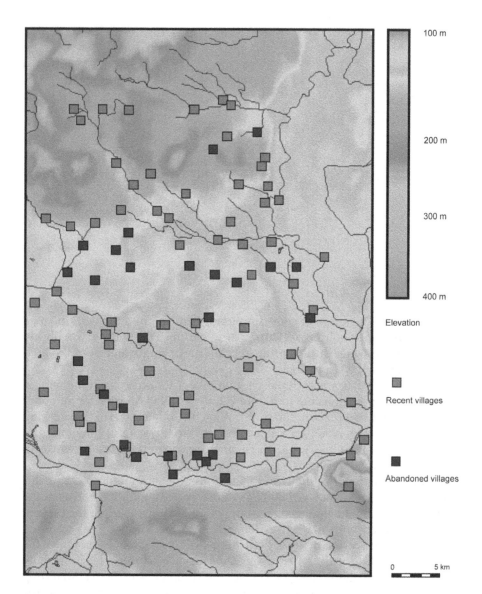

FIGURE 7.2
Recent and abandoned villiages.

connected to the processing of agricultural products. In fact, their impact on
the development of population is still visible in the statistics.

Of course, the modern settlement structure is only the remnant of an older
one going back to the 11th century. In fact, in the working area there were
at least 38 more settlements, most of them abandoned during the 15th century. For 33 settlements the location is known, for 5 more only names are

recorded. Interestingly, the territories of some of these villages survived and are still recorded as cadastral units in the land register.

Existing and abandoned villages are stored as coordinates in a database. Associated with the coordinates, data on landuse, and yield of certain fruits according to the cadastral units of the villages from a survey undertaken in the late 1930s were recorded (Mayer 2001: 274).

7.4.3.1 Data on Landuse and Crop Yield of Recent Settlements

Data on landuse consists of per cadastral unit of:

- size of cadastral unit and population (average of the years 1880-1935).
- acreage of land used for cultivating grain, vine growing, meadow, pasture, for farming including non-grain fruits, forestry, and uncultivated land (Mayer 2001: 274).

Data on yield includes:

- yield of summer and winter barley, summer and winter wheat, oats, summer and winter rye, potatoes, sugar beet, beets for animal fodder, clover, and lucern.

7.4.3.1.1 Exploratory Data Analysis

Data of population, territory size, acreage of land used for cultivating grain, and farmland exhibit lognormal distributions. Exploratory data analysis showed that pasture, vine growing, and meadows are of little importance for the region since the portion of land used for these purposes is very small if it occurs at all. Six factors of a principal components analysis out of nine (population was not included in the analysis) comprise 95% variation within the data, where the first factor alone comprises 29.5% variation. This clearly means that there is no systematic dependency between variables on landuse and the combinations and amount of land used for certain agricultural activities do not follow a discernable rule.

The correlation between population and size of cadastral unit is 0.68 (including outliers) and nearly as high as the correlation between population and tillage. From this follows that the size of a unit satisfies the needs of its inhabitants and therefore meets the requirements of section 7.4.2 to serve as a defined vicinity.

Data on yield showed no special distribution form which is because the data is available only in blocked form. The portions of variation expressed by the factors of a principal components analysis show the presence of factors with a high amount of variation expressed, but all in all there is no hint that data could be reduced to a low number of variables or factors. No preference for a single fruit is visible in the data.

The joint data sets analyzed by a principal components analysis do not show a leading factor. A canonical correlation analysis between data on landuse and yield gave a highly significant value for Λ, proving that landuse is closely tied to the expected yield.

A detailed discussion of factors of the principal components analysis will not be undertaken. Compared to the results of another study (Mayer 2001: 274), results show the homogeneity of the agricultural activities in the study area.

7.4.3.1.2 Trend Phenomena

Trends have been extracted by putting data on a 150 x 150 m grid and performing median polish. Principal components were computed finding three factors for the data set on landuse and three other ones for the data on yield of relevance. The two data sets were additionally compared by canonical correlation analysis. Highly significant values for õ were obtained, proving the close correspondence of the variables. Certainly, the factors of yield can be interpreted as the result of regional fertility, and it is quite obvious that landuse adapts to regional fertility. Trends of population and size of cadastral units have a correlation of 0.7 (not corrected for outliers), underlining that cadastral units are an estimator for the defined vicinity (see section 7.4.2). Similar results have been obtained in an analysis of the same variables in another region (Mayer 2001: 275) confirming the findings from this region.

7.4.3.2 The Distribution of Distances

7.4.3.2.1 Local Intensity Function

To estimate the Local Intensity Function, distances were computed and ordered according to the rank of the neighbor. Median polish was performed for each neighbor of order n. Principal components were extracted from these data showing 13 factors (out of 73) relevant to express 95% of the variation within the data. Of course, one hopes to find a single factor to express all variation, but this would only happen if the underlying pattern is perfectly regular. At least, the low number of factors for the data investigated show a high degree of regularity in the material.

7.4.3.2.2 The Pattern

All statistics, K-functions, L-functions, Paircorrelation Function, and Thompson statistics, show regularity after the removal of systematic effects. Regularity is visible in Figures 7.3 and 7.4 by the graph of detrended data having values higher than 1. The graph of the Paircorrelation Function of detrended data (Figures 7.5 and 7.6) has values of 0 up to a distance of 2.5 km which means there are no neighbors below this distance. The median of the distance to the nearest neighbor suggests this minimum distance at 2.1 km. The sharp

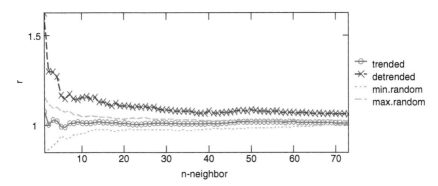

FIGURE 7.3
Recent villages, Thompson statistics.

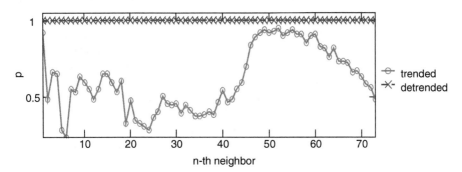

FIGURE 7.4
Empirical probabilities.

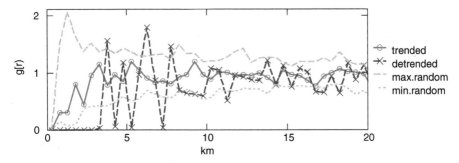

FIGURE 7.5
Recent villages, Paircorrelation.

ups and downs in the graph are result of the regularity of the pattern. Since a regular pattern was detected, the intensity computed from the medians for neighbor 1 to n should equal the actual number of villages. Estimating the intensity by the medians from the 1st to nth neighbor a number of 75 villages was found, only one more than actually exist.

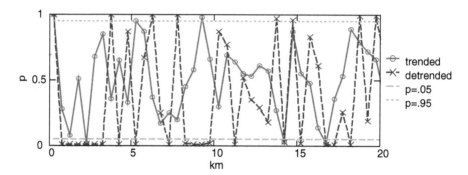

FIGURE 7.6
Empirical probabilities.

7.4.3.3 Estimation of the Local Intensity Function of the Recent Settlement Structure

From the fact that the distance structure of the data cannot be expressed by a single factor (section 7.4.3.2.1), it follows that the Local Intensity Function cannot be estimated directly but only its impact on the distribution for each the distances of order n (Cressie 1993: 579, 654). To check the validity of the model, the iterative approach of model validation in section 7.2.7 was used finding an error of 0.069 ± 0.1703 km. It is obvious that data is reproduced sufficiently. Compared to the distribution of the residuals (-0.008 ± 0.177 km), the measurement error does not differ significantly, therefore the model is valid.

7.4.4 Abandoned villages

The abandoned villages were added to the data set of existing villages and the Local Intensity Function and the pattern were determined. Again a regular pattern was detected after removal of systematic effects (Figures 7.7 and 7.8). A minimum distance is quite obvious in the graphs and is

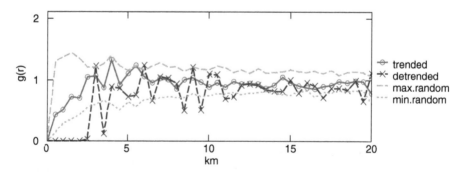

FIGURE 7.7
Recent and abandoned villages, Paircorrelation.

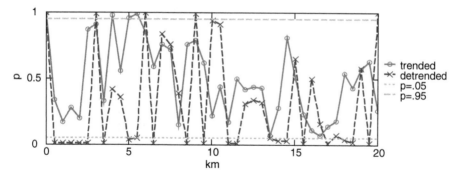

FIGURE 7.8
Empirical probabilities.

about 1.91 km. A principal components analysis was performed on the systematic effects in x- and y- directions finding 15 factors (out of 105) relevant to express 95% variation in the data. Again, variation could not be reduced to a single factor, but the low number of relevant factors underline the regularity found by the statistics on distances. The number of villages in the settlement pattern was estimated to be 107, which is only one more then the observed number. The recomputation of distances yielded an error of 0.0038± 0.13 km (Figure 7.10).

7.4.5 Trend or Local Intensity Function?

Between the results of the estimated Local Intensity Functions of the data sets, with and without abandoned villages, a canonical correlation analysis was performed, yielding highly significant values of Λ without a leading canonical variate. The correlations between the estimated Local Density Functions by order of neighbor n of the two data sets vary between 0.771 and 0.276. From the conception of canonical correlation analysis, one would expect that both data sets are connected by a single canonical variate if the process that generated both data sets has a Local Intensity Function. Since this is not true, the systematic effects in the distance distributions are trend, and therefore not a property of the pattern.

Consequently, a canonical correlation analysis between the data on trends of landuse and yield and the estimated Local Intensity Functions has been performed, finding high values of Λ but no leading canonical variate. Therefore trend components have been fitted to the systematic effects of distances by linear regression, and the results were used to recompute the distance distributions of the data sets with and without abandoned villages, finding a measurement error of 0.069 ± 0.206 km and 0.004 ± 0.122 km, respectively. For comparison, a recomputation of the distances has been undertaken using a randomly distributed trend plotted in Figures 7.9 and 7.10.

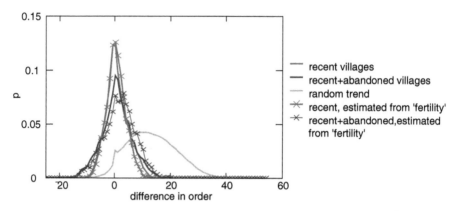

FIGURE 7.9
Difference in observed and estimated order.

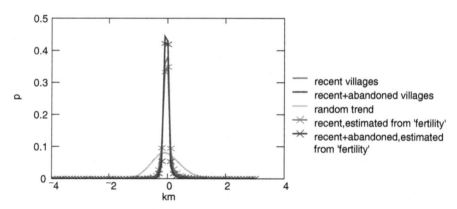

FIGURE 7.10
Difference in observed and estimated distances.

Compared to the recomputation from the extracted x- and y- effects of the distances of 0.0076 ± 0.17 km and 0.017 ± 0.13 km, it is clear that the systematic effects in the distance distributions is an external impact on the settlement pattern. Therefore, the settlement structure in the model consists of a settlement pattern of some regularity that is adopted to the natural conditions in the working area. This conforms to model 2.1, which is validated by the measurement errors presented: the settlement structure can be sufficiently modeled by an interaction of two processes, a point pattern and a spatial process realized by variables to describe natural conditions in a landscape. The point pattern has no Local Intensity Function, all systematic effects in the settlement structure stem from the impact of natural conditions on the pattern.

7.5 Discussion

Of course, the results in section 7.4 rely only on a single realization of the modern settlement structure since only one data set has been considered. In a comparable study, the results are confirmed by using the same model (Mayer 2001). This study includes not only historical and modern data but settlement structures of prehistoric periods as well. Some of these settlement structures do not only involve regular patterns but clustered ones as well. In all cases the model proved to be valid and the estimation of the parameters of systematic effects of the underlying processes have been estimated satisfyingly by the methods in section 7.2.

Stoyan (1988) analyzes a medieval settlement structure consisting of existing and abandoned settlements in eastern Germany by means of spatial statistics. The data was considered to be a realization of a thinned Poisson process without systematic effects. Parameters of the point process were estimated, finding the results not quite satisfying (Stoyan 1988: 55). Under the aspect of the results of section 7.4, the model applied by Stoyan lacked a sufficient specification of parameters to reproduce the settlement structure analyzed. On the other hand, that study shows the abilities of the explicit use of spatial processes as a tool to model settlement structures by including methods to assess the validity of a model. Nevertheless, the analyses of the settlement structure presented in this paper and by another study (Mayer 2001) demonstrate the flexibility of the spatial processes as a general model for human behavior. Additionally, the specification of a suitable set of parameters in a model formulated by spatial processes allows the substitution of results of other sciences in the same methodical framework, opening new aspects of interdisciplinary research.

References

Cressie, N.A.C., *Statistics for Spatial Data*. Wiley, New York, 1993.

Diggle, P.J., Besag, J.E., and Gleaves, J.T., Statistical Analysis of spatial point patterns by means of distance methods, *Biometrics*, 32, 659, 1976.

Hodder, I. and Orton, C., *Spatial Analysis in Archaeology*, Cambridge, Cambridge University Press, U.K., 1979.

Mayer, Ch., Observations on settlement patterns from the Late Bronze Age to end of the Iron Age, *Památky Archeologické* XCII, 254, 2001.

Neustupný, E., Structures and events, in *Space in Prehistoric Bohemia*, Neustupn_, E., Ed., Czech Academy of Science Press, Prague, 1998.

Orton, C., Stochastic process and archaeological mechanisms in spatial analysis, *J. Archaeological Sci.*, 9, 1, 1982.

Stoyan, D., Thinnings of point processes and their use in the statistical analysis of a settlement pattern with deserted villages, *Statistics*, 19, 45, 1988.

Stoyan, D. and Stoyan, H., *Fraktale, Formen, Punktfelder*, Akademie Verlag, Berlin, 1992.

Thompson, H.R. Distribution of distance to n-th neighbour in a population of randomly distributed individuals, *Ecology*, 37, 391, 1956.

8

Integrating Spatial Statistics into Archaeological Data Modeling

Kevin R. Schwarz and Jerry Mount

CONTENTS

ABSTRACT   Spatial analyses of archaeological data typically have suffered from methodological, sampling, and statistical problems. Previous approaches to archaeological inference based on common statistical procedures are briefly discussed. This chapter suggests that wider use of spatial statistical tests, which are designed for spatial inference, can be integrated into geographic information systems (GIS) and similar spatial analyses that are often graphically displayed. A series of examples is presented using archaeological data that demonstrate the utility of the spatial statistical methods.

The problems of previous approaches include lack of proper sampling and violation of the assumptions of normality and independence. The independence assumption underlies many aspatial statistical tests such as the t-test, correlation, ANOVA, and regression methods. It suggests that each case of a particular variable is drawn separately and independently of the influences of other cases of the same variable, as in experimental design. Much spatial data violates this assumption through the existence of spatial autocorrelation.

This concept suggests that values of observations that are in close spatial proximity will be similar, as described by Tobler's first geographical law.

The solution provided is a suite of spatial statistical techniques designed to overcome biases associated with aspatial statistics. The diagnostic use of spatial autocorrelation statistics such as Moran's I and Geary's C allows the researcher to assess the need for more powerful spatial models. In cases where modeling of spatial processes is needed, spatial regression can be used. Ripley's K statistic allows researchers to examine successive neighbors to understand clustering or dispersion of observations. Nearest-neighbor hierarchical clustering provides a means to present clustering as a deviation from spatial randomness. Mapping software such as ArcView GIS and Golden Surfer can be extended to utilize these spatial statistical methods with links to S-Plus and CrimeStat statistical packages.

8.1   Introduction

The use of statistical methods in archaeological analyses has a long and checkered history. Beginning with the Spaulding-Ford debate of the 1950s, archaeologists have argued over the types, methodologies, and utility of various statistical methods and techniques (e.g., see reviews of this era in Thomas 1978; Cowgill 1975). Oftentimes the haphazard application of little-understood statistical methods to poorly sampled data sets creates problems. This chapter discusses a particular subset of statistical issues in archaeology, namely the use of statistics in spatial analysis.

The rise of interest in settlement patterns in archaeology (Willey 1953; Willey et al. 1965; Aldenderfer 1996) and intrasite artifact pattern analysis (Whallon 1973, 1974; Hietala 1983; Carr 1984) created an explicitly spatial component to many statistical problems, though this fact often is not recognized. For example, many statistical treatments of settlement patterns developed in the 1970s dealt with the degree to which human settlement locations (and site characteristics such as size or inferred population) correspond to ecological variables such as land quality or water resources (e.g., Vita-Finzi and Higgs 1970; Brumfiel 1976; Kowalewski 1982). These studies take the approach of applying traditional regression and correlation techniques, in the hopes of identifying statistically significant association between human and land variables. Many of these studies have problematic findings for the statistical reasons described below.

The use of geographic information systems (GIS) and similar mapping software in settlement pattern analysis since the 1980s has greatly increased the tools available to archaeologists (Kvamme 1993; Allen et al. 1990; Lock and Stančič 1995). Not only is map creation, display, and manipulation facilitated, but new analytical capabilities such as buffering and map overlay improve the archaeologist's tool kit for spatial analysis. The interest in predictive modeling, both for its research potential and for archaeological resource management, led to the elaboration of many statistical models for predicting site location from ecological variables (Wood 1978; Carr 1985; Kvamme 1992; Warren 1990a, 1990b). Logistic regression and discriminant function analysis have been the preferred statistical techniques of these modeling efforts.

The problem with many traditional statistical approaches lies in their aspatial nature. Because phenomena distributed in space are related by their proximity to each other, these phenomena may suffer from a problem known as spatial autocorrelation (Odland 1988; Griffith 1987). Autocorrelation creates bias in parameter estimates and makes significance testing difficult (Griffith 1996). Only recently have archaeologists attempted to deal with the spatial nature of many archaeological phenomena by introducing spatial statistics into archaeological data analysis. Spatial statistical tests are based on conventional statistics but incorporate locational data within them, and through this modification provide for more robust findings. As recently as 1998, spatial statistics was the province of only those who could master matrix algebra and code SAS or other statistical macro languages. Currently, however, spatial statistical tests are appearing in user-friendly formats such as S-PLUS SpaceStat and CrimeStat statistical packages and are seamlessly integratable with GIS packages (such as ESRI's ArcView GIS) and usable with other mapping software (such as Golden Surfer). This capability allows archaeologists with some knowledge of the theory and method of spatial statistics to use these improved procedures in their settlement-pattern and intrasite spatial analyses.

In Section 8.2 we briefly review some theoretical considerations involved in the use of spatial statistics. In subsequent sections we present a guide to a few of the more useful statistical procedures, including Moran's I and Geary's C, spatial autoregressive models, Ripley's K, and nearest-neighbor hierarchical clustering. Several examples are given from archaeological settlement-pattern and intrasite artifact-distribution studies. Finally we conclude with a cautionary note on the use of aspatial models in predictive modeling and other spatial contexts.

8.2 Theory and Method of Spatial Statistics

It would be difficult for an archaeologist and a geographer to try to describe the mathematics of spatial statistics as they exist now. It is a specialized, complex, and fast-changing field of endeavor (Griffith and Amrhein 1997). Rather, we will describe some of the conceptual structures involved: specifically why spatial statistics are necessary in the first place and what spatial autocorrelation entails. The preferred methodology is to apply basic diagnostic tests of Moran's I and Geary's C to determine if a spatial statistical estimation is necessary. More specialized spatial tests are readily available in statistical packages for those engaged in spatial analysis.

It is well-known that several formal assumptions are requirements for data sets before traditional inferential statistical techniques are attempted. These include the assumption of randomness, the assumption of normality, and the assumption of independence (Griffith and Amrhein 1997). The first two assumptions are frequently discussed (though just as frequently violated) by nonexperimental social scientists like archaeologists. The assumption of independence is more subtle, however, and many introductory statistics textbooks fail to even mention it (e.g., Pedhazur 1982; Hinkle et al. 1994). The idea of statistical independence requires that the cases of a variable are not intercorrelated. In randomized tests of intelligence or other controlled settings, this idealized requirement may be close to accurate. However, in the natural world, and space and time in particular, most "observations" are part of larger and more continuous phenomena in which characteristics in a given region or interval are at least partially dependent on antecedent events or on neighbors (Odland 1988; Griffith 1984). For the continuum of time, time-series statistics have been developed. Consider the case of a long-distance runner for whom 1-mile splits are measured in a 26.2-mile marathon. These splits are actually part of an autocorrelated continuum. If a runner has a slow split-time on mile 22, then it is likely that he/she will have a slow split-time on mile 23 because of underlying factors of the structure of the race, his/her own physiology, and ultimately time itself.

The same dependence applies to space. There are two types of spatial autocorrelation we can discuss: positive spatial autocorrelation and negative spatial autocorrelation. Positive spatial autocorrelation is described graphically as clustering of similar attributes or values (Figure 8.1). In any region, site, or archaeological excavation block, it is likely that the characteristics of one point in that region, site, or excavation square is related in its attributes or values to that of its neighbors. Thus soils of a particular series may be found in one survey grid, and it may be more likely to be encountered again in a neighboring block than in a more distant block, depending on the type of topography and scale of the blocks. Likewise, a larger settlement may attract smaller settlements to its near periphery for the purposes of trade or alliance, forming a cluster of settlement. Elevated concentrations of artifacts in one excavation square may be due to a broader artifact scatter. All other things being equal, we may expect adjacent squares to also feature relatively high concentrations of artifacts in comparison with a square selected by chance. This characteristic of geographic phenomena was formalized by Waldo Tobler (1979) as Tobler's first law (or the first law of geography). It suggests that: "Everything is related to everything else, but near things are more related than others" (Chrisman 1997). The relationship of dependence is called positive spatial autocorrelation and can be considered as a deviation from randomness (Odland 1988). Negative spatial autocorrelation is described as the interpositioning

Positive Spatial Autocorrelation

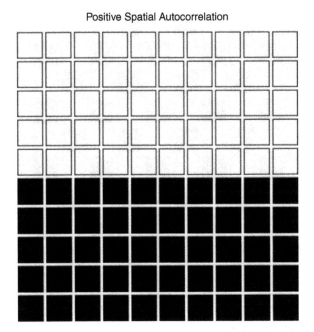

FIGURE 8.1
Positive spatial autocorrelation.

of regions of differing value and is visible as a checkerboard or alternating pattern (Figure 8.2).

Spatial autocorrelation means that using standard statistical procedures on spatialized data sets may result in serious biases because the data may contain undescribed spatial dependencies. For example, regression and correlation coefficients are overestimated (Griffith 1996), significance tests are biased, and estimates of regression-equation parameters (such as the slope) may not be accurate. That said, it should be emphasized that not all spatially distributed data sets are spatially autocorrelated. Test statistics have been developed to gauge the degree of autocorrelation that exists within a data set. Two are reviewed in the following section.

8.3 Tests of Spatial Autocorrelation

Two easy-to-use statistical tests have been designed to diagnose spatial autocorrelation: Moran's I and Geary's C. Both are available in the S-Plus Space-Stat module and CrimeStat program. Moran's I or the Moran coefficient (Moran 1948) measures the covariation of juxtaposed map values of a given variable of interest (Griffith 1987). The interpretation of the statistic is similar

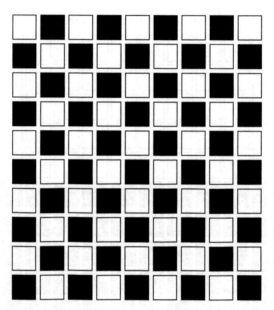

FIGURE 8.2
Negative spatial autocorrelation.

to that of a correlation coefficient. The expected value (of nonassociation) of the Moran coefficient is MC = $-1/n-1$. Positive autocorrelation indicates spatial clustering, with a significant MC approaching 1. Negative autocorrelation approaching -1 indicates a checkerboard distribution of the variates in space. Nonsignificant findings suggest a random relationship (Griffith 1987). Geary's C or the Geary ratio (GR) is similar in concept to the Moran coefficient. The interpretation of the statistic differs in that the expected value of nonassociation is 1. If similar values cluster in space, then GR approaches 2 and positive autocorrelation exists. If dissimilar values tend to cluster, then GR will approach 0 and negative autocorrelation exists (Griffith 1987).

It is possible to test sample distributions with the Moran coefficient and Geary ratio as an end in itself. For example, artifact distributions can be explored with this technique. Consider the following examples. Recent household excavations by Schwarz (2004) of Postclassic (A.D. 1000–1525) Maya sites from the Petén Lakes region of Guatemala have revealed numerous *in situ* artifact scatters. For example, at a domestic site know as Petenxil Group 1, the author discovered scatters and concentrations of chert flakes on and in front of a masonry platform, the remains of an ancient house. Flakes were recorded in 1 × 1-m provenience units in a roughly rectangular excavation block. Flake counts were paired with the northing and easting information for each excavation square, and the data was tabulated in a Microsoft Excel spreadsheet. The output is viewed as an isopleth map of flake concentrations across the floor of the structure and in front of it, prepared in Surfer (Figure 8.3). Given the apparently strong spatial concentration of chert flakes existing to the west

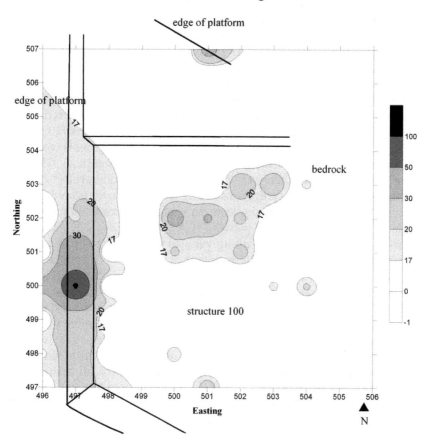

FIGURE 8.3
Isopleth map structure 100, lithic debitage.

of the platform, further exploration of this data was undertaken. The Moran coefficient ($N = 103$; MC $= 0.031$; $Z = 3.01$) indicates a significant finding of positive spatial autocorrelation. Geary's C (GR $= 0.081$; $Z = 3.41$) provides the same information. Interestingly, the relatively weak coefficients suggest that the western flake concentration is not of sufficient size and intensity to create a strongly autocorrelated global distribution and does not overwhelm the intensity of flake concentrations in the surrounding regions.

Global measures of autocorrelation can mask significant amounts of local-level spatial variation. Anselin's local Moran statistic is designed to measure point-to-point (or zone-to-zone) autocorrelation. It adapts Moran's I to assess whether points (or zones) are related in the value of an intensity variable to neighboring points. The adaptation was accomplished by Anselin (1995), who specified the local indicator of spatial association (LISA). The LISA indicates the extent of similarity of the observation or zone in relation to its neighbors. Values are produced for each observation, and the output can be viewed in map form. Levine (2002: 289) states that a significance measure has yet to be worked out for this statistic, but generally, high negative values indicate dissimilarity of values in comparison with neighbors, while high positive values indicate similarities of value.

An example of the local Moran statistic's utility comes from a complex scattering of different artifacts from the vicinity of a small temple in the Quexil Basin in Guatemala. Structure Q1 sits on the Eastern Island in Lake Quexil. A small temple is fronted by a staircase around which three different artifact scatters were found (Figure 8.4). Fragments of ceramic *incensarios* (incense burners), chert flakes, and broken and burnt turtle shell fragments were found in nonoverlapping concentrations in the 6×7-m excavation block. The count data of the three classes then were tabulated with the total artifact count for the squares outside the concentrations. Using the artifact counts for each block, Moran's I and Geary's C both indicate no spatial autocorrelation within the sample ($N = 84$, MC = −0.01; Z = 0.0007; GR = 0.0001, Z = 2.39). The counts were then standardized. The resulting variates were analyzed using the local Moran statistic in CrimeStat. The resultant output of the I statistic for each excavation unit was then remapped in Surfer. The isopleth map (Figure 8.5) demonstrates three interesting local characteristics. The two largest and most intense concentrations (the *incensario* fragments in the northeast and the turtle shell fragments in the south) are visible as large areas of negative value (local MC < −0.90), which result from the neighboring much-lower-value areas outside the scatter. The flake scatters, consisting of elevated levels in single excavation units, are not visible against the background of all artifacts. The surrounding areas, such as the sides of the temple, are of low negative value and are not aggregated, suggesting that the counts of all artifacts in this excavation are not organized on a local scale.

8.4 Spatial Statistical Models

Griffith (1987, 1989, 1996) and Amrhein (Griffith and Amrhein 1997) developed a series of spatial statistical tests based on the regression methodology (generalized least squares) that they adapted to spatial data. In this short chapter, we cannot hope to describe the reasoning and complexity underlying the development of spatial statistics. However, a few points can be emphasized in relation to autocorrelation statistics. The first is that Griffith (1987) recommends regressing two or more variables of interest in a spatialized data set and testing the residuals for spatial autocorrelation with Moran's I or Geary's C. One graphs the regression residuals with the fitted values in a scatter plot. If the plot appears to be random, then the existence of autocorrelation is not likely. However, if a linear pattern appears in the residual plot, then autocorrelation is likely present and needs to be addressed

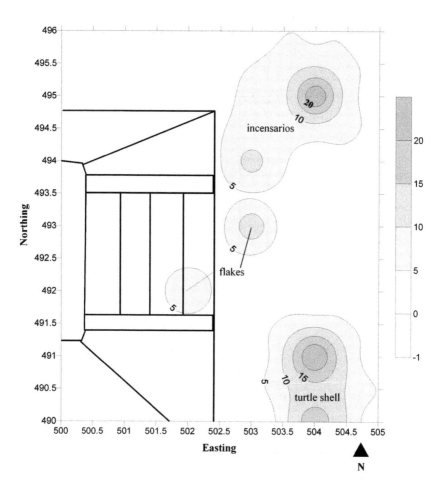

FIGURE 8.4
Isopleth map structure Q1, concentrations.

(this is the linear dependence described above). Significant positive or negative spatial autocorrelation may indicate the need for a spatial autoregressive model as developed by Griffith (1987; Griffith and Amrhein 1997) and others. Readers should consult the original sources for a more in-depth discussion.

When significant spatial autocorrelation exists in a data set, it is often worthwhile to pursue a spatial statistical model. However, it should be noted that sometimes linear dependencies in data sets can be dealt with through simpler means than employing spatial autoregressive models. For example, one author of this paper (Schwarz 1997) encountered this problem in assessing settlement-pattern and ecological data from the pre-Hispanic Basin of

Structure Q1, concentrations

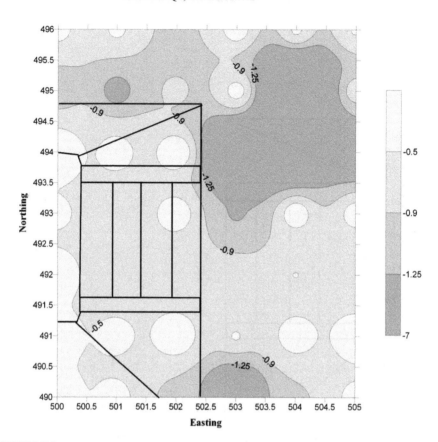

FIGURE 8.5
Isopleth map structure Q1, concentrations, local Moran's I.

Mexico and Valley of Oaxaca (Schwarz 1997; Sanders et al. 1979; Blanton 1982; Kowalewski et al. 1989). For this example, the 17 largest sites (with more than 800 estimated population) were selected for the Valley of Oaxaca Period IIIa (A.D. 200–500; see Figure 8.6). The population estimates were regressed against an environmental index. Site population was estimated by site size and intensity of occupation, and an environmental index of land quality was tabulated from a summation of areas with differing agricultural productivity (alluvium, piedmont, etc.) from within a 3-km-radius catchment. The results of the regression analysis demonstrate a strong positive correlation of the two variables (adj. $r^2 = 0.593$, $p = 0.0001$). However, a puzzling linearity appears in the residual plot (Figure 8.7). It may be that population clustering or autocorrelation derived from the land-productivity zones created this linearity. Upon consideration, the raw variates were normalized and converted to base-10 logarithms and then regressed again (adj. $r^2 = 0.328$, $p = 0.001$; Figure 8.8). At this point, the visible linearity in the

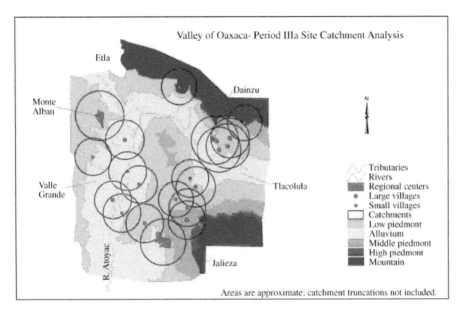

FIGURE 8.6
Site catchment analysis, Valley of Oaxaca, Period IIIa.

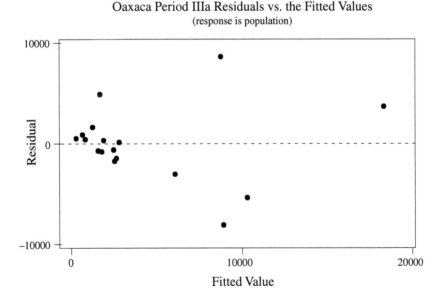

FIGURE 8.7
Residual plot, Oaxaca, Period IIIa, raw variates.

residual plots was reduced (Figure 8.9). A laboriously constructed Moran coefficient test was performed using Minitab macros (as described by Griffith 1989). The Moran test indicated that no significant spatial autocorrelation

Valley of Oaxaca Period IIIa Catchment Analysis

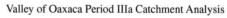

Log10(population) = 1.51 + 0.66[Log10(agric. prod.)]
Adj, R-Sq = 0.328

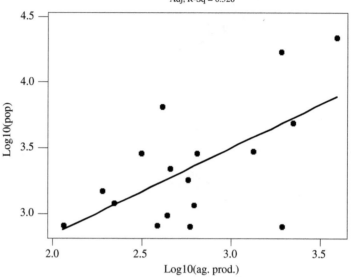

FIGURE 8.8
Scattergram, Oaxaca, Period IIIa.

Oaxaca, Period IIIa Transformed Residuals vs. Fits

(values are log base 10 and normalized)

FIGURE 8.9
Residual plot, Oaxaca IIIa, normalized log-ten variates.

existed in the converted data, so no autoregressive model was needed. It should be noted also that some investigators recommend using standard nonlinear regression techniques to eliminate spatial dependencies in the data (Odland 1988) before proceeding to an autoregressive (or spatial regression) model. Griffith and Amrhein (1997) give an extensive exploration of this complex field focusing on logistic, Poisson, and binomial distribution approaches to estimation.

The last resort of statistical analysis of spatially dependent data sets is the spatial autoregressive model (or spatial regression). The basic necessity of incorporating spatial dependence in the model means that the analyst must account for the spatial relationships among the observations of variates in some way, most often by a spatial connectivity matrix. This device is an $N \times N$ matrix that specifies the contiguity (border relationships) or spatial weights among all observations. The specification of the model incorporates an autoregressive component that relies on a Monte Carlo-type iterative process to estimate parameters of the statistic. The specialists in this procedure stress that it is still experimental in nature and involves methodological and mathematical complexities. As such, discussion of its mathematical basis goes beyond the scope of this chapter.

8.5 Ripley's K Statistic

Ripley's K statistic compares the spatial data input by the analyst with a homogeneous Poisson distribution (which represents complete spatial randomness). A scaled-distance algorithm is introduced, and the output is the K(t) statistic, a measure of global spatial clustering. In practice, the use of L(t) as an estimator has replaced K(t). L(t) can be derived from K(t) via the formula:

$$L(t) = (K(t)\Pi)^{1/2}$$

L(t) is then charted against distance and can be compared with complete spatial randomness (L(csr) = 0). Values of L(t) above 0 indicate spatial clustering, while negative values indicate dispersion. The Ripley's K function is also used to create simple and multivariate spatial models (Ripley 1981; Cressie 1991; Bailey and Gatrell 1995). Ripley's K has the advantage of characterizing spatial relationships at different distance scales, unlike, for example, nearest-neighbor analysis (Dixon 2001).

The graph of L(t) versus distance for an example given above is illustrative. The data from the lithic scatter at structure 100 indicated the existence of positive spatial autocorrelation. Ripley's K function was performed, and the resulting output of L(t) was graphed using a locally weighted smoothing process (Figure 8.10). This reduced to a smooth line the jagged linear

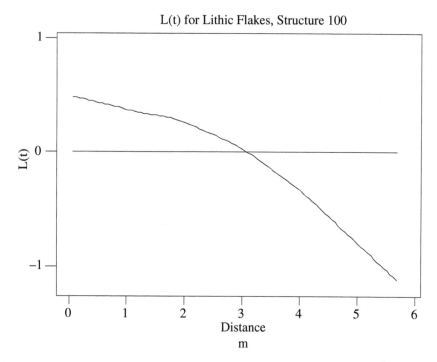

FIGURE 8.10
Ripley's L(t) plot, structure 100.

output that resulted from graphing the gridwork of excavation units. The finished graph depicts spatial clustering at scales of less than 3 m, consistent with the areal breadth of the western flake concentration. At a scale of 3 m, the distribution is completely spatially random. At scales of more than 3 m, the distribution is dispersed. This suggests that the western concentration is the major spatial cluster present within the scatter and that no larger clusters exist.

8.6 Nearest-Neighbor Hierarchical Spatial Clustering

This statistic is a clustering technique that identifies and groups observations that are spatially close. This clustering routine is based on the nearest-neighbor method. It clusters observations into hierarchical groups of nearest neighbors based upon three criteria that the user inputs. The user specifies a threshold distance that is compared with all pairs of points. The second criterion is a minimum number of points to be included in each first-order cluster. For second- and higher-order clusters, the routine is run using the centers of the lower-order clusters as points. A third criterion specifies the magnitude of the standard deviational ellipse that defines the spatial clusters. The clustering routine runs until all observations are united in one high-order cluster or the clustering fails (Levine 2002: 216–221).

An example of nearest-neighbor hierarchical spatial clustering (NNHSC) as an analysis technique derives from the work of one of the authors (Mount) in mapping a multicomponent site in Tennessee (40DV392) as a salvage project. The site includes Archaic, Mississippian, and historic components, though it is the Mississippian period occupation that provides most of the burials and features analyzed here (historic features were excluded from the analysis). The site sits on the floodplain of the Cumberland River and a secondary drainage and extends to a gently sloping hill on the western section of the site. Two discrete cemeteries exist: the western cemetery on the hill and the northern cemetery on a low rise between the river and drainage. The eastern part of the site was an important habitation area and included many features, such as postmold outlines of houses, pit features, and hearths (Figure 8.11).

Given the complexity of feature and burial patterning it was thought that NNHSC analysis would be a fruitful means of identifying clusters of different scales. It was decided also to use a battery of simpler spatial statistical indicators to assess the spatial patterning in this very large data set (N = 1559). These include mapping the geometric mean of features, calculating Moran's I for features and burials, and graphing Ripley's K function for both. Moran's I statistic for burials indicates that no spatial autocorrelation is present in the sample (MC = 0.0001, Z = 0.78), while the Moran's I for features indicates the existence of slightly positive spatial autocorrelation (MC = 0.0006, Z = 1.96).

Nearest-neighbor hierarchical spatial clustering of the burials was run for minimum groups of three, five, and eight burials (Figure 8.12). The interpretation of the output of such clustering routines is impressionistic. It appears that cluster minimums of three create too many insignificant clusters when viewed against the raw data (Figure 8.11). A cluster minimum of three creates seven first-order clusters and one second-order cluster, while a cluster minimum of eight produces sparse output. Therefore five-observation minimum clustering is preferred here. Running a NNHSC routine for features with the same clustering minimums produces quite interesting results (Figure 8.13). Clearly, the distribution of nonburial features across the site is more complex than that of the burials. A plot of the geometric mean of features may indicate the center of site activity. All three iterations of the NNHSC provide valuable insight on the distribution of features, but our impression is again that the five-observation minimum clustering strikes the right balance in identifying clusters and hierarchical groups. The three-observation-minimum analysis presents a more detailed look at the patterning within the site. It is also possible to depict the probabilistic nature of the cluster ellipses, as seen in Figure 8.14. Here three-observation-minimum clusters are drawn as concentric ellipses at distances of 1, 2, and 3 standard deviations. These different

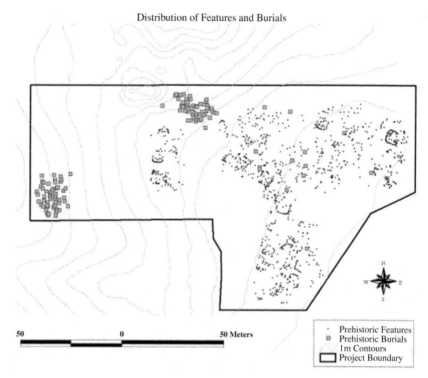

FIGURE 8.11
Distribution of features and burials, Tennessee.

solutions for the boundary of particular clusters are essentially spatial-confidence intervals (Levine 2002: 221).

An analysis of Ripley's K function further clarifies the spatial patterning of the site. A plot of L(t) for burials indicates positive spatial clustering from 0 to 25 m in distance and spatial randomness and dispersion beyond that distance (Figure 8.15). This is approximately the length of both the northern and western cemeteries (25 m). The finding is consistent with the Moran's I analysis of no global spatial autocorrelation (between the two cemeteries) and the NNHSC analysis, which failed to unite the two cemeteries into a global cluster. A graph of L(t) for features reveals clustering from approximately 0 to 35 m in distance and randomness and then dispersion beyond (Figure 8.16). Interestingly, the curvilinear character of L(t) below 35 m suggests increasing clustering with distance from scales of approximately 0 to 15 m and decreasing clustering from 15 to 35 m. This distribution corresponds to the complex patterning observable in the NNHSC clustering maps for features and is consistent with the positive finding from the Moran's I test. The use of NNHSC and the other spatial statistical analyses here demonstrates how researchers can characterize spatial data through multiple complementary means.

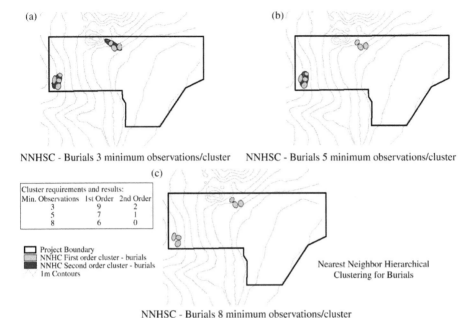

NNHSC - Burials 3 minimum observations/cluster   NNHSC - Burials 5 minimum observations/cluster

NNHSC - Burials 8 minimum observations/cluster

FIGURE 8.12
NNHSC burials, Tennessee.

8.7   Notes on Applications

The foregoing applications of spatial statistics are greatly facilitated by the seamless integration of S-Plus SpaceStat and CrimeStat with mapping software such as ArcView GIS and Golden Surfer. Data can be input from database programs such as Microsoft Access, from spreadsheets such as Microsoft Excel, or as text files. Output for many of these statistical procedures can be viewed graphically in ArcView GIS as *.shp files or in Surfer as contour maps. Data from excavation grids or point data (settlement-pattern data or point-provenience data) can be handled without problem, although it must be kept in mind that particular data-collection methodologies have advantages and disadvantages. For example, excavation block data is subject to edge effects that can distort the findings. Several test statistics, such as Ripley's K function in CrimeStat, allow the researcher to neutralize these border effects.

The ability to integrate georeferenced data from an archaeological database with these spatial statistical tests holds both promise and peril. The promise is that, by giving archaeologists the capability to use these improved statistical tests, they will be able to build more realistic causal

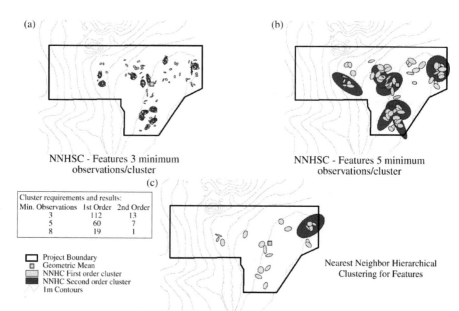

FIGURE 8.13
NNHSC features, Tennessee.

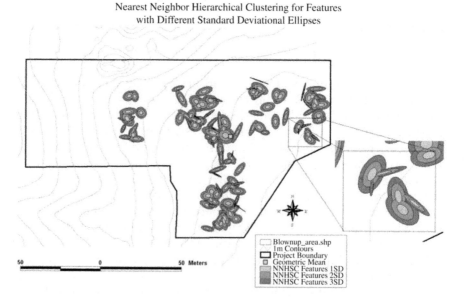

FIGURE 8.14
NNHSC features, different standard deviations (1, 2, and 3), Tennessee.

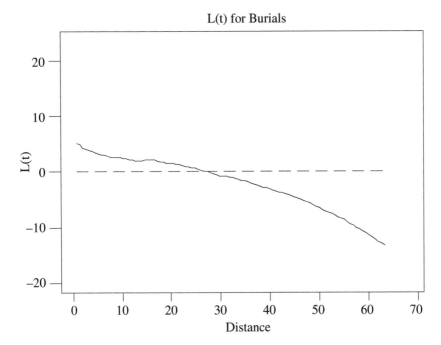

FIGURE 8.15
Ripley's K, burials, Tennessee.

and management-based models that deal with the reality of spatial dependencies. The peril is that archaeologists, as investigators who are generally untrained in the use of these complex statistical tools, will adopt a shotgun approach that will introduce problems of suitability and interpretation (as suggested by Griffith and Amrhein 1997: 322). It is hoped that the latter scenario can be avoided.

8.8   Discussion and Conclusion

The current state of the art in GIS approaches to archaeology involves the development and deployment of predictive models of site location through modeling of environmental variables. This effort can be seen as an outgrowth of the quantitative bent archaeology has taken in recent decades and the cultural ecological framework of the 1960s to the 1980s. As a theoretical goal or a cultural resource management tool, this sort of activity holds much promise. Despite the complaints of some that ecologically oriented modeling ignores aspects of past cultural behavior (Gaffney 1995), the ability to predict site locations sucessfully is a first step in providing plausible causal accounts

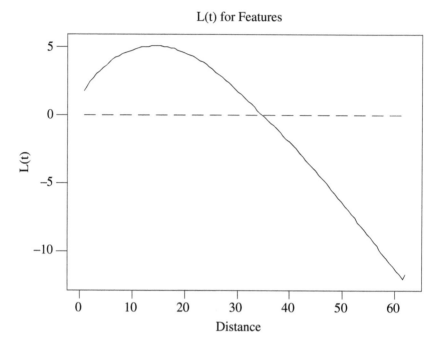

FIGURE 8.16
Ripley's K, features, Tennessee.

of ancient cultures and is an effective tool for resource management. In this sense, this chapter tries to contribute to such an effort by suggesting some sounder footings for spatial modeling. That is, by incorporating spatial auto-correlation, nonlinear and autoregressive models, and scalar and spatialized clustering routines in one's tool kit, a more accurate and valid model of a given archaeological data set can ultimately be developed. Intrasite spatial analyses are also aided in bringing newer, more robust tools to the search for behavioral patterning. It is hoped that this introduction to the topic of spatial statistics spurs further interest in the primary sources (e.g., Odland 1988; Griffith 1987, 1996; Griffith and Amrhein 1997), rather than immediate application, considering the limited description of theory and method that could be presented here.

Two other points emerge from the pioneering work of Griffith (1996; Griffith and Amrhein 1997). The first is that spatial statistical approaches are increasingly being implemented within the GIS context because autocorrelation statistics can be displayed graphically, promoting visualization of spatial relationships previously only described statistically (Vasilev 1996). The second is that for certain types of aspatial statistical procedures, such as logistic regression, there is not yet a well-developed complementary spatial statistical procedure. Griffith and Amrhein (1997: 315) state that a proper spatial statistical companion for logistic regression is virtually impossible; current modeling efforts involve many troubling violations of assumptions and *post hoc* modifications. The authors note that discriminant function analysis is a much easier statistic to adapt to spatial dependencies. Given this finding, it is suggested that logistic regression findings from spatial data sets be treated with a certain skepticism until autocorrelation can be assessed and more-robust spatial statistics can be utilized.

References

Aldenderfer, M., introduction to *Anthropology, Space, and Geographic Information Systems*, Aldenderfer, M. and Maschner, H.D., Eds., Oxford University Press, Oxford, 1996, pp. 1–18.

Allen, K.M., Green, S.W., and Zubrow, E.B., Eds., *Interpreting Space: GIS and Archaeology*, Taylor and Francis, Philadelphia, 1990.

Anselin, L., Local indicators of spatial association: LISA, *Geographical Analysis*, 27, 93–115, 1995.

Bailey, T.C. and Gatrell, A.C., *Interactive Spatial Data Analysis*, Longman Scientific and Technical Publishers, Harlow, U.K., 1995.

Blanton, R.E., Ed., Monte Alban's Hinterlands, Part I: Pre-Hispanic Settlement Patterns of the Central and Southern Parts of the Valley of Oaxaca, Mexico, Memoirs of the Museum of Anthropology, No. 15, University of Michigan, Ann Arbor, 1982.

Brumfiel, E., Regional growth in the Eastern Valley of Mexico: a test of the "population pressure" hypothesis, in *The Early Mesoamerican Village*, Flannery, K.V., Ed., Academic Press, New York, 1976, pp. 234–250.

Carr, C., The nature and organization of intrasite archaeological records and spatial analytic approaches to their investigations, in *Advances in Archaeological Method and Theory*, Vol. 7, Schiffer, M.B., Ed., Academic Press, New York, 1984.

Carr, C., Introductory remarks on regional analysis, in *For Concordance in Archaeological Analysis: Bridging Data Structure, Quantitative Technique, and Theory*, Carr, C., Ed., Westport Publishers, Kansas City, MO, 1985.

Chrisman, N.R., *Exploring Geographic Information Systems*, John Wiley and Sons, New York, 1997.

Cowgill, G.L., A selection of samplers: comments on archaeostatistics, *American Antiquity*, 33, 367–375, 1975.

Cressie, N.A., *Statistics for Spatial Data*, John Wiley and Sons, New York, 1991.

Dixon, P.M., Ripley's K Function, Department of Statistics, Iowa State University, Ames, 2001.

Gaffney, V., Postscript: environmental determinism and archaeology, in *Archaeology and Geographical Information Systems*, Lock, G. and Stančič, Z., Eds., Taylor and Francis, Bristol, PA, 1995, pp. 367–369.

Griffith, D.A., Theory of spatial statistics, in *Spatial Statistics and Models*, Gaile, G.L. and Willmott, C.J., Eds., D. Reidel Publishing, Dordrecht, Holland, 1984, pp. 3–16.

Griffith, D.A., *Spatial Autocorrelation: A Primer*, Association of American Geographers, Washington, DC, 1987.

Griffith, D.A., Spatial Regression Analysis on the PC: Spatial Statistics Using Minitab, Institute of Mathematical Geography, Department of Geography, Syracuse University, Syracuse, NY, 1989.

Griffith, D.A., Introduction: the need for spatial statistics, in *Practical Handbook of Spatial Statistics*, Arlinghaus, S.L., Ed., CRC Press, Boca Raton, FL, 1996, pp. 1–15.

Griffith, D.A. and Amrhein, C.G., *Multivariate Statistical Analysis for Geographers*, Prentice Hall, Upper Saddle River, NJ, 1997.

Hietala, H., Ed., *Intrasite Spatial Analysis in Archaeology*, Cambridge University Press, London, 1983.

Hinkle, D.E., Wiersma, W., and Jurs, S.G., *Applied Statistics for the Behavioral Sciences*, Houghton Mifflin, Boston, 1994.

Kowalewski, S.A., Population and agricultural potential: early I through V, in Monte Alban's Hinterlands, Part I: Pre-Hispanic Settlement Patterns of the Central and Southern Parts of the Valley of Oaxaca, Mexico, Memoirs of the Museum of Anthropology, No. 15, University of Michigan, Ann Arbor, 1982, pp. 149–180.

Kowalewski, S.A., Feinman, G.M., Finsten, L., Blanton, R.E., and Nicholas, L.M., Monte Alban's Hinterlands, Part II: Pre-Hispanic Settlement Patterns in Tlacolula, Etla, and Ocotlan, the Valley of Oaxaca, Mexico, Memoirs of the Museum of Anthropology, No. 23, University of Michigan, Ann Arbor, 1989.

Kvamme, K.L., A predictive site location model on the high plains: an example of an independent test, *Plains Anthropologist*, 37 (138), 19–40, 1992.

Kvamme, K.L., Spatial statistics and GIS: an integrated approach, in *Computing the Past: Computer Applications and Quantitative Methods in Archaeology*, CAA 1992 proceedings, Andresen, J., Madsen, T., and Scollar, I., Eds., Aarhus University Press, Aarhus, Denmark, 1993, pp. 91–102.

Levine, N., CrimeStat: A Spatial Statistics Program for the Analysis of Crime Incident Locations, Ver. 2.0, Ned Levine and Associates, Houston, TX, and the National Institute of Justice, Washington, DC, May 2002.

Lock, G. and Stančič, Z., Eds., *Archaeology and Geographical Information Systems*, Taylor and Francis, Bristol, PA, 1995.

Moran, P., The interpretation of statistical maps, *Journal of the Royal Statistical Society*, 10B, 243–251, 1948.

Odland, J., *Spatial Autocorrelation*, Sage Publications, Newberry Park, CA, 1988.

Pedhazur, E., *Multiple Regression in Behavioral Research*, Holt, Reinhart, Winston, Orlando, FL, 1982.

Ripley, B.D., *Spatial Statistics*, John Wiley and Sons, New York, 1981.

Sanders, W.T., Santley, R.S., and Parsons, J.B., *The Basin of Mexico: Ecological Evolution of a Civilization*, Academic Press, New York, 1979.

Schwarz, K.R., Reconsidering Human-Land Relationships in Pre-Hispanic Central Mexico, unpublished M.A. thesis, Department of Anthropology, Southern Illinois University, Carbondale, 1997.

Schwarz, K.R., Understanding Classic to Postclassic Household and Community Spatial Transformation: The Rural Maya of the Quexil-Petenxil Basins, Guatemala. Unpublished Ph.D. dissertation, Department of Anthropology, Southern Illinois University, Carbondale, 2004.

Thomas, D.H., The awful truth about statistics in archaeology, *American Antiquity*, 43 (2), 231–244, 1978.

Tobler, W.R., Cellular geography, in *Philosophy in Geography*, Gale, S. and Olsson, G., Eds., John Wiley, London, 1979, pp. 379–386.

Vasilev, I.R., Visualization of spatial dependence: an elementary view of spatial autocorrelation, in *Practical Handbook of Spatial Statistics*, Arlinghaus, S.L., Ed., CRC Press, Boca Raton, FL, 1996, pp. 1–15.

Vita-Finzi, C. and Higgs, E.S., Prehistoric economy in the Mount Carmel area of Palestine: site catchment analysis, *Proceedings of the Prehistoric Society*, 36, 1–37, 1970.

Warren, R.E., Predictive modeling in archaeology: a primer, in *Interpreting Space: GIS and Archaeology*, Allen, K.M., Green, S.W., and Zubrow, E.B., Eds., Taylor and Francis, Philadelphia, 1990a, pp. 90–111.

Warren, R.E., Predictive modeling of archaeological site location: a case study in the Midwest, in *Interpreting Space: GIS and Archaeology*, Allen, K.M., Green, S.W., and Zubrow, E.B., Eds., Taylor and Francis, Philadelphia, 1990b, pp. 201–215.

Whallon, R.R., Spatial analysis of occupational floors I: applications of dimensional analysis of variance, *American Antiquity*, 38 (3), 266–278, 1973.

Whallon, R.R., Spatial analysis of occupational floors II: nearest neighbor analysis, *American Antiquity*, 39 (1), 16–34, 1974.

Willey, G.R., Prehistoric Settlement Patterns in the Viru Valley, Peru, Bureau of American Ethnology, Bulletin 155, Washington, DC, 1953.

Willey, G.R., Bullard, W.R., Jr., Glass, J.B., and Gifford, J.C., Prehistoric Maya Settlements in the Belize Valley, Papers of the Peabody Museum of Archaeology and Ethnology, Harvard, Vol. LIV, Peabody Museum, Cambridge, 1965.

Wood, J.J., Optimal location in settlement space: a model for describing location strategies, *American Antiquity*, 43 (2), 258–270, 1978.

9

Quantifying the Qualified: The Use of Multicriteria Methods and Bayesian Statistics for the Development of Archaeological Predictive Models

Philip Verhagen

CONTENTS

ABSTRACT   Over the past ten years, archaeological predictive modeling for cultural resource management (CRM) in the Netherlands has experienced a shift from the dominant use of quantitative methods to a more qualitative approach, where expert judgment plays an important role in the definition of zones of high and low archaeological value. However, the use of expert judgment poses a problem for the development of predictive models. The process leading to the establishment?/specification? of expert judgment is often vague; the criteria for inclusion of areas in certain archaeological zones are poorly specified; and as a consequence, mapping between regions is often incompatible. In many cases, the process is vague; the criteria for inclusion of areas in certain archaeological zones are poorly specified; and mapping between regions is frequently incompatible. This chapter demonstrates the use of a multicriteria decision-making framework that increases the transparency of "expert judgment" in predictive mapping. The chapter also evaluates the potential of Bayesian statistical methods for use in developing weights for the criteria involved. Hopefully, this will lead to the development of models that, although subjective to a certain extent, are at least consistent and comparable between regions. The proposed methodological framework is demonstrated using a case study for the municipality of Ede in the central part of the Netherlands.

9.1 Introduction

Over the past ten years, archaeological predictive modeling in the Netherlands has been the subject of a sometimes-heated debate (see Verhagen et al. 2000). After seminal publications by Wansleeben (1988), Ankum and Groenewoudt (1990), and Soonius and Ankum (1990), a number of predictive maps have been produced by public archaeological institutions in the Netherlands (RAAP[1] and ROB[2]). At the same time, academic archaeologists have studied the methodological and theoretical aspects of predictive modeling, and some have criticized the modeling concepts used in public archaeology in several publications (Wansleeben and Verhart 1997; van Leusen 1993, 1995, 1996; Kamermans and Rensink 1999).

The inferential or inductive approach, already criticized by Brandt et al. (1992) for its inability to cope with the low quality of many archaeological data sets, has gradually been replaced by a more intuitive way of model development that tries to make the best of both worlds by including quantitative data when they are available. Coupled to this development toward more-deductive mapping, it is notable that the multivariate approach has been replaced by the use of a reduced number of variables that are supposed to have the strongest predictive power for a particular region or archaeological period. These models can best be characterized as hybrids and are essentially descriptions of existing knowledge rather than extrapolations. A consequence of this approach is that error margins and uncertainties are never specified, and on the whole the models lack a clear formalized methodology for including both "hard" and "soft" knowledge.

The current chapter discusses the possibilities of improving the applied methodology by focusing on the formalization of the inclusion of "expert judgment" or subjective knowledge into the mapping. Two methodological innovations are suggested for this: the application of a multicriteria decision-making framework to the modeling, and the use of Bayesian statistics for developing the knowledge base needed for the model. This methodological framework is tested on a case study in the municipality of Ede (Figure 9.1), where the model of Soonius and Ankum (1990) was originally applied.

9.2 Multicriteria Decision Making and Its Relevance to Predictive Modeling

Multicriteria decision making (MCDM) is a set of systematic procedures for analyzing complex decision problems. By dividing the decision problem into small, understandable parts and then analyzing these parts and integrating them in a logical manner, a meaningful solution to the problem can be achieved. Decision making includes any choice among alternative courses of action and is therefore of importance in many fields in both the natural and social sciences. These types of decisions usually involve a large set of feasible alternatives and multiple, often conflicting and incommensurate evaluation criteria. Archaeological predictive modeling fits into this framework, as it is a way to evaluate the archaeological potential of an area and provide the basis for decision making in prospection design as well as in

FIGURE 9.1
Location of Ede.

planning procedures. In fact, it is recognized as such by Kvamme (1990): "A [predictive] model is a *decision rule* conditional on other nonarchaeological features of locations" [emphasis added].

Many archaeologists may not be familiar with the concepts and terminology used in MCDM, so a condensed description of its core notions follows here. This description closely follows the outline presented by Malczewski (1999); a slightly different terminology can be found in other publications, such as Nijkamp et al. (1990). MCDM can be broken down into the following components (Malczewski 1999: 82):

- The definition of a goal that the decision maker attempts to achieve
- The selection of a set of evaluation criteria (called objectives or attributes)
- The decision maker and his or her preferences with respect to the evaluation criteria
- The definition of a set of decision alternatives
- The calculation of a set of outcomes associated with each attribute/ alternative pair

9.3 Defining Goals

A goal is a desired state of affairs. In a predictive modeling context, the goal can for example be defined as minimizing the impact of planning measures on the archaeological record. A similar goal could be a maximum reduction of the costs associated with archaeological investigations in the area under consideration. The defined goal can be broken down into several *objectives*, which can best be thought of as intermediate goals. This conceptual framework forms the basis of the analytical hierarchy process (Saaty 1980), a widely used method for MCDM. By defining objectives, a hierarchical structure of decision making can be developed. Decisions can then be made by comparing objectives or, at the lowest level of the hierarchy, by comparing *attributes*. Attributes are measurable quantities or qualities of a geographical entity or of a relationship between geographical entities. These are the basic information sources available to the decision maker for formulating and achieving the objectives. An attribute is a concrete descriptive variable; an objective is a more abstract variable with a specification of the relative desirability of that variable. The attributes used for archaeological predictive modeling are usually a limited number of environmental variables; they can be supplemented with expert knowledge on less measurable variables.

Given the goal of minimizing the impact of planning measures on archaeology, one of the objectives under consideration might be the selection of areas of minimal site density, and another the selection of building methods that are not damaging to the archaeological remains. In this example, predictive modeling is only a way of defining an objective, rather than a way of achieving an immediate goal. Archaeological predictive modeling is a typical example of multiattribute decision making (MADM) that obtains preferences directly for the attributes in the form of functions and weights. MADM problems are those that have a predetermined, limited number of alternatives, as is common in environmental impact assessments where, for example, a limited number of railway alignments must be compared. However, in a GIS-context, each single raster cell or polygon can be seen as a decision alternative, as outcomes are calculated for each entity.

When applying the MCDM framework to a predictive-modeling study, such as was done in Ede by Soonius and Ankum (1990), it is easy to break down the model into a hierarchical structure of objectives (Figure 9.2). The final model presented is aimed at predicting the potential of every single raster cell for finding undisturbed remains of prehistoric settlements. It has two objectives at the highest level: an assessment of the chance of survival of archaeological remains, and an assessment of overall site density. The assessment of overall site density is achieved by combining five intermediate objectives: determining site density for five separate archaeological periods. Each of these subobjectives is in turn evaluated using six attributes, selected by means of a χ^2 test.

9.4 Selection of the Evaluation Criteria

In this phase of MCDM, the attributes to be used are specified, and a measurement scale is established. This is equivalent to the selection of the predictor variables and establishing their values, for example, in terms of site density or probability. Attributes may be used for different objectives, possibly with other values attached.

Attributes should be *complete, operational, decomposable, nonredundant,* and *minimal.* A set of attributes is complete if it covers all relevant aspects of the decision problem and indicates the degree to which the overall objective is achieved. One of the enduring criticisms of archaeological predictive modeling is that the set of attributes used is not complete; especially social and cultural variables are assumed to be missing from the full set of attributes. A set of attributes is operational if it can be used meaningfully in the analysis (it is understandable), so that decision makers can understand the consequences associated with alternative decisions. Furthermore, an attribute should be *comprehensive* (have a direct relation to the decision problem) and *measurable.* This last condition in many cases implies choosing a proxy attribute; so instead of "a dry piece of land to live on," the more tractable variable "groundwater table" might be employed. The use of proxy attributes implies that there is a missing link between the information available and the information necessary (Beinat 1995).

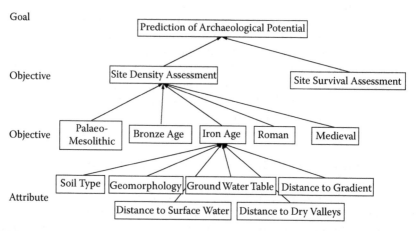

FIGURE 9.2
Hierarchical structure of objectives.

In the model created by Soonius and Ankum (1990), the environmental variables used were selected by means of a χ^2 test using the known archaeological sites. As has been shown by van Leusen (1996), the actual application of the test was not done correctly. The violations of the statistical assumptions for the Ede model are, however, not too grave as long as the analysis is not considered per period. Even with Yates's correction applied, the values of χ^2 are high enough to justify the selection of the analyzed variables for the predictive mapping on methodological grounds. However, this does not mean that the set of attributes analyzed is complete, although quite a number of environmental descriptors were analyzed.

Furthermore, it may be suspected that some of the variables used are redundant, as no independence check was performed. Spatial cross-correlation indices (Goodchild 1986) can be calculated for pairs of raster maps and used as a first measure of the independence of the attributes used. The calculation should be based on the attribute values associated with the different maps. Correlation values range from +1 to −1. A positive correlation indicates a direct relationship between two layers, such as when the cell values of one layer increase, the cell values of another layer are also likely to increase. A negative correlation means that one variable changes inversely to the other. A correlation of 0 means that two layers are independent of each other.

$$c = \Sigma c_{ij} / \{\sqrt{[\Sigma (z_i - z_i)^2]} \cdot \sqrt{[\Sigma (z_j - z_j)^2]}\}$$

where

$c_{ij} = (z_i - z_i) \cdot (z_j - z_j)$, reflecting the similarity of i and j attributes

z = attribute value

i = any cell in grid 1

j = any cell in grid 2 (on the same location)

TABLE 9.1

Correlation Matrix for Six Variables

	1	2	3	4	5	6
1	1.00000	0.17658	0.10074	0.18295	0.33574	−0.11646
2	0.17658	1.00000	0.55118	0.43743	−0.01326	−0.08668
3	0.10074	0.55118	1.00000	0.36713	−0.15656	−0.09246
4	0.18295	0.43743	0.36713	1.00000	0.16368	−0.14212
5	0.33574	−0.01326	−0.15656	0.16368	1.00000	−0.07790
6	−0.11646	−0.08668	−0.09246	−0.14212	−0.07790	1.00000

Note: 1 = soil type; 2 = geomorphological unit; 3 = groundwater table; 4 = distance to dry valley; 5 = distance to ecological gradient; 6 = distance to surface water.

Source: Soonius, C.M. and Ankum, L.A., Ede, II: Archeologische Potentiekaart, RAAP rapport 49, Stichting RAAP, Amsterdam, 1990.

From Table 9.1 it is clear that at least two variables should be regarded with suspicion for use in the predictive model, i.e., groundwater table and distance to dry valleys, both of which seem to be somewhat correlated to the geomorphological map. This is not very surprising, as the groundwater table is related to elevation, and the map for the distance to the dry valleys was obtained by extracting the dry valleys from the geomorphological map. A third possible correlation can be observed between the soil map and the distance to ecological gradient. This seems to indicate that certain soil types are related to the presence or absence of an ecological gradient.

9.5   Defining Measurement Scales

Measurement scales can be obtained by using normalization, value (or utility) functions (Keeney and Raiffa 1976); probabilistic methods; and fuzzy-set membership (e.g., see Burrough 1989 and Burrough et al. 1992). A distinction can be made between objective probability (like site-density measures) and subjective probability. The latter is also known as "prior belief" in the context of Bayesian statistics.

   The objective probability approach received a lot of attention in archaeological predictive modeling in the 1980s, and in particular the use of multivariate statistical techniques like logistic regression that can be used to obtain objective weights (e.g., see Warren 1990). However, in many cases the weights obtained by multivariate statistics are not as objective as one would like them to be. As was noted at the outset, the use of quantitative, multivariate methods for predictive modeling in the Netherlands has been replaced over the past 10 years by qualitative or semiquantitative methods. The main reason for this is the fact that frequency statistics have not been able to deliver their promise of objective predictions because of the poor quality of the archaeological data sets used. Many archaeological data sets are biased toward certain site types that have been recorded under specific terrain conditions, so there will be many situations where the available archaeological data set cannot be used as a representative sample of the target population. In those cases, expert judgment or subjective belief may be used to estimate map weights.

9.6   Defining Preferences

The decision maker's preferences with respect to the evaluation criteria should be incorporated into the decision model; they express the importance of each criterion relative to other criteria. The decision maker is simply the person(s) involved in trying to achieve the defined goal. Ideally, experts provide facts and decision makers values (Beinat 1995). In predictive modeling, however, the decision makers are often also the experts. They might judge that, for the prediction of Bronze Age graves, other evaluation criteria should be used than for Medieval settlement locations. Similarly, different preferences can be specified for the Bronze Age graves and Medieval settlements when making a decision on what to investigate under the constraints of the available budget.

A number of methods are available to obtain a numerical representation of preferences. Of these methods, pairwise comparison (Saaty 1980) seems to be the most effective, but ranking methods and rating methods are easier to apply. The important element is that these methods are all meant for the comparison of value judgments, and as such involve expert opinion and subjective reasoning.

The definition of the preferences, in fact, takes place at *all* hierarchical levels of the decision-making process: attributes can be compared, but objectives can be as well. For example, the decision maker might decide that groundwater table is more important for site location than soil texture. After finishing this evaluation, he or she might decide that Neolithic sites are not as important as Roman sites. And after this evaluation, he or she might decide that site density is a less important criterion than site preservation. In this way, a nested hierarchy of decision making is created, in which all decisions can be subjected to the same cycle of criterion selection, establishment of measurement scales, definition of alternatives, and preference definition (the analytic hierarchy process). It should be noted that the definition of preferences can be avoided only when the criteria used are truly independent and can be measured in terms of objective probability (for example in a logistic regression equation when all statistical requirements have been fulfilled). In all other cases, value judgments will be necessary to weigh the evaluation criteria.

The procedures of establishing measurement scales and criterion preferences, as for example applied by Dalla Bona (1994, 2000), who refers to it as the "weighted value method," and even of nesting objectives are of course not new in archaeological predictive modeling (e.g., Kohler and Parker 1986, who distinguish a Hierarchical Decision Criteria Model). However, they are usually not recognized as belonging to the more generic class of MCDM methods. Van Leusen (1993) for example suggested that "translating" archaeological intuition into weighting schemes and other types of classification rules and embodying them in an expert system would at least make "intuitive" approaches reproducible, which seems an adequate description of applying MCDM methods. It should be noted, however, that this methodology, which heavily relies on expert judgment, has not been the dominant one in (especially American) literature on the subject up to today, as is illustrated by most of the applications found in Wescott and Brandon (2000). An example of the rather suspicious attitude toward subjective weighting can be found in the paper by Brandt et al. (1992), who rejected the use of purely deductively derived weights for map layers and features, basically because the current state of archaeological theory would not be able to give more than rough notions about human locational behavior. However, at the same time, these authors could not achieve complete reliance on inductive methods either (van Leusen 1996), leading to the "hybrid approach" that was also used by Soonius and Ankum (1990). Recently, some interest can be observed in the use of land-evaluation methods as a purely deductive technique for predictive modeling (Kamermans 2000), an approach that has been applied previously in archaeological studies of prehistoric land use (Kamermans 1993; Finke et al. 1994). However, one should not necessarily equate deductive modeling with expert judgment weighting: the experts usually arrive at their judgment through a combination of deductive and inductive arguments.

9.7 Establishing the Decision Rules

This phase brings together the preceding three steps for the overall assessment of the alternatives (ranking of alternatives). The most commonly applied method is simple additive weighting, also known as weighted linear combination. Similarly, probabilistic additive weighting can be used to obtain a ranking of alternatives. The prerequisite of all addition methods is that the attributes used be conditionally independent.

Kohler and Parker (1986) review four different types of decision rules that can be applied to archaeological predictive models. The Fatal-Flaw Decision Criteria Model is the most constraining of these; it results in a binary response (yes or no). In MCDM, this decision rule is known as a noncompensatory method called *conjunctive screening*. Under conjunctive screening, an alternative is accepted if it meets specified standards or thresholds for *all* evaluation criteria (Boolean AND). The Hierarchical Decision Criteria Model is a variant of this; it performs a conjunctive screening as well, but each time for a different objective. Conjunctive screening is also found in the application of land evaluation, where a land unit can only be classified as being suitable for certain kinds of cultivation if it meets all evaluation criteria. *Disjunctive screening*, on the other hand, accepts the alternative if it scores sufficiently high on *at least one* of the criteria (Boolean OR). This is a method not usually found in archaeological predictive modeling.

If no direct binary response is desired, compensatory methods are applied. They require a value judgment for the combination of (possibly conflicting) criteria, and can be applied only when all the evaluation criteria are measured in the same units. Afterward, a constraint can be placed on the outcome of the decision rule, for example, to distinguish crisp zones of high and low probability. Kohler and Parker (1986) distinguish the unweighted Decision Criteria Model and the Weighted Additive Decision Criteria Model. The unweighted model is in fact a special case of additive weighting, as all weights are equal — which is a value judgment in itself.

The combination of nonindependent attributes can be done by means of multiplication. This is to be avoided in most cases, as it implies that the interactions between the attributes are known, and these must then serve as the input for the multiplication equation.

9.8   Bayesian Statistics and Predictive Mapping

9.8.1   Combining Objective and Subjective Weights

It is important to note that between the extremes of purely objective and purely subjective weighting, compromises of the two can be used. Two routes can then be followed: the first starts with an objective weighting, and the weights are adapted afterward by consulting experts. This is basically the method employed by Deeben et al. (1997). This form of combination lacks a set of formal rules for application and therefore does not produce a transparent model unless all modifications to the objective weighting are clearly specified. The second route, which is further explored in this chapter, starts with a subjective weighting and uses any quantitative data available to modify the weights, but only when the data are considered to be a representative sample. In essence, this is the concept of Bayesian statistics, where a subjective prior belief is modified using quantitative data to obtain a posterior belief:

posterior belief = conditional belief × prior belief

The few published applications of Bayesian statistics in predictive modeling inside (van Dalen 1999) and outside archaeology (Aspinall 1992; Bonham-Carter 1994) have one thing in common: the assumption of a uniform prior probability for all map categories. This assumption is the simplest possible form of formulating prior beliefs, and it equates to a situation where no prior information is available. In the studies mentioned, most attention is paid to the establishment of the conditional beliefs. It can be shown that the p_s/p_a ratio[4] (a commonplace indicator of site density that is also used by Deeben et al. [1997]) is equivalent in a Bayesian context to the ratio of prior to conditional probabilities under the assumption of a uniform prior probability (see Buck et al. 1996). This is the reason that Bonham-Carter (1994) uses p_s/p_a ratios as well for the development of a Bayesian geological predictive model. Similarly Aspinall (1992), in an ecological application of Bayesian statistics, comments that conditional probabilities can be expressed as relative frequencies of occurrence. If the condition of independence of the variables is met, the calculation of posterior probabilities is then simply a question of multiplying the p_s/p_a ratios per variable with the prior probabilities. When using a logarithmic normalization, the posterior probabilities can even be calculated by simple addition instead of multiplication (Bonham-Carter 1994).

The uniform prior probabilities are based on the currently found site density per area unit. When combined with p_s/p_a ratios used as conditional probabilities, the sum of the predicted posterior probabilities per area unit should equal the number of observed sites, as p_s/p_a is a dimensionless number indicating relative site densities. Bonham-Carter (1994) notes that the ratio of observed to predicted sites can therefore serve as a measure of violation of the assumption of conditional independence of the variables. However, it may be desirable not to predict an absolute number of sites, as the size of the target population is not known. In that case, the prior probabilities should be normalized on a scale of 0 to 1. When using a uniform prior probability, this implies that any Bayesian predictive modeling exercise equates to calculating the p_s/p_a ratios per map category.

Apart from the uniform probability distribution, Buck et al. (1996) show a number of other distributions, either discrete or continuous, that can be used to model both prior (Orton 2000) and posterior probabilities. In the case of categorical maps, a binomial distribution can be implemented by breaking down the nominal variables into binary ones (Bonham-Carter 1994). The prior and posterior belief for each map category can subsequently be modeled by means of a (continuous) Beta distribution (Buck et al. 1996), allowing for the establishment of standard deviations around the mean of the posterior belief. The form of the Beta distribution is directly dependent on the sample size and the conditional probability derived from it. The larger the sample size, the smaller the standard deviations will become, and the closer the mean of the distribution will be to the conditional probability found. A useful aspect of Beta distributions is that they will also yield a mean and a standard deviation in cases where no sites have been found on the map unit of interest. They can be used when the available sample is small.

9.8.2 Formulating the Priors

The following question to be answered is: If we do not want to use an assumption of uniform probability, how can we establish prior probabilities based on expert judgment? This brings us back into the realm of multicriteria decision making and, precisely, the issue of specifying preferences by the decision maker. Three basic methods can easily be applied:

1. *Ranking methods*: The decision maker expresses a ranking of the criteria under consideration, and the following equations can then be used to obtain numerical weights from the rank-order informa- tion (rank sum weights):

   $$w_j = n - r_j + 1/\Sigma(n - r_k + 1)$$

   or (rank reciprocal weights)

   $$w_j = (1/r_j)/\Sigma(1/r_k)$$

   where

   w_j = weight for criterion j
   n = number of criteria under consideration
   r = rank position of the criterion

   In general, the larger the number of criteria to be ranked, the less useful the method will become.

2. *Rating methods*: The decision maker tries to estimate weights on a 1 to 100 scale (or any other conceivable numerical scale). This is the most widely applied method in archaeological predictive mapping (e.g., Dalla Bona 1994, 2000).

3. *Pairwise comparison method* (as part of the analytical hierarchy process; Saaty 1980): Criteria are compared in pairs, and intensities of impor- tance are attributed to each pair. It can only be performed if all criteria are measured on the same scale. It is suggested by Malczewski (1999) that the method is the most effective technique for spatial decision making, and as such should receive more attention. A good introduc- tion to the technique is given in Eastman et al. (1993). The number of comparisons involved can become very large if the number of criteria increases. Comparisons are made in linguistic terms (Table 9.2). A maximum number of nine comparison levels is given that can be related to an intensity number. An intensity number of 1 implies equal importance (the diagonal in the comparison matrix), an intensity of 9 is translated as extreme importance. This means that when a zone of high archaeological value is judged to be "extremely more important" than a zone of low archaeological value, a value of 9 is given to the

TABLE 9.2

The Scale Used for Pairwise Comparison

Intensity of Importance	Definition
1	Equal importance
2	Equal to moderate importance
3	Moderate importance
4	Moderate to strong importance
5	Strong importance
6	Strong to very strong importance
7	Very strong importance
8	Very to extremely strong importance
9	Extreme importance

Source: Malczewski, J., *GIS and Multicriteria Decision Analysis*, John Wiley and Sons, New York, 1999.

"high compared to low" cell in the comparison matrix. A value of 1/ 9 should then be given to the "low compared to high" cell, resulting in a reciprocal matrix.

In essence, the procedures described above solve the problem of formalizing the use of subjective information in a predictive map.

9.9 Bayesian Statistics and Inductive Learning

The available methods for preference specification demand that the expert express his or her preferences in a numerical manner, or at least be able to provide a preference ranking. One might therefore ask: Is it of any use to specify preferences in a numerical way, apart from specific cases where a numerical answer is desired? Two arguments can be given in favor of using numerical preferences:

1. The measurement scales should be comparable when applying decision rules; some form of quantification is necessary to judge the outcome of any multicriteria decision-making process.
2. The fact that predictive maps will be nothing but a representation of current knowledge makes it necessary that the models produced be amenable to improvement, i.e., the models should be able to learn. To achieve this, the modeling procedure should be transparent and reproducible; this is more easily done using numerical preferences than by using linguistic terms.

The second point may need some clarification. Let us take the case where a map unit has been qualified as having a low archaeological value. The definition of "low value" in this case may imply a relatively low density of archaeological finds. However, the linguistic definition does not include an assessment of the actual quantities or probabilities involved. Suppose that, in the course of several years, a number of new settlement sites are found within this particular unit, e.g., because of the use of new prospection methods. When do we decide that the archaeological value of this particular unit no longer is low but should be intermediate or high? Without a numerical decision rule, the interpretation of the archaeological value of the unit is neither transparent nor reproducible.

Advocates of Bayesian statistics are always very confident that it offers a way of "inductive learning." Venneker (1996), for example, states that "it constitutes a computationally efficient recursive process in which the entire data stream is captured in the posterior belief of a hypothesis and need not be recalculated each time new or additional independent data become available." The bottleneck in this statement is the fact that the new data should be independent from the old data in order to be able to adapt the posterior belief; otherwise there is the very real danger of self-fulfilling prophecies. This is precisely the problem of using archaeological predictive maps for guiding surveys: there will be a natural tendency to select those areas where high site densities are predicted, and this may lead to an ever-increasing amount of biased-sample data. Even though the Bayesian approach has the appeal of being a formal method applicable in a rather straightforward way to data that are far from optimal, the approach itself does not solve the problem of using biased data. One advantage of Bayesian statistics, however, is that if the new data are collected independently, there is no need for random sampling (Orton 2000); representative sampling — a task that may be difficult enough in itself — is good enough.

9.10 Application: The Predictive Map of Ede

To illustrate the approach outlined above, a case study was performed using the new predictive map of the municipality of Ede that was recently made by RAAP (Heunks 2001) (Figure 9.3). This map was commissioned by the municipality to replace the 1990 map, which had become outdated. The new map is primarily based on a qualitative interpretation of the 1:50,000 soil map of the area. Map units were combined to arrive at what can best be described as "archaeological land units" that were subsequently evaluated for their archaeological value (Table 9.3). In essence, we are dealing with a single-attribute map that can be evaluated for different criteria, such as site density per period or site-preservation conditions.

The prior probabilities were obtained by contacting four experts on the archaeology of the region. Of these, one refused to cooperate on the grounds that the land-units map alone could not be used for a predictive

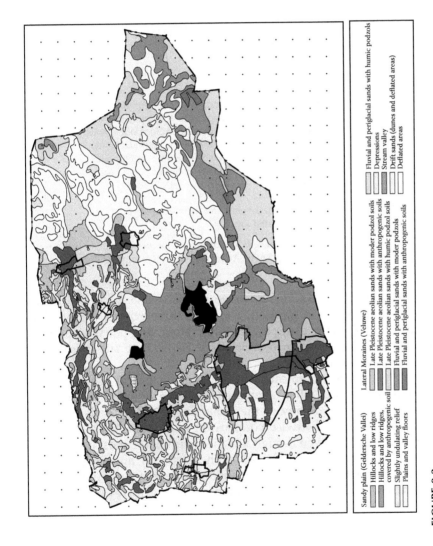

Sandy plain (Geldersche Vallei)

☐ Hillocks and low ridges
☐ Hillocks and low ridges, covered by anthropogenic soil
☐ Slightly undulating relief
☐ Plains and valley floors

Lateral Moraines (Veluwe)

☐ Late Pleistocene aeolian sands with moder podzol soils
☐ Late Pleistocene aeolian sands with anthropogenic soils
☐ Late Pleistocene aeolian sands with humic podzol soils
☐ Fluvial and periglacial sands with moder podzols
☐ Fluvial and periglacial sands with anthropogenic soils

☐ Fluvial and periglacial sands with humic podzols
☐ Depressions
☐ Stream valley
☐ Drift sands (dunes and deflated areas)
☐ Deflated areas

FIGURE 9.3
Predictive map of the municipality of Ede.

TABLE 9.3

Archaeological Land Units Used as the Basis for the Ede
Predictive Map

Number	Land Unit

Low-Lying Sandy Plain (Geldersche Vallei)

1	Low ridges and hillocks
2	Low ridges and hillocks with anthropogenic soils
3	Undulating plains
4	Valleys and depressions

Lateral Morainic Hills (Veluwe)

5	Late Pleistocene aeolian sands with moder podzol soils
6	Late Pleistocene aeolian sands with anthropogenic soils
7	Late Pleistocene aeolian sands with humic podzol soils
8	Fluvial and periglacial sands with moder podzol soils
9	Fluvial and periglacial sands with anthropogenic soils
10	Fluvial and periglacial sands with humic podzol soils
11	Depressions
12	Stream valley
13	Drift sands, both covered and deflated areas
14	Drift sand, deflated areas

Source: Heunks, E., Gemeente Ede: archeologische verwachtingskaart, RAAP-rapport 654, RAAP Archeologisch Adviesbureau BV,
Amsterdam, 2001.

model as it does not try to take into account social and cultural factors that
might have influenced site location. The other three experts were willing
to evaluate the map units using the ranking, rating, and pairwise comparison methods that were outlined in Section 9.8.2 for both site-density and
site-preservation potential.

The responses received indicated that the rank reciprocal method results
in inconsistent weights when compared with the other three methods.
There is no evidence that any of the remaining three methods performs
better than the others. Pairwise comparison is without any doubt a lot
more time consuming, even though it is theoretically the most appropriate
and efficient (Malczewski 1999). The main reason for applying pairwise
comparison is to obtain weights that are independent of the overall weighting obtained with the rating and ranking methods. However, when confronted with all three methods, the experts took the exercise as a test of
maintaining consistency between methods, resulting in a relatively consistent weighting between methods. Apart from that, the experts were hesitant
to indicate differences between units that they were uncertain about, pre-
ferring to attribute equal weights instead (or in Bayesian terms, specifying
uninformative priors).

Comparison of weights between experts showed that some disagreement
exists on the importance of the land units, both for site-density as well as

for site-preservation potential. Whereas one expert took soil type as the most important factor influencing site density, a second one evidently believed that geomorphology was more important. There was also a strong disagreement on the importance of the partly deflated drift-sand areas for site preservation. This may have been the consequence of not fully explaining the map legend to the experts. This particular unit was thought to be important in terms of site preservation, as drift sand may have covered the previously existing surface. In fact, site-preservation potential is influenced by two factors — the presence of a soil cover and the groundwater table — and one expert clearly believed that a high groundwater table was much more important, possibly reflecting a preference for well-conserved organic remains. Of course, this brings us to the question of who to believe. As it will in most cases be impossible to weigh the experts' opinions on a scale of reliability of response, the obvious solution is to take the mean weight of the experts' responses as the prior belief (Beinat 1995). This also implies that it is possible to calculate the variance of the experts' responses. An alternative option is to consider each expert's opinion as an independent sample.

9.11 Site Density

For the site-density mapping, a Beta distribution was used to model the prior and posterior belief. The Beta distribution has the following form (for a more detailed description, see Buck et al. 1996):

$$C \cdot p^{a-1} \cdot (1-p)^{b-1}$$

where

C = a normalizing constant dependent on a and b

p = the proportion of sites in unit X

$1 - p$ = the proportion of sites not in unit X

$a - 1$ = the number of "successful draws" from a sample of size n

$b - 1$ = the number of "unsuccessful draws" from a sample of size n, where $(b = n + 2 - a)$

The prior form of the Beta distribution is obtained by setting a to 1, as no sample data are available (Orton 2000). The value of b and the corresponding standard deviation can then be calculated directly, as the mean of the distribution (the weight attributed by the expert to the map category) is given by $a/(a + b)$, and the variance by $ab/[(a + b)^2 (a + b + 1)]$ (Buck et al. 1996). Once sample data become available, the β distribution can be updated by simply adding the number of "successful" and "unsuccessful" draws to $a - 1$ and $b - 1$, respectively. Eventually, the mean of the distribution will move closer toward the mean of the actual sample, and standard deviations will decrease with increasing sample size. In this particular case we are dealing with three experts. If each of these experts' opinions is treated as an independent sample, the total value of a increases to 3, and the value of b to the sum of b for each independent sample.

Alternatively, by taking the mean of all experts' opinions and calculating the variance of the responses, one can obtain values for a and b using the equations given in Robertson (1999).[3] This will result in a radically different outcome for the prior-probability distribution, and consequently will have a profound effect on the influence of the conditional probabilities on the posterior belief. This method will attribute much more weight to the experts' opinions (especially when they are in agreement), whereas the first method will tend to emphasize the importance of the incoming data for a new site.

When setting a to 1 for formulating the prior probabilities, the relatively large number of 235 known sites in the municipality of Ede leads to posteriors that are very close to the conditional probabilities. The experts' judgment will become quickly less important once considerable amounts of data become available. Unfortunately, it seems improbable that the data set can be considered a representative sample of the total area. The most important reason for this seems to be a research bias of the registered archaeological data set toward areas that are open to field survey. However, large areas of the municipality are covered in woodland or grassland, and even though no quantitative data are currently available to estimate the importance of this effect, it can be suspected that certain land-use types are related to specific land units.

The use of the mean and standard deviation of the experts' opinions to define prior probabilities leads to posteriors that are closer to the prior probabilities in cases where the experts more or less agree. Where strong disagreement exists (larger standard deviations), the posterior probabilities are closer to the conditional probabilities. Whether this alternative method of prior formulation is preferable is probably dependent on the importance attributed to the available site data. Given the fact that the existing site sample is viewed with suspicion, in this particular case it may be the best method for formulating priors.

It is possible to determine confidence intervals around the prior and posterior means, but as we are not dealing with a normal distribution, the standard deviations cannot be used to create confidence intervals; these must be calculated in a statistical package. Plotting the prior distribution together with the posterior distribution shows how far the posterior beliefs are removed from the prior beliefs. Tables 9.4a and 9.4b show how the model develops if all available site data in the municipality of Ede are included to obtain posterior beliefs for both methods of prior formulation.

TABLE 9.4a

Development of Weights for Site Density Using Beta Distributions for All Units on the Ede Predictive Map

Number [a]	Prior Mean (%)	Conditional Mean (%)	Posterior Mean (%)	Posterior Std. Dev. (%)	95% Confidence Interval (%)	
					Lower Bound	Upper Bound
1	9.99	6.94	7.06	1.63	4.22	10.56
2	12.31	28.95	28.40	2.87	22.94	34.20
3	3.00	3.03	3.03	1.04	1.34	5.38
4	1.38	1.98	1.84	0.76	0.66	3.61
5	12.59	11.37	11.41	2.03	7.75	15.67
6	12.79	10.17	10.25	1.93	6.78	14.34
7	7.40	2.29	2.57	1.00	0.99	4.85
8	9.44	3.57	3.82	1.22	1.81	6.53
9	8.64	0.42	0.80	0.57	0.10	2.23
10	4.20	0.63	0.96	0.60	0.16	2.44
11	0.73	0.42	0.53	0.38	0.06	1.48
12	9.99	29.97	29.16	2.89	23.67	34.97
13	4.07	2.28	2.45	0.95	0.95	4.64
14	3.46	3.03	3.08	1.06	1.36	5.46

Note: Based on all available site data ($n = 235$). The prior Beta distribution was calculated by assuming $a = 1$ and taking the mean of the experts' opinions to calculate b. The conditional means were corrected for the effect of land-unit size. The resulting posterior weights should therefore not be interpreted as the percentage of sites to be found on each land unit. The 95% confidence interval is shown in the last two columns.

[a] As defined in Table 9.3.

9.12 Site-Preservation Potential

The initial goal of finding the areas with a minimal number of well-preserved sites was broken down into two objectives: mapping of site-density and site-preservation potential. The same attribute (the land-units map) was used for both criteria, but different weights were applied. However, when the two objectives are combined, it is clear that site-density and site-preservation potential are not independent objectives that can simply be added to arrive at a final weight. Indeed, a multiplicative weighting is needed. This is simple in the case of complete destruction (e.g., in quarries) or in the case of perfect preservation (e.g., under 2 m of drift sand), but a more subtle approach is needed in the intermediate situations where sites may have been partially destroyed. The experts' weighting can provide a first estimation in this respect, and these weights can be normalized to reflect the range from optimal to minimal preservation. However, when these weights are multiplied with the site-density estimates, the resulting weights highly favor the areas where some form of protection is present (Table 9.5).

TABLE 9.4b

Development of Weights for Site Density Using Beta Distributions for All Units on the Ede Predictive Map

Number [a]	Prior Mean (%)	Conditional Mean (%)	Posterior Mean (%)	Posterior Std. Dev. (%)	95% Confidence Interval (%)	
					Lower Bound	Upper Bound
1	9.99	6.94	7.85	1.41	5.31	10.83
2	12.31	28.95	23.51	2.24	19.26	28.03
3	3.00	3.03	3.47	0.69	2.24	4.94
4	1.38	1.98	2.07	0.69	0.95	3.61
5	12.59	11.37	11.65	1.35	9.14	14.42
6	12.79	10.17	12.47	1.34	9.95	15.22
7	7.40	2.29	3.90	1.04	2.13	6.17
8	9.44	3.57	4.01	1.22	1.98	6.71
9	8.64	0.42	2.20	0.84	0.87	4.12
10	4.20	0.63	4.07	0.32	3.47	4.72
11	0.73	0.42	0.94	0.09	0.78	1.12
12	9.99	29.97	15.34	1.26	12.95	17.88
13	4.07	2.28	2.59	0.82	1.23	4.42
14	3.46	3.03	3.28	0.88	1.78	5.22

Note: The same as Table 9.4a, but the prior Beta distribution was calculated by using the mean and standard deviation of the experts' opinions to obtain values for *a* and *b*.

[a] As defined in Table 9.3.

TABLE 9.5

Combined Weighting of Land Units for Site-Density and Site-Preservation Potential

Numbers [a]	Site-Density Weight (%)	Site-Preservation Weight (%)	Normalized Site-Preservation Weight (%)	Product of Site Density and Preservation (%)	Normalized Product (%)
1	9.99	4.11	29.78	2.98	5.27
2	12.31	13.81	100.00	12.31	21.78
3	3.00	4.83	34.94	1.05	1.86
4	1.38	7.64	55.30	0.76	1.35
5	12.59	5.60	40.57	5.11	9.04
6	12.79	13.57	98.21	12.56	22.22
7	7.40	4.45	32.24	2.39	4.22
8	9.44	4.45	32.24	3.04	5.38
9	8.64	13.39	96.95	8.38	14.83
10	4.20	4.60	33.30	1.40	2.48
11	0.73	10.94	79.19	0.58	1.02
12	9.99	5.33	38.61	3.86	6.83
13	4.07	6.67	48.25	1.96	3.48
14	3.46	0.59	4.29	0.15	0.26

Note: Site-density and -preservation potential are both based on the mean of the experts' opinions. Preservation was normalized to reflect the assumption that the highest ranking unit equates to perfect preservation. The final weighting was again normalized to a 100% scale.

[a] As defined in Table 9.3.

TABLE 9.6

Development of Weights for Site Density Using Beta Distributions for All Units on the Ede Predictive Map

Numbers [a]	Prior Mean (%)	Conditional Mean (%)	Posterior Mean (%)	Posterior Std. Dev. (%)	95% Confidence Interval (%)	
					Lower Bound	Upper Bound
1	4.11	6.25	5.44	2.81	1.35	12.11
2	13.81	6.25	7.41	3.77	1.86	16.31
3	4.83	6.25	5.76	2.97	1.44	12.80
4	7.64	5.00	5.65	3.14	1.20	13.19
5	5.60	15.00	12.10	4.25	5.10	21.57
6	13.57	11.25	11.61	4.61	4.24	22.03
7	4.45	5.00	4.80	2.68	1.02	11.27
8	4.45	25.00	17.61	4.78	9.29	27.90
9	13.39	3.75	5.27	3.21	0.91	13.10
10	4.60	12.50	9.72	3.74	3.71	18.17
11	10.94	2.50	4.07	2.79	0.51	11.04
12	5.33	12.50	10.21	3.92	3.91	19.06
13	6.67	15.00	12.73	4.45	5.37	22.63
14	0.59	3.75	1.20	0.75	0.20	3.05

Note: Based on data from preserved sites ($n = 38$). The prior Beta distribution was calculated by assuming $a = 1$ and taking the mean of the experts' opinions to calculate b. The 95% confidence interval is shown in the last two columns.

[a] As defined in Table 9.3.

It is important to note that quantitative data on the preservation aspect can actually be obtained. For example, find spots in the Dutch national archaeological database ARCHIS, maintained by the ROB, can be registered as being intact, partially disturbed, or fully disturbed, and as such could be used to get some idea of the actual condition of sites in the various land units. However, the majority of find spots turn out not to have the relevant information registered. Of the 585 find spots registered in the territory of Ede as of April 2001, only 90 (15.4%) had information concerning the state of conservation at the time of discovery. Of these, 35 were completely disturbed, 45 were partially disturbed, and only 10 remained intact. For a traditional inferential model, these numbers are far too small to justify a statistical analysis. However, in a Bayesian model it is perfectly acceptable to add this information to the expressed priors. By counting the partially disturbed sites as 50% intact, the ratio of disturbed to intact sites becomes 52/38. As was done for the site densities, a Beta distribution was also used to model both the prior and posterior beliefs. Table 9.6 shows the results of including the site-disturbance data.

9.13 Conclusions

The case study presented here is a first effort at incorporating multicriteria decision making and Bayesian statistics into predictive modeling. Obviously, a number of potential problems still need to be addressed:

1. The "objective" data used in this case study are not likely to be a representative sample. Bayesian statistical methods are not dependent on the strict sampling conditions that must be observed with frequency statistics, but the sample should in some way reflect the target population.

2. Modeling by means of Beta distributions becomes considerably more complex when multivariate modeling is undertaken. The interplay of various distributions leads to increasingly complex mathematical models that are difficult to interpret. However, in the context of this chapter — modeling Dutch cultural resource management (CRM), which usually deals with univariate models — this is (at least at the moment) not a major obstacle.

3. The Beta distribution itself may not be the most appropriate statistical distribution, as it is a form of the binomial distribution, which is applicable to very large target populations. However, in this particular case, the total number of sites in the area may not be extremely large, and a considerable proportion of it will already have been sampled. Furthermore, this sampling is done without replacement, unlike in the binomial situation. In such cases, the hypergeometric distribution is more appropriate (Davis 1986), but its density function cannot be calculated when the size of the total population is unknown. This means that the area under consideration should be divided into equal-area sampling units (Nance 1981). The actual size of these units will then highly influence the outcome of any statistical modeling.

4. Sometimes experts are consulted only *after* a quantitative model has been constructed (e.g., Deeben et al. 1997). Bayesian statistics assumes that any new information that becomes available is in the form of a sample comparable with the one used for the first calculation. A potential solution is to treat the "conditional" experts' opinions similarly, as an independent sample, with corresponding means and standard deviations.

5. The attributes chosen may be ill-considered. In theory, if an attribute is not important, it should — in the course of Bayesian model development — "average out" and become unimportant. However, it will be harder to add a previously neglected attribute or to incorporate a "revised" attribute (e.g., a new edition of a soil map or an improved

DEM) into the model without going back to the basics of model construction.

   Having said this, it is nonetheless clear that the methodological framework outlined is promising and can readily be applied to archaeological predictive modeling. Indeed, apart from the use of Beta distributions, the proposed methodology does not use any techniques that are drastically different from earlier studies, although it does place predictive modeling in the wider context of multicriteria decision making. As such, this approach offers a more structured approach to the modeling, promises to be useful for the formalized inclusion of expert judgment in the model, and provides an easy way to develop the model once new data become available. The use of the Beta distributions can also tell us something about the strength or subjectivity of the model: if the posteriors are not backed up by quantitative data, then the standard deviations will become larger. The Beta distributions should also provide insight into the sample size needed to restore uncertainties to an acceptable level.

Acknowledgments

The author would like to thank the three experts who were kind enough to specify the numerical preferences for the establishment of subjective probability: Jos Deeben (ROB, Amersfoort), Eckhart Heunks (RAAP), and Huub Scholte Lubberink (RAAP). The author would also like to thank RAAP Archeologisch Adviesbureau BV for providing the opportunity, resources, and time necessary to pursue this line of research.

Notes

1. RAAP is a private archaeological consultancy firm, specialized in archaeological survey and predictive modeling.
2. Rijksdienst voor het Oudheidkundig Bodemonderzoek (ROB) is the Dutch national archaeological service.
3. $a = \mu\{[\mu\,(1-\mu)\,/\,\sigma^2] - 1\}$; $b = (1-\mu)\,\{[\mu\,(1-\mu)\,/\,\sigma^2] - 1\}$; from Iversen (1984).
4. p_s = proportion of sites found in map unit X.
   p_a = proportion of area taken up by map unit X.

References

Ankum, L.A. and Groenewoudt, B.J., De situering van archeologische vindplaatsen, RAAP-rapport 42, Stichting RAAP, Amsterdam, 1990.

Aspinall, R.J., An inductive modeling procedure based on Bayes' theorem for analysis of pattern in spatial data, *International Journal of Geographical Information Systems*, 6, 105–121, 1992.

Beinat, E., Multiattribute Value Functions for Environmental Management, Ph.D. thesis, Research Series 103, Tinbergen Institute, Amsterdam, 1995.

Bonham-Carter, G.F., *Geographic Information Systems for Geoscientists: Modelling with GIS*, Vol. 13, *Computer Methods in the Geosciences*, Pergamon, Oxford, 1994.

Brandt, R.W., Groenewoudt, B.J., and Kvamme, K.L., An experiment in archaeological site location: modelling in the Netherlands using GIS techniques, *World Archaeology*, 24, 268–282, 1992.

Buck, C.E., Cavanagh, W.G., and Litton, C.D., *Bayesian Approach to Interpreting Archaeological Data*, John Wiley and Sons, Chichester, U.K., 1996.

Burrough, P.A., Fuzzy mathematical methods for soil survey and land evaluation, *Journal of Soil Science*, 40, 477–492, 1989.

Burrough, P.A., MacMillan, R.A., and van Deursen, W., Fuzzy classification methods for determining land suitability from soil profile observations and topography, *Journal of Soil Science*, 43, 193–210, 1992.

Dalla Bona, L., Cultural Heritage Resources Predictive Modelling Project, report prepared for Ontario Ministry of Natural Resources, Center for Archaeological Resource Prediction, Lakehead University, Thunder Bay, ON, 1994.

Dalla Bona, L., Protecting cultural resources through forest management planning in Ontario using archaeological predictive modeling, in *Practical Applications of GIS for Archaeologists: A Predictive Modeling Toolkit*, Wescott, K.L. and Brandon, R.J., Eds., Taylor and Francis, London, 2000, pp. 73–99.

Davis, J.C., *Statistics and Data Analysis in Geology*, 2nd ed., John Wiley and Sons, New York, 1986.

Deeben, J., Hallewas, D., Kolen, J., and Wiemer, R., Beyond the crystal ball: predictive modelling as a tool in archaeological heritage management and occupation history, in *Archaeological Heritage Management in the Netherlands, Fifty Years State Service for Archaeological Investigations*, Willems, W., Kars, H., and Hallewas, D., Eds., ROB, Amersfoort, Netherlands, 1997, pp. 76–118.

Eastman, J.R., Kyem, P.A.K., Toledano, J., and Jin, W., *GIS and Decision Making*, Vol. 4, *Explorations in Geographic Information Systems Technology*, United Nations Institute for Training and Research (European Office), Geneva, 1993.

Finke, P., Hardink, J., Sevink, J., Sewuster, R., and Stoddart, S., The dissection of a Bronze and early Iron Age landscape, in *Territory, Time and State: The Archaeological Development of the Gubbio Basin*, Malone, C. and Stoddart, S., Eds., Cambridge University Press, Cambridge, 1994.

Goodchild, M.F., *Spatial Autocorrelation*, Catmog. 47, Geo Books, Norwich, 1986.

Heunks, E., Gemeente Ede: archeologische verwachtingskaart, RAAP-rapport 654, RAAP Archeologisch Adviesbureau BV, Amsterdam, 2001.

Iversen, G.R., *Bayesian Statistical Inference*, Thousand Oaks, CA, Sage Pub., 1984.

Kamermans, H., Archeologie en landevaluatie in de Agro Pontino (Lazio, Italië), Ph.D. thesis, Universiteit van Amsterdam, Amsterdam, 1993.

Kamermans, H., Land evaluation as predictive modelling: a deductive approach, in *Beyond the Map: Archaeology and Spatial Technologies*, Lock, G., Ed., NATO Science Series, Series A: Life Sciences, Vol. 321, IOS Press, Amsterdam, 2000, pp. 124–146.

Kamermans, H. and Rensink, E., GIS in palaeolithic archaeology: a case study from the southern Netherlands, in *Archaeology in the Age of the Internet: CAA97, Computer Applications and Quantitative Methods in Archaeology*, Dingwall, L., Exon, S., Gaffney, V., Laflin, S., and van Leusen, M., Eds., BAR International Series 750, Archaeopress, Oxford, 1999.

Keeney, R.L. and Raiffa, H., *Decisions with Multiple Objectives: Preferences and Value Trade-Offs*, John Wiley and Sons, New York, 1976.

Kohler, T.A. and Parker, S.C., Predictive models for archaeological resource location in: Schiffer, M.B. Ed.: *Advances in Archaeological Method and Theory*, 9, Academic Press, New York, 1986, pp. 397–452.

Kvamme, K.L., The fundamental principles and practice of predictive archaeological modeling, in *Mathematics and Information Science in Archaeology: A Flexible Framework*, Voorrips, A., Ed., Vol. 3 in *Studies in Modern Archaeology*, Holos-Verlag, Bonn, 1990, pp. 257–294.

Malczewski, J., *GIS and Multicriteria Decision Analysis*, John Wiley and Sons, New York, 1999.

Nance, J.D., Statistical fact and archaeological faith: two models in small-sites sampling, *Journal of Field Archaeology*, 8, 151–165, 1981.

Nijkamp, P., Rietveld, P., and Voogd, H., Multicriteria Evaluation in Physical Planning, Contributions to Economic Analysis 185, North-Holland, Amsterdam, 1990.

Orton, C., A Bayesian approach to a problem of archaeological site evaluation, in *CAA 96, Computer Applications and Quantitative Methods in Archaeology*, Lockyear, K., Sly, T., and Mihailescu-Bîrliba, V., Eds., BAR International Series 845, Archaeopress, Oxford, 2000, pp. 1–7.

Robertson, I.G., Spatial and multivariate analysis, random sampling error, and analytical noise: empirical Bayesian methods at Teotihuacan, Mexico, *American Antiquity*, 64 (1), 137–152, 1999.

Saaty, T., *The Analytical Hierarchy Process*, McGraw-Hill, New York, 1980.

Soonius, C.M. and Ankum, L.A., Ede, II: Archeologische Potentiekaart, RAAP rapport 49, Stichting RAAP, Amsterdam, 1990.

Van Dalen, J., Probability modelling: a Bayesian and a geometric example, in *Geographical Information Systems and Landscape Archaeology*, Gillings, M., Mattingley, D., and van Dalen, J., Eds., Vol. 3, *The Archaeology of Mediterranean Landscape*, Oxbow Books, Oxford, 1999, pp. 117–124.

Van Leusen, P.M., Cartographic modelling in a cell-based GIS, in *Computer Applications and Quantitative Methods in Archaeology 1992*, Andresen, J., Madsen, T., and Scollar, I., Eds., Aarhus University Press, Aarhus, Denmark, 1993, pp.105–123.

Van Leusen, P.M., GIS and archaeological resource management: a European agenda, in *Archaeology and Geographical Information Systems*, Lock, G. and Stančič, Z., Eds., Taylor and Francis, London, 1995, pp. 27–41.

Van Leusen, P.M., Locational modelling in Dutch archaeology, in *New Methods, Old Problems: Geographic Information Systems in Modern Archaeological Research*, Maschner, H.D.G., Ed., Occasional Paper 23, Center for Archaeological Investigations, Southern Illinois University, Carbondale, 1996, pp. 177–197.

Venneker, R.G.W., A Distributed Hydrological Modelling Concept for Alpine Environments, Ph.D. thesis, Vrije Universiteit, Amsterdam, 1996.

Verhagen, P., Wansleeben, M., and van Leusen, M., Predictive modelling in the Netherlands: the prediction of archaeological values in cultural resource management and academic research, in *Workshop 4 Archäologie und Computer 1999*, Hartl, O. and Strohschneider-Laue, S., Eds., Forschungsgesellschaft Wiener Stadarchäologie, Vienna, 2000.

Wansleeben, M., Applications of geographical information systems in archaeological research, in *Computer Applications and Quantitative Methods 1988*, Rahtz, S.P.Q., Ed., BAR International Series 466(ii), Tempus Reparatum, Oxford, 1988, pp. 435–451.

Wansleeben, M. and L.B.M. Verhart, Geographical Information Systems. Methodical progress and theoretical decline? *Archaeological Dialogues* 4–1, pp. 53–70, 1997.

Warren, R.E., Predictive modeling in archaeology: a primer, in *Interpreting Space: GIS and Archaeology*, Allen, K.M.S., Green, S.W., and Zubrow, E.B.W., Eds., Taylor and Francis, New York, 1990, pp. 90–111.

Wescott, K.L. and Brandon, R.J., *Practical Applications of GIS for Archaeologists: A Predictive Modeling Toolkit*, Taylor and Francis, London, 2000.

Section 5:

Large Databases and CRM

10

Points vs. Polygons: A Test Case Using a Statewide Geographic Information System

Philip B. Mink, II, B. Jo Stokes, and David Pollack

CONTENTS

Since 1997, the Kentucky Heritage Council, the University of Kentucky Department of Anthropology, and the Kentucky Transportation Cabinet have been working together to develop geographic information systems (GIS) coverages for the cultural resources of the state of Kentucky. The project is funded, in part, by ISTEA and TEA-21 Transportation Enhancement funds provided by the Kentucky Transportation Cabinet. The goal of the project is to link the existing textual databases of the Kentucky Office of State Archaeology (KOSA) and Kentucky Heritage Council (KHC) with GIS coverages that were developed by digitizing archaeological site areas, survey areas, and historic resources. This project was designed to provide a GIS that could be used as a planning tool by the Kentucky Transportation Cabinet and other land-managing agencies and by researchers to learn more about prehistoric and historic settlement patterns.

To gain a better understanding of prehistoric and historic settlement patterns, archaeologists are often interested in interpreting and predicting the spatial distribution of archaeological sites on the landscape. To accomplish this task they have begun to use GIS to develop locational models (Lock and Stančič 1995; Maschner 1996; Wescott and Brandon 2000). In archaeology, these models have two major objectives: (1) to generate maps that can be used by land managers to manage cultural resources more effectively and (2) to gain insights into prehistoric and historic landscapes and settlement patterns (Judge and Sebastian 1988).

Locational models are only as dependable as the data used to create them (Altschul 1988), and in the current study the question is whether points or polygons produce more reliable models. This question was answered by using a sample of Archaic sites from Henderson County, Kentucky, to develop test models using both polygon and point data and then comparing the results of each test model. Before we describe the results of our evaluation of the different models, a brief discussion is presented concerning the benefits to cultural resources managers of digitizing site locations from maps and for using polygon rather than point data to manage cultural resources.

10.1   Digitizing Site Locations and Using Polygons to Manage Archaeological Sites

In Kentucky, when an archaeological site is recorded, a site inventory form is submitted to KOSA, and the information on the form is entered into the site inventory database. These forms require the submission of both written UTM coordinates and a copy of the appropriate part of a USGS Quadrangle map with the site boundaries drawn upon it. During the creation of the Kentucky cultural resources GIS, the original UTM coordinates on the site forms were compared with the calculated GIS-derived coordinates from the hand-drawn map locations provided with the site forms. As a result of this comparison, numerous instances were found where the coordinates recorded on site forms were hundreds of meters and sometimes several kilometers from the locations drawn on a USGS Quadrangle map (Figure 10.1). In developing the GIS coverages, it thus was assumed that the locations digitized from maps were more accurate than the recorded UTM coordinates recorded on the site survey forms and included in the site inventory database. Site recorders appear to be able to visually locate and draw the site on a topographic map more reliably than they can calculate the actual geographic coordinates. As a consequence, the choice to digitize the sites from site areas drawn on topographic maps has resulted in greater accuracy than if sites had simply been plotted as points using the coordinates recorded in the site inventory database.

The question of how to represent archaeological sites in a GIS is one faced by many archaeologists. The norm in most statewide systems, where the archaeological site is the unit of analysis, is to use a point to represent the site. However, the decision to digitize archaeological sites as polygons rather than points was made early in the process of creating the Kentucky GIS. The advantage of polygon data over point data is that polygon data provide a more accurate representation of the size and shape of a site and provide a

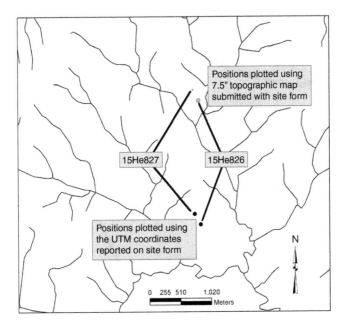

FIGURE 10.1
Example of sites plotted by UTM coordinates vs. corrected with reference-map polygon.

more complete representation of the spatial distribution of archaeological resources within a project or study area. In addition, once the polygons were digitized, calculating a centroid (site-point representation) for each polygonal site area could easily be done. Therefore, in Kentucky, where polygons were digitized, both types of data are available to cultural resource managers and researchers.

As previously stated, digitizing sites as polygons allows differences in site size and shape to be depicted on maps. While a single point may work well for showing the location of an Archaic rock shelter or a Woodland burial mound, it does not work as well for characterizing the spatial limits of a Late Prehistoric village that covers 14 ha. For the latter, a polygon that shows the actual site limits does a better job of depicting the size of this site and distinguishing it from smaller sites. An example of the range in variation in archaeological site size and shape is illustrated in Figure 10.2, which shows archaeological sites along a portion of the Ohio River. In this area, many sites are extremely long, yet narrow. A point does not adequately reflect the size and shape of these sites.

In addition to more adequately representing the area encompassed by a site, polygons provide a more complete representation of the spatial distribution of archaeological resources on the landscape. This is a particularly important point for site management and planning. For example, a fictional scoping study of a highway corridor demonstrates that polygon data provide a much better sense of the project impacts than a map using point data (Figure 10.3). These two maps illustrate, that when sites are represented as

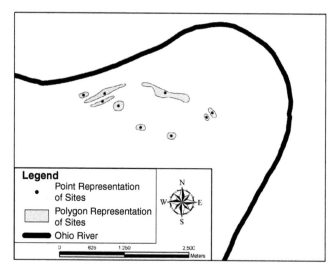

FIGURE 10.2
Differences in site-size representation between points and polygons.

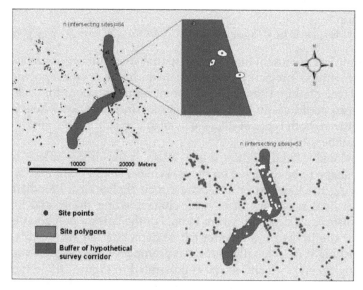

FIGURE 10.3
Highway corridor with polygons and points.

points, there are 53 located within the corridor, but when polygons are used, the project impacts 64 sites. In other words, there is an 11-site increase in sites potentially affected by the project when using polygon data to represent sites rather than point data. From a cultural resource management perspective, this type of locational information is invaluable.

10.2 Locational Models: Points vs. Polygons

While there are some advantages to using polygon information rather than points to manage cultural resources, the question arises as to whether the spatial information associated with polygon data is useful in other ways. After all of the archaeological site and survey coverages for Kentucky were digitized into a GIS, the decision was made to explore the usefulness of polygon data in developing locational models. A review of predictive modeling in archaeology (Cordell and Green 1983; DeBloois 1975; Stevenson 1993; Weed et al. 1996; Wescott and Brandon 2000) showed that point data are the site representation used for most if not all modeling procedures. However, because the Kentucky GIS allows for the use of either point or polygon data, it presented an excellent opportunity to evaluate the respective benefits of either data source in predictive modeling. To accomplish this task, a sample archaeological site from the Kentucky site inventory was used to generate models using point and polygon data to represent the location of archaeological sites.

10.2.1 Archaeological Data

The archaeological sample used to develop the models consisted of Archaic period (8000–1000 B.C.) sites from Henderson County, Kentucky. Henderson County was chosen because of the large number of Archaic sites recorded in the county (*n* = 123) and the high percentage of the county previously surveyed (7%) (Figure 10.4).

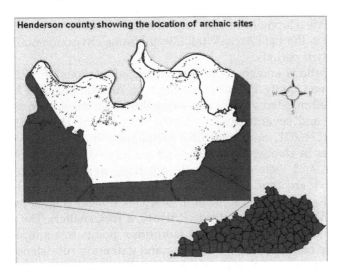

FIGURE 10.4
Archaic sites in Henderson County.

The Archaic period is traditionally divided into three subperiods: the Early, Middle, and Late Archaic (Jefferies 1996; Lewis 1996; Stoltman 1978). The Early Archaic (8000–6000 B.C.) is seen as a time of cultural transition between the Archaic and the preceding Paleo-Indian period. Early Archaic groups exploited local flora (such as nut-bearing trees like oak, hickory, and walnut) and fauna (such as white-tailed deer and turkey) and followed a highly mobile subsistence pattern similar to their Paleo-Indian forebears (Jefferies 1996: 40). In Henderson County, there are 48 sites with an Early Archaic component located in both lowland (e.g., Ohio and Green River floodplains) and upland environments (e.g., bluff tops and interior ridge tops).

The Middle Archaic (6000–3000 B.C.) corresponds to the Hypsithermal, which resulted in drier, warmer conditions that may have restricted the distribution of subsistence resources and encouraged more-intensive exploitation of more-local food items (Jefferies 1996). In addition, there seems to be increased cultural regionalization expressed through the development of varied projectile point types (Cook 1976; Fowler 1959; Nance 1986). The Middle Archaic is underrepresented in Henderson County, with only 13 sites designated as having a Middle Archaic component. A majority of Middle Archaic sites are located in the northwest portion of the county. These sites are associated with the same types of environments as Early Archaic sites.

An emphasis on hunting and gathering continued into the Late Archaic (3000–1000 B.C.), with some important changes. People became more dispersed on the landscape and became increasingly reliant on starchy seeds (such as goosefoot and marsh elder) and freshwater mussels (Jefferies 1996: 54). This period also corresponds to distinct social changes identified by the development of large shell middens along the Green River (Conaty 1985; Fowler 1959; Nance 1985; Rothschild 1979). In Henderson County, the number of sites increases to 51, and the increased utilization of mussels can be seen in the development of sites with thick shell middens along the Green River. Sites in the Late Archaic fall into the same environmental settings as the earlier two periods.

Overall, little is known about the specific adaptations of Archaic people in Henderson County. Primarily, attention on Archaic sites here has focused on large shell-midden sites located along the Green River, such as Bluff City (15He160), and several smaller Archaic camp sites (15He580, 15He589, 15He635, 15He631, and 15He638) (Hockensmith 1984). However, most Archaic sites in Henderson County that are located along the Green River, Ohio River, and their tributaries do not have shell-midden components. Archaic sites are found in various environmental zones in Henderson, including floodplains (15He537 and 15He538), upland ridge tops (15He804, 15He805, 15He820), and a hillside (15He10, a rock shelter). The association of Archaic sites with a variety of landforms points to multiple zones of resource use consistent with a hunting and gathering subsistence base.

While we recognize that there were changes in sedentism, mobility patterns, and resource use during the long span of time covered by the Archaic, all of the Archaic period sites were grouped together for the purpose of modeling a general hunter-gatherer settlement pattern. It was hoped that by choosing a time period characterized by a general pattern of high mobility and a subsistence strategy reliant upon multiple environmental zones, such as floodplain and upland ridges for hunting game and collecting nuts and seeds, a model could be developed using a wide range of environmental variables.

10.2.2   Points vs. Polygons

As suggested by Kvamme (1990), the 123 Henderson County Archaic sites used in this study were divided into two groups (a training group and a test group) in order to "train" the decision rules on which the models were to be built. The training group consisted of 63 archaeological sites and 125 nonsite locations, and the test group consisted of the other 60 archaeological sites. The training group was analyzed to create the model, and sites in the test group were used to test the validity of the model.

The modeling approach chosen, empirical correlation, is an inductive modeling strategy that fits mathematical functions to a training data set (Kvamme 1990). In these models, the underlying assumption is that environmental factors influenced the location of prehistoric settlements. Kohler and Parker (1986) note that the long history of this approach can be traced back to the works of Steward (1938) and Willey (1953). Examples of environmentally based GIS models include those developed by Duncan and Beckman (2000) for West Virginia and Pennsylvania as well as the model developed by Warren and Asch (2000) for the Eastern Prairie Peninsula of Illinois. While it is recognized that culture plays a role in determining site location (Hill 1971; Wobst 1976), the difficulties in extracting quantifiable cultural variables for a locational model led us to rely upon available environmental data.

When quantifying the environmental data, it is also important to choose the right unit of analysis. In most archaeological predictive models, a parcel of land is preferred (Kvamme 1990), represented in a GIS by a uniformly sized grid cell. Warren (1990) indicates that high-resolution data (e.g., small grid-cell size) are more desirable than low-resolution data (large grid-cell size). The readily available North American data provided by the United States Geological Survey (utilized by most North American model builders) are high-resolution data typically provided at a 30×30 m grid-cell resolution (900 m^2), which is potentially smaller than an average site size (e.g., in this study the average site area is 14,338 m^2). Therefore, it is hypothesized that an archaeological site represented by a point that is then used to extract environmental data for model construction will likely result in a high potential for underrepresentation of certain environmental variables and the diversity of environmental zones associated with a particular site. Conversely, a site represented by a polygon that is then used to extract environmental data for model construction should result in a more precise representation of the environmental variables that may have influenced site placement.

To test this hypothesis, relatively simple correlative models were used to assess the applicability of using polygons or points to represent sites in archaeological locational modeling. The nature and availability of existing data necessitated this approach. While more-robust statistical techniques such as logistic regression have been used to develop models (Kvamme 1990), the data for this study were not appropriate for inclusion into regression analysis. It was felt that, because the purpose of this study was to compare and contrast models developed using points and polygons, a simpler approach would be best.

One of the most common strategies followed in creating locational models is to develop a decision rule that is then used to create a probability surface by calculating the intersection of a set of nonarchaeological variables that are indicators of archaeological site presence (Kvamme 1990). The following method was used to create and test the models.

1. Primary data-set collection
2. Secondary data-set derivation
3. Independent-variable sampling for both training and test data sets for both points and polygons
4. Statistical analysis of the data sets
5. Creation of decision rule
6. Development of decision surfaces
7. Testing of the models

Primary data sets were assembled from a variety of sources, including the KOSA, Kentucky Geological Survey, Kentucky Environmental Resources Cabinet, and the Governors Office of Technology, Office of GIS. The data acquired from these sources include: archaeological site boundaries, digital elevation models (DEMs), stream data, soil data, and landform data. ESRI's ArcView 3.2 GIS software, along with the Spatial Analyst Extension and the Grid Analyst Extension Version 1.1 developed by Dr. Arun K. Saraf (2000), was used to develop the models. Many of the primary data sets had to be imported into ArcView and converted to the appropriate formats. The landform, soil, and stream data were all provided as ArcInfo export files (*.e00) and were imported into ArcView and converted to shapefiles. The DEMs were provided in the standard USGS STDS format and imported into ArcView as grids. The archaeological data were provided in shapefile format as polygons representing the site boundaries and did not need to be modified.

Derivation of secondary data sets included modifying the archaeological, hydrological, and elevation data sets. The archaeological data provided by KOSA were transformed into two data sets: (1) the original polygonal site boundaries and (2) a calculated centroid point. To complete the training data set, 125 random points were created within the study area using an ArcView Avenue script (Cederholm 2000). These randomly created points were joined with 63 randomly selected Henderson County Archaic sites to complete the training point data set used to develop the point models. To create a random set of polygons for the polygon models, the average Henderson County Archaic site size was calculated (n = 14,338 m²). That number was used to calculate the appropriate buffer size (n = 64 m) around the randomly generated points created for the point data set. Ultimately, this created a random set of polygons matching the random set of points. The newly created buffer polygons were then merged with the polygon representation of the same 63 sites as in the point training data set to create a polygon training data set.

Secondary data sets also were derived for the independent environmental data, including slope, elevation, aspect, and distances to the various rank-ordered streams. The DEMs were used to create elevation, slope, and aspect grids. These surfaces were created using the embedded script in the ArcView Spatial Analyst extension. The hydrological data were provided in polyline form with attribute data as to name and Strahler (1957) rank order. The stream data were divided into three classes: rank-order 8 (the Ohio and Green Rivers), rank-order 4–6 (there are no rank-order 7 streams in Henderson County), and rank-order 1–3 streams. Once the hydrologic data were divided into these classes, an ArcView Spatial Analyst–extension-embedded script was used to create three distance grids showing the Euclidian distance to each of the stream classes.

To sample the independent (environmental) variables of both the point and polygon training data sets, a third-party software extension for ArcView 3.2, Grid Analyst Ver. 1.1 (Saraf 2000), was used. Continuous surfaces (elevation, slope, distance to rank-order x streams) were sampled for the point training set using the "Extract X, Y and Z Values for Point Theme from Grid Theme" routine. This routine extracts the z-value for every x-, y-location within a point data set. Noncontinuous categorical surfaces (soil, aspect, and landform) were sampled for the point training data set by utilizing a spatial join. This procedure allows the user to join attributes of features in separate surfaces based on their relative locations (Kennedy 2001). Thus, by performing a spatial join on the polygon soil data set and the point training data, the soil type located at that point is recorded. The advantage of using a GIS to develop locational models is that this type of operation could be performed on all of the training sample data at the same time in a matter of seconds, compared with the tedious methodology of deriving this data by hand, like that required less than two decades ago (Judge and Sebastian 1988).

Sampling the environmental data for the polygon training data set also was accomplished using the Grid Analyst Ver. 1.1 (Saraf 2000) extension. Utilizing two routines — "Extract Grid Theme Using Polygon" and "Convert Grid Theme to X, Y, Z Text File" — the values for all of the grid cells under the polygons were obtained. This methodology resulted in an increased sample size for the polygon training data, as many sites crossed into multiple grid cells versus the single-cell point training data. The actual increase in sample size depended on the individual environmental variables and how fractured their surfaces were. This increase in sample size is a particularly important consideration, as many modelers lament the small sample size provided when using archaeological sites for developing predictive models (Stevenson 1993; Weed et al. 1996). Handling polygons in this manner also seems to create a more accurate representation of the preferable variables for archaeological site placement. If the land parcel is truly the unit of investigation, then every parcel intersected by an archaeological site should be counted.

Once the archaeological data sets were chosen, variables were selected based on the availability of information and the statistical likelihood that they actually contributed to determining site location. Possible variables for this analysis included: landform, soil, aspect, slope, elevation, and distance to rank-order 1–3, 4–6, and 8 streams. Each potential noncontinuous environmental variable considered for the model was subjected to a χ^2 goodness-of-fit analysis.

$$\chi^2 = \sum (O_i - E_i)^2 / E_i$$

The goodness-of-fit χ^2 test questions how well a model fits the observed data. Small observed significance levels (i.e., less than 0.10) indicate that the model does not fit well (Shennan 1997). Variables subjected to this procedure included landform, soil, and aspect variables. For one of these to be included in the model, the null hypothesis — that sites are randomly located on the landscape — had to be rejected. For both the point and polygon models, the noncontinuous variables included in the model building were landform, aspect, and soil classification. Within these variables, categories were chosen based on a positive difference between the number of observed occurrences in each category versus the expected number of occurrences in a random distribution (Table 10.1).

Continuous variables available for use in the point and polygon models included elevation, slope, distance to rank-order 1–3 streams, distance to rank-order 4–6 streams, and distance to rank-order 8 streams. In these cases, the χ^2 analysis cannot be used effectively for choosing viable variables. Instead, each of these variables was subjected to a Mann-Whitney test. An independent, random sample of 125 sites was generated for comparison with

TABLE 10.1

Noncontinuous (Categorical) Variables and Their Values Included in Both the Point and Polygon Models

Environmental Variable	Initial Point Model Value(s)	Extra Point Model Value(s)	Polygon Model Value(s)
Landform	22, 82	22, 82	22, 82
Soil	10, 30, 100, 190	10, 30, 100, 190	100, 130, 180
Aspect	N, NW, SW, E	N, NW, SW, E	N, NW

the point model, and another independent, random sample was generated for comparison with the polygon model. The Mann-Whitney test is the most popular and powerful of the nonparametric, two-independent-samples tests. It is equivalent to the Wilcoxon rank sum test and the Kruskal-Wallis test for two groups (Shennan 1997: 66).

Mann-Whitney tests whether two populations are equivalent. This test is used in place of a two-sample *t* test when the populations being compared are not normal. It requires independent random samples of sizes n_1 and n_2. The test is very simple and consists of combining the two samples into one sample of size $n_1 + n_2$, sorting the result, assigning ranks to the sorted values (giving the average rank to any "tied" observations), and then letting *T* be the sum of the ranks for the observations in the first sample. If the two populations have the same distribution, then the sum of the ranks of the first sample and those in the second sample should be close to the same value. The *p* value for the null hypothesis is that the two distributions are the same. Small significance values (<0.05) indicate significant variance between the locations of each set (Shennan 1997: 66–67). In this case, if a variable showed a small significance value, it was included in the model. Table 10.2 highlights the continuous variables chosen for the point and polygon models.

For the point model, only distance to rank-order 8 streams was shown to be significant in the Mann-Whitney test. Interestingly, the polygon set produced significant results for not only the rank-order 8 streams, but for the elevation and the distance to rank-order 4–6 streams as well. These additional data sets were therefore added to the polygon predictive models. Ranges were based on the standard deviation for each variable. The upper and lower bounds for the range are one standard deviation above and below the mean for a variable.[1]

The resultant models produced in ArcView give a range of grid-cell values where variables overlap. For the point model, a total of four variables was used to define the model parameters. As such, the result produced five possible combinations of output: 0, 1, 2, 3, 4. This included all combinations

TABLE 10.2
Continuous Variables and the Value Ranges Entered into Both the Point and Polygon Models

| Environmental Variable | Value Ranges | | |
	Initial Point Model	Extra Point Model	Polygon Model
Elevation	None	107–144 m	107–144 m
Slope	None	None	None
Distance to rank-order 1–3 streams	None	None	None
Distance to rank-order 4–6 streams	None	67–10,290 m	67–10,290 m
Distance to rank-order 8 streams	0–856 m	0–856 m	60–8,184 m

of variables overlapping (or none in the case of the 0 return). Consequently, the polygon model included two additional variables (distance to rank-order 4–6 streams and elevation), so the range of outputs increased to seven: 0, 1, 2, 3, 4, 5, 6.

Because GIS-based locational models are actually just sets of decision rules that assign a grid cell to an event class (i.e., site present or site absent) based on characteristics of that location (Kvamme 1990), results from the various statistical analyses were used to develop decision rules used to construct the models. Given that the statistical analysis of both the polygon and point training data sets resulted in different outcomes, two sets of decision rules had to be developed.

In the following decision rules,

M = (model indicates archaeological site present)

M' = (model indicates archaeological site absent)

the decision rule for the point model is as follows:

M = {(landform = floodplain or side-slope) ∩ (soil group # = 10 or 30 or 100 or 190) ∩ (aspect = N or NW or SW or E) ∩ (354 m ≤ distance rank order 8 streams ≤ 6,832 m)}

M' = {(landform ≠ floodplain or side-slope) ∩ (soil group # ≠ 10 or 30 or 100 or 190) ∩ (aspect ≠ N or NW or SW or E) ∩ (354 m > distance to rank order 8 streams > 6,832 m)}

and the decision rule for the polygon model is as follows:

M = {(landform = floodplain or side-slope) ∩ (soil group # = 100 or 130 or 180) ∩ (aspect = N or NW) ∩ (60 m ≤ distance to rank order 8 streams ≤ 8,184 m) ∩ (67 m ≤ distance to rank order 4–6 streams ≤ 10,290 m) ∩ (107 m ≤ elevation ≤ 144 m)}

M' = {(landform ≠ floodplain or side-slope) ∩ (soil group # ≠ 100 or 130 or 180) ∩ (aspect ≠ N or NW) ∩ (60 m > distance to rank order 8 streams > 8,184 m) ∩ (67 m > distance to rank order 4–6 streams > 10,290 m) ∩ (107 m > elevation > 144 m)}

Additionally, to ensure that the lesser number of variables included in the point model did not masquerade as a legitimate difference between using points and polygons, a third set of decision rules was developed and used to compensate for this. To this end, a second point model was developed that included, along with the variables used in the initial point model, elevation and the distance to rank-order 4–6 streams (i.e., all of the same variables used in the polygon model), even though statistically they should have been left out of the model. The following is the decision rule for the extra point model:

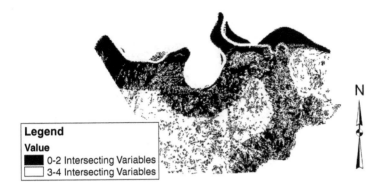

N

FIGURE 10.5
Model developed using points to represent site area.

M = {(landform = floodplain or side-slope) ∩ (soil group # = 10 or 30 or 100 or 190) ∩ (aspect = N or NW or SW or E) ∩ (354 m ≤ distance rank order 8 streams ≤ 6,832 m) ∩ (67 m ≤ distance to rank order 4–6 streams ≤ 10,290 m) ∩ (107 m ≤ elevation ≤ 144 m)}

M′ = {(landform ≠ floodplain or side-slope) ∩ (soil group# ≠ 10 or 30 or 100 or 190) ∩ (aspect ≠ N or NW or SW or E) ∩ (354 m > distance to rank order 8 streams > 6,832 m) ∩ (67 m > distance to rank order 4–6 streams > 10,290 m) ∩ (107 m > elevation > 144 m)}

Once the decision rules for each model were created, the process of developing the models began. An identical procedure was followed to develop each model, including converting all of the data into binary grids using the "Map Calculator" function in ArcView 3.2 Spatial Analyst extension:

1 = favorable value for site presence, e.g., (landform = 22 or 82)
0 = unfavorable value for site presence, e.g., (landform ≠ 22 or 82)

After all of the binary grids were created, they were added together (once again using "Map Calculator") to create a new surface containing a range of values of 0 to 4 for the initial point model and of 0 to 6 for the polygon and extra point model. The values are representative of the number of favorable archaeological variables intersecting at that grid cell. Thus, a value of 2 would indicate the intersection of two variables (e.g., soil and aspect); a value of 3 would indicate the intersection of three variables (e.g., soil, aspect, landform), and so forth. Not surprisingly, a higher number of intersecting variables (three or four for the point models and five or six for the polygon models) indicated a high probability for site presence. The resulting probability surfaces are shown in Figures 10.5 through 10.7.

The effectiveness of the probability surfaces had to be tested after their creation. Kvamme (1988, 1990) discusses various methodologies for assess-

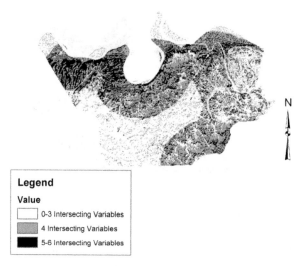

FIGURE 10.6
Model developed using polygons to represent site area.

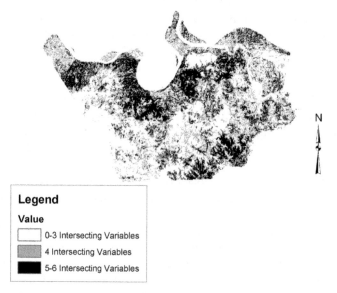

FIGURE 10.7
Extra model developed using points to represent site area but including the same environmental variables used in the polygon model.

ing the effectiveness of predictive models, including the first assessment method chosen for this investigation. This method focused on comparing the percentage of correctly predicted training sites (percent-correct statistic) with the percentage of the study area considered to have a high probability of containing sites (study area percent-occupied statistic). The second assessment is a more robust test of model effectiveness: the gain statistic (Kvamme

TABLE 10.3

Comparison between the Study Area Percent-Occupied Statistic (Listed by Number of Intersecting Variables, Value) and the Percent-Correct Statistic of Both the Training and Testing Data Sets

Number of Intersecting Variables	Study Area Percent-Occupied Statistic	Training Sites Percent-Correct Statistic	Testing Sites Percent-Correct Statistic
Initial Point Model			
0	5.70	3.17	5.00
1	16.01	7.94	10.00
2	32.11	17.46	31.67
3	32.37	42.86	43.33
4	13.81	28.57	10.00
Extra Point Model			
0	3.29	0.00	0.00
1	4.64	3.17	6.67
2	13.02	11.12	5.00
3	27.92	14.29	20.00
4	30.52	26.98	38.33
5	16.62	28.57	26.67
6	3.99	15.87	3.33
Polygon Model			
0	0.48	0.00	0.00
1	2.88	0.00	0.00
2	8.91	0.50	0.50
3	29.14	10.80	7.60
4	41.03	41.60	41.70
5	15.63	39.80	40.20
6	1.93	7.30	10.00

1988). This statistic provides an effective means for comparing models and is discussed in more detail below.

Table 10.3 presents the results for comparing the percent-correct statistics of the three models. The grid-class column shows the cell values (number of intersecting variables) for each predictive model. The other three columns show the percentages of the study area occupied by the various values and the percentages of the training and testing sites that occur in each individual class. When comparing the results of the three models, one can see that the polygon models seem to have a smaller portion of the study area classified as high probability and a larger percentage of sites occurring within those high-probability areas than the point models. The point models, in many cases, have an almost equal percentage of the study area classified as high probability in comparison with the percent-correct statistics.

The initial point model has a total of five potential cell values (0 to 4). Cell values of either 3 or 4 (cells representing the intersection or either three or four of the variables) are considered high-probability cells, i.e., they have a high likelihood of containing sites. The combination of those two cell values results in the high-probability zone. The high-probability zone in this model covers 46.18% of the study area. The percentage of training and testing sites occurring in that high-probability zone is 71.43 and 53.33%, respectively. While the ratio between study-area percentage and training-site percentage seems to indicate some predictive utility for the model, the almost equal ratio between study-area percentage and testing sites seems to indicate a model with very little predictive value. When only a cell value of 4 is considered to be the high-probability zone, then the ratio between study-area percentage, training-site percentage, and testing-site percentage (13.81, 28.57, and 10.00%, respectively) shows a further decline in the predictive value of the model. The ratio between study-area percentage and the percentage of testing sites in the high-probability zone is indicative of a model with absolutely no predictive value.

The extra point model — developed to ensure that no biases were introduced into the point versus polygon comparison due to the difference in the number of variables used to develop the model — has seven potential cell values (0 to 6). If high probability is considered to be the intersection of four or more environmental variables, then the percentage of the study area considered high probability is 51.3%, with 71.42% of the training sites and 68.33% of the testing sites occurring in the high-probability zone. Comparing these percent-correct statistics indicates that the extra point model's performance is equal to or possibly slightly worse than the initial point model. This pattern holds even if the criterion for high probability is increased to a higher number (five or six) of intersecting environmental variables.

The polygon model also has seven potential cell values (0 to 6) representing the intersection of environmental variables. If the intersection of four or more variables is considered high probability, then the comparison between study-area percentage considered high probability (58.59%) with the percent-correct statistic for training sites (89.77%) and testing sites (91.9%) indicates a model with a positive predictive value. The fact that there is a higher percentage of test sites falling within the high-probability zone than the initial training sample is also a good indication of the predictive value of the polygon model. Both of these patterns continue when the number of environmental variables intersecting to create the high-probability zone is increased.

Overall, neither point model seems to have any predictive utility based on a comparison of the percent-correct statistics. The percentage of the study area considered to be high probability is too high, no matter how many intersecting variables are used, when compared with the percent-correct statistic of both the training and testing data sets. On the other hand, the polygon model does seem to have some predictive utility. The percent-correct statistic of both the training and testing data sets is far higher than the percentage of the study area considered high probability.

Comparing the ratio between the percentage of total study area considered to be high probability with the percent-correct statistic of both the training

TABLE 10.4

Gain Statistic and Predictive Utility Comparison for the Three Models

Number of Intersecting Variables	Data Set	Gain Statistic	Predictive Utility (gain)
Initial Point Model			
3,4	Training	0.36	None
3,4	Testing	0.13	None
4	Training	0.52	Positive
4	Testing	−0.38	Negative
Extra Point Model			
4,5,6	Training	0.29	None
4,5,6	Testing	0.25	None
5,6	Training	0.54	Positive
5,6	Testing	0.31	None
6	Training	0.75	Positive
6	Testing	−0.20	Negative
Polygon Model			
4,5,6	Training	0.34	None
4,5,6	Testing	0.36	None
5,6	Training	0.63	Positive
5,6	Testing	0.65	Positive
6	Training	0.74	Positive
6	Testing	0.81	Positive

and testing data sets is a good way to begin assessing the effectiveness of a model. However, to actually compare two or more models, a measure of gain must be computed. Kvamme (1988: 320) proposes the use of the gain statistic to measure gain.

Gain = 1 − [(percentage of total area covered by model)/(percentage of total sites within model area)]

When using this statistic a value close to 1 indicates a model with predictive utility; a value closer to 0 suggests the model has no predictive function. A gain calculated as a negative value implies a model with reverse predictive utility (Kvamme 1988: 320). The gain statistic was calculated for all of our models, and the results are shown in Table 10.4. An examination of the results indicates that the polygon models result in greater predictive utility than the point models.

The following parameters were used to interpret the gain statistic (gs) of our models:

gs > 0.5 = positive predictive utility

gs > 0 but < 0.5 = no predictive utility

gs < 0 = negative predictive utility

The gain statistic calculated for the initial point model shows no predictive utility when the intersection of three or more variables is considered the high-probability surface and tested with the percent correct of both the training and testing data sets. Furthermore, when the intersection of all four relevant environmental variables is indicated as the high-probability surface, a negative predictive utility is indicated when the statistic is calculated with the testing data set. Clearly, the gain statistic indicates a model with no predictive utility for the combination of environmental variables when their ranges are acquired by using a point to represent archaeological site location.

When the gain statistic was calculated for the polygon model, the result indicated a model with an increased predictive utility as well as one demonstrating an increase in the number of intersecting variables creating the high-probability surface. When the quantity of intersecting variables is reduced to a minimum of four intersecting variables, the predictive utility of the polygon model becomes null. However, when the number of environmental variables composing the high-probability surfaces is set at five or greater intersecting variables or at six intersecting variables (all of the pertinent environmental variables), the predictive utility of the model increases. This increase in predictive utility again indicates a benefit of using polygons to develop locational models.

10.3 Conclusions

In this study, the applicability of using points and polygons to represent archaeological sites in the development of locational models was examined. A sample of archaeological data (Archaic sites from Henderson County, Kentucky) derived from the Kentucky statewide site inventory was used to develop and test several models. The results of our study suggest that there are advantages to using polygons rather than points to develop archaeological locational models.

When a point is used to extract the environmental data, only one value for each variable can be selected for that site, regardless of its size and the number of environmental zones it is associated with. This sampling strategy can limit the development of the model by not adequately reflecting the relationship between archaeological sites and environmental factors. Though models derived using point data are slightly less labor intensive, in regard to digitizing site boundaries and extracting environmental data, they appear to perform poorly when compared with polygon models derived using the same data set.

On the other hand, the use of polygons allows for a more accurate sampling of environmental variables that may affect the placement of archaeological sites on the landscape. By utilizing a multiparcel sampling strategy, afforded by using polygonal site boundaries, even a limited sample of archaeological sites and environmental variables allows for the creation of a model with a gain in predictive utility over a purely random or a point-generated model. While polygon models are more labor intensive to develop, they do seem to perform better than point models.

The results of the polygon models also have the potential to contribute to studies of Archaic site-distribution patterns. Association with a variety of resource areas, with a focus on the Ohio and Green Rivers (rank-order 8 streams) appears to reflect the highly mobile hunter-gatherer settlement pattern that has been suggested for the Archaic period in the Ohio Valley. Additional avenues of research into Archaic settlement patterns in Kentucky would be to compare the pattern identified within Henderson County with the distribution of Archaic settlements in other parts of the state or within specific drainages.

In conclusion, there seem to be advantages of utilizing polygons rather than points in a statewide GIS. From a resource-management perspective, polygons provide a more accurate representation than points of the size and shape of a site and allow for a more complete representation of the spatial distribution of archaeological resources within a project or study area. Cultural resource managers need to know the full extent of the sites to ensure that they are adequately protected and, if they cannot be protected, that impacts to archaeological sites are taken into account in project planning. From a locational-modeling viewpoint, polygons allow for a better sampling of environmental variables in the development of a model. This increased sampling ability allows for the development of a model with greater predictive utility than an analogous model based on point data.

Notes

1. This standard was applied even though large ranges resulted for the rank-order 4–6 and rank-order 8 stream variables.

References

Altschul, J.H., Models and the modeling process, in *Quantifying the Present and Predicting the Past: Theory, Method, and Application of Archaeological Predictive Modeling*, Judge, W.J. and Sebastian, L., Eds., U.S. Government Printing Office, Washington, DC, 1988, pp. 61–96.

Cederholm, M., Random Points in a Polygon, 2000; available on-line at http://www.pierssen.com, 2000.

Cook, T.G., Koster: An Artifact Analysis of Two Archaic Phases in West Central Illinois, Koster Research Reports 3, Northwestern University Archaeological Program, Prehistoric Records, Evanston, IL, 1976.

Conaty, G.T., Middle and Late Archaic Mobility Strategies in Western Kentucky, unpublished Ph.D. dissertation, Department of Archaeology, Simon Fraser University, Burnaby, BC, 1985.

Cordell, L.S. and Green, D.F., Theory and Model Building: Defining Survey Strategies for Locating Prehistoric Heritage Resources, Cultural Resources Document 3, USDA Forest Service, Southwestern Regional Office, Albuquerque, NM, 1983.

DeBloois, E.I., The Elk Ridge Archaeological Project: A Test of Random Sampling in Archaeological Survey, Cultural Resources Report 2, USDA Forest Service, Intermountain Region, Denver, CO, 1975.

Duncan, R.B. and Beckman, K.A., The application of GIS predictive site location models within Pennsylvania and West Virginia, in *Practical Applications of GIS for Archaeologists: A Predictive Modeling Toolkit*, Wescott, K.L. and Brandon, R.J., Eds., Taylor and Francis, Philadelphia, 2000, pp. 33–58.

Fowler, M.L., Summary Report of Modoc Rock Shelter, 1952, 1953, 1955, 1956, reports of Investigation 6, Illinois State Museum, Springfield, 1959.

Hill, J.N., Research propositions for consideration: Southwestern Anthropological Research Group, in *The Distribution of Prehistoric Population Aggregates*, Gumerman, G.J., Ed., Prescott College Press, Prescott, 1971, pp. 55–62.

Hockensmith, C.D., Reassessment of the Archaeological Resources in the Proposed Campground Expansions at the Green River State Park, Taylor County, Kentucky, research report on file with Kentucky Office of State Archaeology, Lexington, 1984.

Jefferies, R., Archaic Period, in The Archaeology of Kentucky: Past Accomplishments and Future Directions, Pollack, D., Ed., State Historic Preservation Comprehensive Plan, Report 1, Kentucky Heritage Council, Frankfort, 1966, pp. 143–246.

Judge, W.J. and Sebastian, L., Eds., *Quantifying the Present and Predicting the Past: Theory, Method, and Application of Archaeological Predictive Modeling*, U.S. Government Printing Office, Washington, DC, 1988.

Kennedy, H., Ed., *The ESRI Press Dictionary of GIS Terminology*, ESRI Press, Redlands, CA, 2001.

Kohler, T.A. and Parker, S.C., Predictive models for archaeological resource location, in *Advances in Archaeological Method and Theory*, Vol. 9, Schiffer, M.B., Ed., Academic Press, Orlando, FL, 1986, pp. 397–452.

Kvamme, K.L., Development and testing of quantitative models, in *Quantifying the Present and Predicting the Past: Theory, Method, and Application of Archaeological Predictive Modeling*, Judge, W.J. and Sebastian, L., Eds., U.S. Government Printing Office, Washington, DC, 1988, pp. 325–428.

Kvamme, K.L., Fundamental principles and practice of predictive archaeological modeling, in *Mathematics and Information Science in Archaeology: A Flexible Framework*, Vol. 3, Voorrips, A., Ed., Verlag-Holos, Bonn, 1990, pp. 257–295.

Lewis, R.B., Ed., *Kentucky Archaeology*, University of Kentucky Press, Lexington, 1996.

Lock, G. and Stančič, Z., Eds., *Archaeology and Geographic Information Systems: A European Perspective*, Taylor and Francis, London, 1995.

Maschner, H.D.G., Ed., *New Methods and Old Problems: Geographic Information Systems in Modern Archaeological Research*, Occasional Paper 23, Center for Archaeological Investigations, Southern Illinois University, Carbondale, 1996.

Nance, J.D., The Archaic Sequence in the Lower Tennessee-Cumberland-Ohio Region, *Southeastern Archaeology,* 6: 129–140, 1986.

Nance, J.D., The Morrisroe site: projectile point types and radiocarbon dates from the Lower Tennessee Valley, *Midcontinental Journal of Archaeology,* 11, 11–50, 1983.

Rothschild, N.A., Mortuary behavior and social organization at Indian Knoll and Dickson Mounds, *American Antiquity,* 44, 658–675, 1979.

Saraf, A.K., Grid Analyst Extension 1.1, Department of Earth Sciences, University of Roorkee, Roorkee, India, 2000.

Shennan, S., *Quantifying Archaeology,* 2nd ed., University of Iowa Press, Iowa City, 1997.

Steward, J.H., Basin-Plateau Aboriginal Sociopolitical Groups, Bulletin 150, Bureau of American Ethnology, Smithsonian Institution, Washington, DC, 1938.

Stevenson, C.M., A Predictive Model of Prehistoric Site Location for the Route 30 Study Area, Wyandot and Crawford Counties, Ohio, report submitted by ASC Inc. to the Ohio Department of Transportation, Columbus, Ohio, 1993.

Stoltman, J., Temporal models in archaeology: an example from eastern North America, *Current Anthropology,* 19, 703–746, 1978.

Strahler, A.N., Quantitative analysis of watershed geomorphology, *American Geophysical Union, Transactions,* 38, 913–920, 1957.

Warren, R.E., Predictive modeling in archaeology: a primer, in *Interpreting Space: GIS and Archaeology,* Allen, K.M.S., Green, S.W., and Zubrow, E.B.W., Eds., Taylor and Francis, Philadelphia, 1990, pp. 90–111.

Warren, R.E. and Asch, D.L., A predictive model of archaeological site location in the Eastern Prairie Peninsula, in *Practical Applications of GIS for Archaeologists: A Predictive Modeling Toolkit,* Wescott, K.L. and Brandon, R.J., Eds., Taylor and Francis, Philadelphia, 2000, pp. 5–32.

Weed, C.S., Tuttle, E.H., Harris, E.J., and Meyers, R.G., Phase I Cultural Resources Investigations of Proposed Support Areas for the Route 30 STA/COL-30-18.35/ 0.00 Alternatives Eastern Segment, Carroll and Columbiana Counties, Ohio, Vol. I (revised), report submitted by Gray and Pape Inc. to the Ohio Department of Transportation, Columbus, Ohio, 1996.

Wescott, K.L. and Brandon, R.J., Eds., *Practical Applications of GIS for Archaeologists: A Predictive Modeling Toolkit,* Taylor and Francis, Philadelphia, 2000.

Wobst, H.M., Locational relationships in Paleolithic society, *Journal of Human Evolution,* 5, 49–58, 1976.

Willey, G.R., Prehistoric Settlement Patterns in the Viru Valley, Peru, Bulletin 155, Bureau of American Ethnology, Smithsonian Institution, Washington, DC, 1953.

11

Relating Cultural Resources to Their Natural Environment Using the IEDROK GIS: A Cultural Resources Management Tool for the Republic of Korea1

Bruce Verhaaren, James Levenson, and James Kuiper

CONTENTS

1 This chapter addresses the development and use of the IEDROK GIS. The development of IEDROK required the collection and integration of different types of data from many unrelated sources. Its use requires viewing cultural resources from the perspective of a culture having somewhat different priorities and perspectives. The development of IEDROK, Integrated Ecosystems Databases for the Republic of Korea, was supported by U.S. Department of Defense Legacy Resource Management Program, project no. 991888. Work was supported under a military interdepartmental purchase request (W31RY091653834) from the U.S. Department of Defense, through U.S. Department of Energy contract W-31-109-Eng-38. IEDROK was developed by Argonne National Laboratory personnel in collaboration with several Korean agencies and organizations, including the Korean Association for Conservation of Nature, a research arm of the Ministry of Environment, and the Cultural Properties Administration, an arm of the Ministry of Culture and Tourism. Assistance, information, and support were provided by both the 8th U.S. Army and the U.S. Air Force, Pacific Air Forces.

ABSTRACT   Under the Status of Forces Agreement (SOFA) negotiated between the United States and the Republic of Korea, U.S. forces stationed in Korea are required to take into account the effects of their actions on Korean cultural properties. In addition to this legal requirment, cultural resource management is important in maintaining good relationships between U.S. forces and the local population. The Integrated Ecosystem Databases for the Republic of Korea (IEDROK) geographic information system (GIS) provides information on natural and cultural resources to U.S. Forces Korea (USFK) resource managers. Focused on areas surrounding USFK installations, it is a rapid-assessment tool that alerts managers to the presence of protected resources that could be affected by USFK undertakings.

Geographic information systems are ideal tools for the management of archaeological and other cultural resources. A graphic display of cultural resources against the backdrop of significant natural geographic features permits a rapid visual assessment of the relationship of the cultural world to the natural world. Displaying cultural resources of a functional or temporal class in relation to such features as hydrology, soil type, slope, vegetation cover type, and the contemporary built environment provides the cultural resource manager with the necessary tools to locate and manage known cultural resources. This capability can also enable the manager to predict the likelihood of encountering significant cultural resources in similar but not-yet-investigated areas. IEDROK is an example of such a tool. It was designed for the cultural resource managers at USFK installations. Those managers have stewardship responsibility for significant natural and cultural resources, even though their primary mission is not resource management.

11.1 Introduction

Korea has spent much of the last century under foreign domination, and Koreans place great value on preserving things that are distinctively Korean and that highlight the accomplishments of the Korean people. The Status of Forces Agreement (SOFA) between the governments of the United States and the Republic of Korea establishes environmental governing standards (EGS) for Korea (USFK 1997) that U.S. forces are committed to follow. These include the inventory and preservation of natural and cultural resources in areas under their control. However, the Cold War never ended on the Korean Peninsula, and the need for security and military preparedness is intense. In this context, the Integrated Ecosystem Databases for the Republic of Korea (IEDROK) geographic information system (GIS) is a tool designed to aid U.S. Forces Korea (USFK) planners and resource managers in meeting their SOFA obligations in support of their military mission.

Typically, military bases by their very nature provide important protections to both natural and cultural resources. Because access to USFK-managed lands is limited, these lands have been protected from the effects of urban growth and industrialization that are frequently characteristic of the civilian society surrounding them. USFK installations include extensive buffer zones, where access is limited even for military personnel. These areas often become refuges for wild species. Archaeological sites and other cultural resources in these areas also remain relatively undisturbed. For USFK, cultural resources include any structures, natural features, artifacts, or archaeological remains important to Korean or American history or culture. They form a part of the legacy that the U.S. Department of Defense (DoD) seeks to exercise responsible stewardship over.

The primary mission of cultural and natural resource managers of USFK installations is to support the military defense of the Republic of Korea. However, as resource managers, they manage the cultural resources present on USFK installations on a daily basis. Many managers are young military officers trained in engineering, without extensive training or experience in archaeology, anthropology, history, or the natural sciences. For the most part, they first encounter the Korean culture during their military tour. IEDROK was designed for them. It is not solely or even primarily a database of cultural information. Rather, it graphically presents both the natural and cultural features of the Korean Peninsula, placing cultural resources in their broader environmental contexts. IEDROK provides basic information about Korea's natural and cultural environment, information that was previously available only in disparate sources and formats accessible only to local specialists. It is a flexible framework that can be customized by the end user to focus on a particular installation by adding additional detail or by modifying installation configurations as conditions change.

11.2 IEDROK and the Legacy Resource Management Program

The development of IEDROK was funded by the DoD Legacy Resource Management Program. The Legacy Resource Management Program was established by Congress in 1990 to provide for the stewardship of all DoD-controlled or -managed air, land, and water resources and establish inventories of all scientifically significant biological, geophysical, cultural, and historical assets on DoD lands. Such inventories catalog the attributes of these resources, their scientific and cultural significance, and their interrelationship to the surrounding environment, including the military mission carried out on the land upon which they are found.

Legacy Program managers have developed the following statement of purpose (DoD 2001):

1. Legacy will determine how to better integrate the conservation of irreplaceable biological, cultural, and geophysical resources with the dynamic requirements of military missions. To achieve this goal, the DoD will give high priority to inventorying, conserving, and restoring biological, cultural, and geophysical resources in a comprehensive, cost-effective, state-of-the-art manner, in partnership with federal and local agencies and private groups.

2. Legacy activities will help ensure that the DoD better understands the need for protection and conservation of natural resources and that the management of these resources will be fully integrated with, and support, DoD mission activities and the public interest.

3. Legacy, through the combined efforts of DoD components, will achieve the legislative purposes of Legacy with cooperation, industry, and creativity to make the DoD the federal environmental leader.

These Legacy Program goals mesh well with the commitments made by USFK in the EGS (USFK 1997). Chapter 12 of the EGS includes the following cultural resources management criteria:

Criterion 12-3b: "If financially and otherwise practical, installations will inventory cultural property and resources in areas under USFK control."

Criterion 12-3d: "Installations will establish measures sufficient to protect known cultural property or resources identified on the installation inventory and for mitigation of adverse effects."

Criterion 12-3e: "Installations will establish measures to prevent excavation of cultural properties. Areas known to contain buried or submerged historic properties shall not be excavated or disturbed."

Criterion 12-3i: "Installations will ensure that planning for major actions includes consideration of possible effects on cultural or archeological property or resources."

EGS Chapter 13 deals with the management of natural resources and commits USFK installations to "develop a program for conserving, managing, and protecting natural resources," "maintain a current list of endangered species and specified wild species," "initiate surveys for endangered species and specified wild species," and "take reasonable steps to protect and enhance known endangered or specified wild species and their habitat."

In other words, USFK managers must be able to take cultural and natural resources into account in their planning and mitigate any projected adverse effects on these resources. IEDROK assists USKF resource managers in this task. IEDROK presents the locations of protected resources in their environmental context. By overlaying the area or potential effect of a proposed action on this underlying base data, a manager can rapidly identify the potential impacts to known natural and cultural resources likely to result from planned USFK actions, thus providing a framework for analyzing new information generated during environmental reviews. Resource managers can use IEDROK in their environmental review process and in the development of the installation's master plans.

IEDROK is a flexible GIS tool that assists in achieving the first two Legacy Program goals, thereby furthering the third. IEDROK brings together natural- and cultural resource inventory information and provides a tool for integrating them into the USFK planning process. As guests in a foreign country, USFK planners and resource managers are often presented with new issues regarding natural and cultural resources. IEDROK is designed to aid the managers by locating and identifying those resources on and in the vicinity of their installations. The GIS accommodates the entry of more-precise, recently recorded information as it is encountered. Current DoD guidance emphasizes an interdisciplinary and ecosystems approach to resource management. IEDROK integrates natural and cultural resources information in support of this approach.

11.2.1   IEDROK Structure: The Data Layers

IEDROK is broad in scope and includes the best available data (see Table 11.1). The data sets included come from a variety of sources and vary in scale and detail. DoD standard National Imagery and Mapping Agency (NIMA) products were used for the baseline layers. Natural and cultural resource data were obtained from Korean sources augmented by field observations. This section briefly identifies each of the data sources and discusses the reliability of the data from each source. Table 11.1 lists the GIS data layers in the system and summarizes the nature of each layer and the data sources.

IEDROK should be thought of as a screening tool intended for use early in the planning process. It alerts managers to the potential of natural and cultural resource issues rather than presenting a detailed representation of resource locations. The earlier in the planning process it is used, the more time will be available to assess and mitigate any adverse impacts. For Korean security reasons, none of the source maps include the locations or boundaries of any military facilities. It is assumed that users can integrate IEDROK with the GIS at their own installations.[1]

TABLE 11.1

Sources for the IEDROK GIS

Shapefile Legend Name	Data Description	Date	Source	Format	Data Type	Number of Records	Projection	Scale	Conversion Function	Caveats
Baseline Layers										
See Table 11.2	Coastlines and political boundaries	1973–1989	*Vector Map Level 1 (VMap1)* (NIMA 1998)	Vector Product Format (VPF) on CD and tape	Polygon and line	2 themes	UTM datum WGS84	1:250,000	ArcView VPF extension	Some sources as old as 1966
See Table 11.3	Land use, topography, vegetation, soils, hydrology	1995–2000	*Vector Product Interim Terrain Data (VITD)* (NIMA 1996)	VPF on CD and tape	Polygon and line	10 themes	UTM datum WGS84	1:50,000	ArcView VPF extension	Projection more accurate than VMap1
Natural Resources — Points										
natural_point.shp Protected species	Provincial protected species	1980	*Rare and Endangered Wildlife in Gangweon Do, Korea* (KACN 1980)	Book	Point	466	UTM zone 52	≈1:8,000,000	Direct construction of shapefile	Source scale smaller than baseline; points approximate
natural_point.shp Protected species	Rare and endangered	1990	*Current Condition of Korea's Rare and Endangered Flora and Fauna* (KACN 1990)	Book	Point	1,588	UTM zone 52	≈1:8,000,000	Direct construction of shapefile	Source scale smaller than baseline; points approximate
natural_point.shp Protected species	Freshwater fishes	1990	*Coloured Illustrations of the Freshwater Fishes of Korea* (Choi et al. 1990)	Book	Point	486	UTM zone 52	≈1:8,000,000	Direct construction of shapefile	Source scale smaller than baseline; points approximate

Layer	Feature	Year	Source	Format	Type	Count	Projection/datum	Scale	Section	Notes
natural_point.shp Protected species	Endangered and reserved species	2000	*Endangered Wild Species of Korea* (KACN 2000a)	Intergraph GeoMedia and ESRI ArcView shapefiles on CD ROM	Point (1 polygon)	916	Geographic, Korean datum	≈1:8,000,000	Section 4.2.1.2	Source scale smaller than baseline; points approximate
natural_point.shp Protected species	Field observations	2000	Field observations at Kunsan AB and Osan AB by J. Levenson	Sheet map	Point	20	UTM zone 52	1:30,000	Direct construction of shapefile	Source scale larger than baseline
natural_mon_ point.shp Natural monuments	Natural monuments	1997	*The Management Atlas of State Designated Cultural Properties* (KCPA 1997)	Sheet maps	Point	112	TMX — transverse Mercator with five different Korean zones and datums	1:5,000 1:25,000 1:50,000	Section 4.2.1.3	Source scale larger than baseline; some displacement possible
Natural Resources — Polygons										
natural_mon_poly. shp Natural monuments	Natural monuments	1997	*The Management Atlas of State Designated Cultural Properties* (KCPA 1997)	Sheet maps	Polygon	137	TMX — transverse Mercator with five different Korean zones and datums	1:5,000 1:25,000 1:50,000	Section 4.2.1.3	Source scale larger than baseline; some displacement possible
Cultural Resources — Points										
cultural_point.shp Cultural properties	National treasures	1997	*The Management Atlas of State Designated Cultural Properties* (KCPA 1997)	Sheet maps	Point	37	TMX — transverse Mercator with five different Korean zones and datums	1:5,000 1:25,000 1:50,000	Section 4.3.2	Source scale larger than baseline; some displacement possible

TABLE 11.1

Sources for the IEDROK GIS (continued)

Shapefile Legend Name	Data Description	Date	Source	Format	Data Type	Number of Records	Projection	Scale	Conversion Function	Caveats
cultural_point.shp Cultural properties	Other national cultural properties	1997	The Management Atlas of State Designated Cultural Properties (KCPA 1997)	Sheet maps	Point	22	TMX — transverse Mercator with five different Korean zones and datums	1:5,000 1:25,000 1:50,000	Section 4.3.2	Source scale larger than baseline; some displacement possible
cultural_point.shp Cultural properties	Seoul properties	1997	The Management Atlas of State Designated Cultural Properties (KCPA 1997)	Sheet maps	Point	38	TMX — transverse Mercator with five different Korean zones and datums	1:5,000 1:25,000 1:50,000	Section 4.3.2	Source scale larger than baseline; some displacement possible
cultural_point.shp Cultural properties	Local cultural properties	1995–1997	The Compendium of Cultural Relics (NRICP 1995–1997)	CD	Point	194	UTM zone 52	1:30,000– 1:100,000	Direct construction of shapefile	Source maps general; some displacement likely
cultural_point.shp Cultural properties	Local recorded cultural properties	1995–1997	The Compendium of Cultural Relics (NRICP 1995–1997)	CD	Point	495	UTM zone 52	1:30,000– 1:100,000	Direct construction of shapefile	Source maps general; some displacement likely
Cultural Resources — Polygons										
cultural_poly.shp Cultural properties	National treasures	1997	The Management Atlas of State Designated Cultural Properties (KCPA 1997)	Sheet maps	Polygon	17	TMX — transverse Mercator with five different Korean zones and datums	1:5,000 1:25,000 1:50,000	Section 4.3.2	Source scale larger than baseline; some displacement possible
cultural_poly.shp Cultural properties	Other national cultural properties	1997	The Management Atlas of State Designated Cultural Properties (KCPA 1997)	Sheet maps	Polygon	163	TMX — transverse Mercator with five different Korean zones and datums	1:5,000 1:25,000 1:50,000	Section 4.3.2	Source scale larger than baseline; some displacement possible

11.2.1.1 Base Layers

The best available base map data for the Korean Peninsula came from NIMA. While NIMA data sets include important environmental information, they are designed to meet the tactical needs of the DoD (e.g., factors affecting the movement of troops), not its environmental stewardship requirements. NIMA layers indicate slope (topography), soil type, water resources, and vegetation cover. However, because these same layers depict significant factors affecting the range and distribution of both natural and cultural resources, they also provide a sound baseline for the construction of a useful management tool. Soil, vegetation, and water clearly constrain the kinds of plants and animals that are able to live in a given area. All of these factors constrain human settlement and land use. For centuries, Koreans have considered the physical environment when selecting a site for the construction of a dwelling, shrine, or grave. The principles of chŏnghyŏlbŏp or geomancy, relate the cultural to the natural environment, and a propitious site location is based on spatial relationships with mountain ridges and water (Choi 1997; Feuchtwang 1974).

Two NIMA products with multiple components were used to construct the baseline data set: Vector Product Interim Terrain Data (VITD) and Vector Map Level 1 (VMap1). Because access to these products is restricted, data layers of IEDROK cannot be depicted in this report, and generic illustrative background layers have been substituted in the illustrations presented. There are no such restrictions on the IEDROK natural and cultural resource layers.

For IEDROK, the themes most relevant to cultural and ecological analyses were first extracted from the VITD and VMap1 data sets. Although there is some overlap between the themes available in VITD and VMap1 formats, the two data sets are not identical. VMap1 data include a simplified rendition of Korea's coastlines and internal political boundaries and are used for small-scale displays. VITD source maps are at a larger scale, provide more detail, and form the heart of the baseline data set. VITD coastlines are used for larger scale views.

The VITD data set was designed principally for military unit mobility analysis. A full suite of data includes themes such as slope, vegetation, surface materials (soils), surface drainage, transportation, and obstacles. Source data for the VITD is typically taken from 1:50,000-scale hard-copy maps (NIMA 1996). The NIMA VITD data for this region were produced during 1996, but the dates of the source materials are not specified. VMap1 is a vector-based electronic data set that provides basic cartographic data for larger areas. The VITD data are believed to be more precise than the VMap1 data, which are horizontally less accurate in absolute terms than VITD data sets as the distance from Seoul increases.

11.2.1.2 Cultural Resource Layers

IEDROK displays the locations of nationally and locally designated cultural properties. Icons are used to indicate the type of resource (Figure 11.1), and

Cultural Properties (point)

Archaeological Site

Fortification

Historic Building

Monument

Other Historic Place

Pagoda

School

Shrine

Temple

Tomb

FIGURE 11.1
IEDROK icons.

more-detailed information is available by clicking on the symbol. The Republic of Korea divides cultural properties into four categories. The most significant properties are designated "national treasures." Therefore, IEDROK incorporates the locations of all national treasures throughout South Korea. Resources falling in other categories — treasures, historic sites, and important folk materials — are included in IEDROK only when they are located relatively close to a major USFK facility.

The locations of cultural resources were obtained from source maps supplied by the Korean Cultural Properties Administration (KCPA), located in Taejon, and the National Research Institute for Cultural Properties (NRICP), located in Seoul. The locations of the nationally designated properties were digitized from map sheets found in *The Management Atlas of State-Designated Cultural Properties* (KCPA 1997). The positions of the locally designated cultural properties and locally recorded cultural resources were derived from maps provided electronically by the National Research Institute for Cultural Properties in *The Compendium of Cultural Relics* (NRICP 1995–1997). The source maps varied in scale and precision. The precise locations of military installations are not shown on the maps. Screening for cultural resources near or on USFK facilities was based on the authors' knowledge of the installation locations. The most precise locations are determined by the individual user when the IEDROK layers are incorporated into an installation's GIS, or vice versa.

Cultural resources are abundant throughout the Korean Peninsula. However, due to constraints of time and resources, with the exception of national treasures, the basic IEDROK includes only those resources in relatively close proximity to major USFK installations. Three classes of cultural resources are included in IEDROK: (1) nationally designated cultural properties, (2) locally designated cultural properties, and (3) locally recorded cultural resources. Nationally and locally designated cultural properties have special status under Korean law and are protected against damage or destruction. Designation of cultural properties on the national level occurs under the authority of the Ministry of Culture and Tourism upon the advice of the Cultural Properties Committee. Provincial and local departments of culture may designate locally important properties. Once designated, properties are listed by the Cultural Properties Administration by category and receive a unique number (e.g., Treasure 43 or Monument 88). Locally recorded cultural resources receive no designation number and have fewer legal protections. They remain locally important, however, and are included in IEDROK so they will not be inadvertently harmed. Some are currently in the designation process.

11.2.1.3   Natural Resource Layers

IEDROK natural resource layers were compiled from four types of data resources, including: (1) distribution maps in published works on Korea's flora and fauna, (2) hard-copy maps of designated Korean natural monuments found in *The Management Atlas of State-Designated Cultural Properties* (KCPA 1997), (3) electronic distribution point data generated for this project by the Korean Association for the Conservation of Nature (KACN), and (4) field observations made at USFK facilities sources (see Table 11.1). For the most part, these sources indicate the distribution of threatened or endangered species using points to indicate where the species were observed or reported. Because of discrepancies in scale between these sources and the baseline sources, points should be interpreted to indicate the presence of a species in the general area, but not at the exact location of the point. The location information based on field observations by Argonne personnel at U.S. Air Force bases in Korea, while also displayed as points, is more precise because the exact locations of the observations are known.

The Ministry of Culture and Tourism designates specific species, wildlife refuges, forests, groves, or even individual trees as nationally protected natural monuments. Designated natural monuments are considered cultural properties in the Republic of Korea. Cultural Properties Administration data often refer to specific places with known addresses or legal descriptions. The Cultural Properties Administration provided data in the form of larger-scale maps. Data sets included point and polygon data. Because of the larger scale, the Cultural Properties Administration data sets could be projected with more precision than the published and electronic point data. IEDROK includes all of the natural monuments on the topographic maps in the Cultural Properties Administration's *Atlas of State-Designated Cultural Properties* (KCPA 1997).

11.2.2   Creating the Layers: Source Data

One of the major challenges in constructing IEDROK was blending data derived from a variety of sources. As an integrated database, IEDROK includes previously existing GIS baseline data; data sets derived from maps that were scanned, screen digitized, and georeferenced; data from maps that were digitized on a tablet and georeferenced; and point data from published sources that were entered directly into the GIS.

An ArcView®²-structured GIS data set was constructed to form the baseline data set for IEDROK. The baseline data set was derived from two NIMA products with multiple components (VITD and VMap1). NIMA products use World Geodetic System 1984 (WGS84) as the horizontal datum. The vertical datum on the topographic layer is mean sea level.

Vector Product Interim Terrain Data (VITD) is a vector-based electronic data set designed principally for military-unit mobility analysis. Sources used by NIMA in the compilation of VITD may include aerial photography, topographic maps, soil surveys, hydrographic studies, land-use inventories, and transportation reports. Information may also have been derived from hard-copy tactical terrain analysis data bases (TTADB) at a scale of 1:50,000 or conversion from interim terrain data (ITD) in standard linear format (SLF). The data sets for the VITD were produced during 1996; however, the specific dates of the data sources are not identified in the release notes of the VITD product (NIMA 1998).

VITD includes physical features such as vegetation type, surface materials (soils), degree of slope, and surface drainage. These themes are of ecological significance. Other useful VITD parameters include transportation corridors and obstacles. Of these, the themes shown in Table 11.2 were selected as most

TABLE 11.2

VITD Themes Included in IEDROK

Theme	Description
Slope (polygon)	Slope
Soils (polygon)	Soils
Land use (polygon)	General vegetation
Forest vegetation (polygon)	Forest vegetation
Open water and built-up areas (polygon)	Open water and built-up areas
Major rivers (polygon)	Surface drainage: areas (rivers and canals wide enough to be represented as area features)
Rivers and streams (line)	Surface drainage: lines (rivers and canals wide enough to be represented as line features)
Dams and canals (line)	Surface drainage: lines (dams and canals wide enough to be represented as line features)
Roads (line)	Roads and cart tracks
Railroads (line)	Railroad tracks

appropriate for the current application. A formal horizontal accuracy has not been defined for the VITD product at this time. However, because the sources of VITD data are typically 1:50,000 source scale products, they are likely more accurate than the VMap1 data described next.

Vector Map Level 1 (VMap1) is a vector-based electronic data set that provides basic cartographic data for larger areas on the ground. The VMap1 data set was derived from NIMA air navigation map products at the 1:250,000 source scale. Most of the source maps were produced from 1973 to 1989, with a few being as old as 1966. Additional electronic data were used to update particular themes in the VMap1 data during the digital production cycle (NIMA 1998).

Comparison of VMap1 with VITD themes indicated that the data sets match closely in and around Seoul, but there was a discrepancy between the two in outlying regions. That discrepancy increased with the distance from the capital. Data tests and NIMA sources indicate that discrepancy appears to be the result of systematic errors in the conversion of the original Tokyo datum sources to the more recent WGS84 datum for the VMap1 data set.

The data themes included by NIMA in VMap1 are political boundaries, data quality, elevation, hydrography, industry, physiography, population, transportation, utilities, and vegetation. For IEDROK, only coastline and boundary data were used to supplement the VITD data (Table 11.3).

The Korea baseline data set was constructed by extracting an overview area of boundaries and coastlines from the VMap1 data source (Table 11.3). This provided a guideline to fill in areas with the VITD and VMap1 vector data. The VITD data were extracted into shapefiles[3] to provide the detailed baseline or background data shown in Table 11.1. VITD is distributed by NIMA in quarter-degree cells. This arrangement is difficult to work with for larger regions. Merging the cells to form larger, contiguous areas of GIS data allowed for easier data manipulation in exchange for slightly slower performance in some operations.

The VMap1 data themes were also converted into the shapefile format and merged into one shapefile for each layer. Some VITD and VMap1 data sets resulted in extremely large files. The final database was divided into five parts, corresponding to USFK regions, to fit them onto a set of five CDs.

This data compilation effort resulted in the set of data themes identified in Tables 11.2 and 11.3, forming a base upon which the cultural and natural layers of IEDROK have been built. All data are registered to the WGS84 datum.

TABLE 11.3

VMap1 Themes Included in IEDROK

Theme	Description
Coastline (line)	Coastal linework
Political boundaries (polygon)	Political boundaries: area

Some data are more accurate than others. For example, the VITD information is more accurate and detailed than the information from the VMap1 data source because of the differing source scales and the placement-error-probability accuracies defined in the product specifications. However, coastlines and political boundaries are available only in the VMap1 data sets, and were therefore included by default. The imperfect synthesis of data from these two sources should have little effect on the practical use of the GIS. Depending on the source of the thematic layers, the discrepancies are likely to be trivial. The differences will be apparent only at a very large scale, and then mostly in the southern end of the Korean Peninsula. In general, when a layer in a GIS is displayed at a larger scale than the source data scale, limits to accuracy should be taken into account and are sometimes clearly visible.

11.2.2.1 Cultural Resource Data

Cultural resource data were extracted from two sources: the *Management Atlas of State-Designated Cultural Properties* (KCPA 1997) and *The Compendium of Cultural Relics* (NRICP 1995–1997).

The *Management Atlas of State-Designated Cultural Properties* is a 21-volume set primarily comprised of 1:5,000, 1:25,000, and 1:50,000 topographic base maps produced by the Korean National Geography Institute, with cultural-property locations added by the KCPA. Depending on the property size and the scale of the map, the property may be shown as either a point or a polygon. Each of the map quadrangles is projected in the Korean TMX transverse Mercator system based on a corrected Tokyo datum. It divides Korea into five areas with five different Korean points of origin. Latitude and longitude are indicated at the corners of each quadrangle sheet. The majority of the maps in the atlas are at the 1:5,000 scale. However, for locations that include military installations, the atlas includes only smaller scale maps covering a broader area. In keeping with South Korean practice and security restrictions, none of these maps indicates the presence of any military installations. Almost 1,000 black-and-white photocopies of the atlas maps were obtained directly from the KCPA. KCPA personnel from the Monuments Division highlighted the location of each cultural property in color on the maps. Limitations of time and resources did not allow all sheets to be processed, and priorities were established in consultation with USFK. All topographic sheets showing national treasures, the most important Korean cultural properties, were identified and processed. USFK identified the U.S. Army facilities with the highest priority for identifying natural and cultural resources. With the exception of facilities in Seoul, all quadrangles showing any cultural properties within a 15-km radius of U.S. Army facilities were processed. Quadrangles showing cultural resources within the same radius of the two active U.S. air bases were also processed. Cultural properties of all categories found on these maps have been included in the database. Attribute tables were generated for the cultural properties that

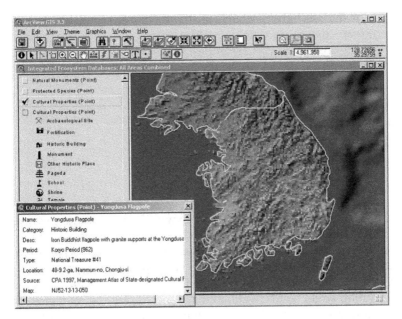

FIGURE 11.2
Selecting an icon displays attribute information.

include the name of the site, its designation, its address, its date, and a brief description (Figure 11.2).

The KCPA 1:5,000 maps for Seoul have no positional markings, such as latitude/longitude or UTM tick marks. Therefore, locations were entered as points directly into a shapefile using a backdrop of the most detailed VITD street-grid layer to determine point location. As with other areas of the country, no detailed maps were available for locations including military installations, thus most of Yongsan-gun did not appear on any KCPA quadrangle. Therefore, those quadrangles directly adjacent to Yongsan-gun were used. This resulted in data extraction from the area within about 5 km of the military base.

The remaining KCPA maps had geographic coordinates at their corners. For point data, the most efficient method of determining the latitude and longitude was by measurement and linear interpolation from known coordinates at map corners. The coordinates, map number, and site were recorded on a spreadsheet; then the data were transferred to ArcView and a point layer was produced from the coordinates. After the point layer was made, the coordinates were converted from the map coordinate system, which used the Korean datum, to the final coordinate system having a WGS84 datum. Comparing the projection against the printed maps indicated that points were projected with horizontal accuracy within 20 to 25 m.

The longitude/latitude pairs for the tabular data were obtained via hand measurements from the base maps. Four measurements were taken: width of the map through the data point, height of the map through the data point, width from the western map edge to the data point, and height from the southern map edge to the data point. The estimation of the geographic coordinates of the data point was accomplished using the following equations:

$$\lambda_{est} = \left(\Delta\lambda_{map} \frac{h_i}{h_{map}} \right) + \lambda_{ref}$$

$$\phi_{est} = \left(\Delta\phi_{map} \frac{w_i}{w_{map}} \right) + \phi_{ref}$$

where λ_{est} and ϕ_{est} are the latitude and longitude of the data point, respectively; $\Delta\lambda_{map}$ and $\Delta\phi_{map}$ are the latitude and longitude ranges of the map, respectively; h_i and h_{map} are the heights from the southern edge of the map to the data point and the northern map edge, respectively; w_i and w_{map} are the widths from the southern edge of the map to the data point and the northern map edge, respectively; and λ_{ref} and ϕ_{ref} are the reference latitude and longitude values obtained from the southwestern corner of the map, respectively. The data set was converted to shapefiles by loading the ASCII data into the program and creating an event theme.

Whenever a natural monument was large enough to appear as a polygon rather than a point on the topographic sheets, the shape of the polygon was captured by using one of two methods. Some maps were scanned as TIFF files using a 36-in.-wide drum scanner. These files were then brought into ArcView and viewed on a monitor. The position and orientation of each map were established from the coordinates of the southwestern corner, the scale of the map, and by digitizing the map's extent. ArcView tools were then used to create digitized polygons matching those shown on the maps, yielding a shapefile output. Each feature was coded with map and site numbers. The shapefile was then converted into an Arc/Info® coverage[4] for georeferencing. The georeferencing process used the latitude and longitude of the southwest corner and the map scale. Using a batch process, the three remaining corners of the quadrangle were calculated, and the resulting rectangle was projected into the proper Korean projection. A least-squares fit was then used to fit the quadrangle to a known coordinate system and transform the digitized information to the Korean projection used on the map. Then the data were projected from the Korean projection to the final coordinate system. This process places the polygon accurately relative to the underlying baseline data set.

Other maps were digitized directly using a digitizing tablet. In this method, the map was attached to a digitizing tablet; then a digitizing tool was used to mark the known position of the four corners of the quadrangle, and the boundaries of each polygon were traced. Again, each feature was coded with the source map number and the number of the site being plotted.

This process resulted in digitized shapefiles. Then, the same process was used to georeference these files as was used for the digitized scanned images. All maps were merged after georeferencing. A single natural monument was often split across more than one map sheet. In such cases, the polygon segments from each sheet were combined into one feature. Completed polygon features were compared against the paper originals to verify the accuracy and completeness of the position and form and were corrected as needed.

The locations of locally designated cultural properties came from the NRICP database. The scale of these maps is usually not indicated, but it is estimated to range from about 1:30,000 to 1:100,000. Site locations are shown as points. There are no polygon or line data from these sources. This database, received on four CD-ROMs, was produced in Korean (*Hangul*). Extracting these data therefore required the extra step of translating attribute information into English. Because local cultural resources are numerous, coverage was once again restricted to the priority areas discussed above. The NRCIP data set was divided according to local political boundaries. Data sets from governmental units, including priority military bases and the jurisdictions immediately adjoining them, were included because these are the areas where cultural resources are most likely to be affected by USFK activities. The maps included with the NRCIP data show cultural-property locations as points on a background consisting of a simplified street grid, coastlines, and political boundaries. NRCIP maps were compared with the VITD road grid and 1:100,000 road maps of the local area to determine their position. A GIS layer was created directly from these sources. The attributes attached to the points are the same as those used for nationally designated properties, except that the local cultural-property categories of "monument," "tangible cultural property," "cultural properties material," and "folk material" replace the national categories. The majority of the recorded sites are undesignated properties and are categorized in the attribute table as "miscellaneous."

11.2.2.2 Natural Resource Data

Constructing the natural resource layers involved obtaining data from additional source types. Natural-monument data were derived from the *Management Atlas of State-Designated Cultural Properties* as described above. Natural resource data were also derived from hardbound texts and electronic data in a different format.

Three bound volumes provided the maps that were used to construct the initial natural resources data set. Each volume describes Korean species of concern, accompanied by a small-scale, point distribution map. Points were screen-digitized over a base map similar to the published maps, resulting in separate shapefiles for each volume. A set of attributes was assigned to each point, including the genus and species, common name, protected status, and taxon. Reference information, including page number, was also included.

Data were extracted from the following three sources: Choi et al. (1990) and KACN (1980, 1990).

These layers were later combined with other point data to form a single natural resources point layer. The attribute fields were first standardized according to their names and features, and then joined. To enhance the ease of use, display points are represented with icons depicting the different taxa: mammals, birds, reptiles, amphibians, insects, or plants.

The horizontal accuracy of point placement is constrained by the accuracy of the source data. The scale of the published sources is approximately 1:8,000,000. The scale of the VMap1 data is 1:250,000. Even though ArcView software allows the user to zoom in to large-scale views of the data, such actions will provide a false impression of local precision. Because the location can be no more precise than the original source map, the best use of this GIS is to screen for possible presence. IEDROK alerts the user to the likelihood of a particular species in the area. The utility of IEDROK is as an initial screening tool that provides a level of information not previously available.

The Korean Association for the Conservation of Nature (KACN) provided species point distributions for inclusion in IEDROK (KACN 2000a, 2000b). Additional species point plots were created by KACN specifically for this database. These plots updated and expanded the distribution plots available from published sources. KACN provided the point distributions for protected species as GIS layers. Important identifying species attributes, such as species and common name, were extracted from the KACN reports (KACN 2000a, 2000b).

While there seems to be a relatively good fit between the KACN projections and the VMap1 baseline data, KACN did not document their projection, so the projection had to be deduced from the delivered product. There may be some displacement of points as a result. However, the degree of displacement is insignificant compared with the existing accuracy limitations of the source data. The most valuable contribution of material supplied by KACN was the textual descriptions. Brief narratives for each species include distributional information. These narrative distributions can be transformed into polygons and entered into IEDROK.

11.3 Using IEDROK: System Requirements

IEDROK was developed for use with the Environmental Systems Research Institute's (ESRI) ArcView software, and is provided on five CD-ROMs. IEDROK can be used on any personal computer with ArcView 3.1 to 3.3 installed.[5] The Republic of Korea is divided into four geographic areas by USFK. Because of the large volume of data, IEDROK files based on those areas are contained on five separate CD-ROMs. (Areas I and II were combined on a single CD, while the larger Areas III and IV each had to be divided in two.) Provided on each CD is a simple set of base map layers for the entire Republic of Korea and a list of all of the reference documents.

Microsoft's Internet Explorer and a means of accessing the Internet are necessary to use the Korean Cultural Properties Administration's Web site. Documents provided on the CDs require Adobe Reader® for viewing and printing. The minimum system requirements are listed below:

Software:

ESRI ArcView 3.1

Microsoft Internet Explorer 5

Adobe Reader 6.0

Computer:

Personal computer with at least a Pentium or higher Intel-based or compatible microprocessor, a hard disk, and a CD-ROM drive

Operating System:

Microsoft Windows 95, 98, NT 4.0, 2000, or XP

Memory:

32 MB RAM (64 MB or more is recommended)

Display:

256 colors at 800 × 600 resolution (higher resolution and color depth, and at least 32 MB of display RAM are strongly recommended)

11.4 Finding a Location of Interest

IEDROK provides a wealth of base map, natural, and cultural information that can be examined relative to the location of an existing installation or planned activity. There are several ways to identify that location in IEDROK. The simplest is to use the zoom and pan controls to locate a site of interest based on recognizable features on the maps.

A second method is available if the latitude and longitude coordinate of a location is known. The coordinate should be in the World Geodetic System, 1984 (WGS84), datum to match the projection used in IEDROK. The location-finder window was added to the interface for this purpose. Figure 11.3 shows the location-finder window with coordinate entered. After the user enters the coordinate, the map is then panned to the entered position and a red dot is placed at the entered location. If the coordinate entered falls in an area beyond the extent of the current map, the system will indicate which CD to use for that location, or whether the point falls beyond the limits of IEDROK data layers.

FIGURE 11.3
Location-finder window with coordinates entered.

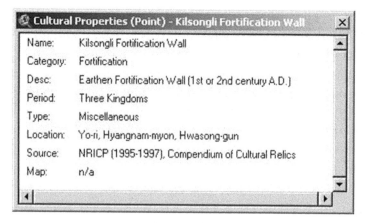

FIGURE 11.4
Information retrieved by the custom-identify tool for a selected cultural resource.

A third method depends upon the availability of a GIS layer for the location of interest. If the available layer is in a data format supported by ArcView and in a geographic coordinate system using the WGS84 datum, it can be easily used by adding it to the map and zooming to the extent of the layer.

11.4.1 Looking Up Information around a Location

Another key activity is to examine and query the IEDROK to answer questions about the resources depicted on the map. The first and sometimes most effective means is to use either ArcView's standard "identify" tool or the IEDROK "custom identify" tool. Figure 11.4 shows the window displayed by the custom-identify tool for a selected cultural resource.

A second method is to select a group of features around a location and then examine the selected records in the associated GIS database table. The interface contains a variety of tools for selecting groups of features. For example, a circle-selection tool allows features to be selected within a particular distance from a location of interest. This approach is useful for identifying resources that may be affected by actions at a particular location (Figure 11.5).

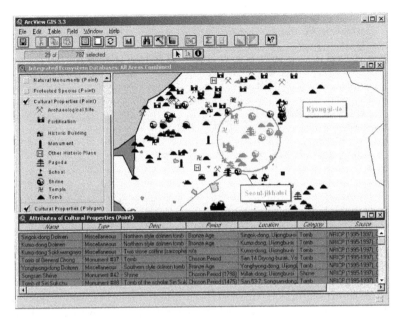

FIGURE 11.5
Map showing selection of cultural properties within a 10-km circle north of the Seoul area, with table showing selected records.

11.4.2 Examining the Distribution of a Set of Similar Features

In evaluating the consequences of a planned action on an identified cultural resource, it is often necessary to understand how common the resource is and the extent of its distribution. For example, suppose an action were planned near a Bronze Age tumulus. The resource manager might want to know if tombs of a similar type are common in the area and might be affected if the venue of the action were changed. An approach opposite to that described in the previous section can be used to focus on these questions. This involves selecting resources with characteristics of interest from a database table, and then examining the spatial distribution of the selected features on the map. Figure 11.6 shows a query operation searching the cultural-properties point layer for Bronze Age tombs, which would yield 30 matches in the USFK Areas I and II.

11.5 Further GIS Analysis

The previous examples describe simple yet powerful operations that the IEDROK GIS can perform. For an experienced ArcView user, many other

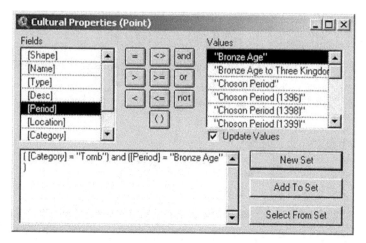

FIGURE 11.6
Query-builder window to search for Bronze Age tombs in the cultural-properties point layer.

analysis and modeling activities can be implemented. For example, relationships between layers can be examined, such as which physical features are associated with a particular resource and how common or rare those features are. A predictive model might be formulated if strong associations between GIS layers are seen, such as cultural resources of a particular type or period in association with such factors as hydrology, soil type, and slope. These are the very factors traditionally employed by Korean shamans, or geomancers, in recommending burial and construction sites. Resources can be visualized on the map according to their database characteristics, such as grouping cultural resources by their age or grouping natural resources according to their protection status. Distributions of cultural resources and examination of their characteristics could lead to a better understanding of the spatial and temporal ranges of the cultures that produced them.

11.6 Conclusions and Future Directions

IEDROK was constructed to provide USFK managers tasked with stewardship responsibility over Korea's cultural and natural resources with an initial screening tool. This tool allows them to determine the reported presence of protected resources in the vicinity of a proposed action and to predict the likelihood of such resources in areas where none has been previously reported. Disparate data sets from a variety of sources were brought together and combined. The process highlighted the wealth of data that exist concerning these resources, but it also revealed the challenges in making the data available in a format that resource managers could readily use. The effort exposed the difficulties of using data from different sources and at different scales. Even the two NIMA data sets did not prove to be perfectly congruent.

IEDROK includes field observations only for those USFK facilities visited by Argonne scientists. As additional field data become available for other USFK facilities, IEDROK provides a framework into which these data can easily be added. The exact locations of military facilities must be added to IEDROK by each user, or IEDROK layers can be incorporated into an installation's existing GIS.

In conclusion, IEDROK has brought together for the first time Korean natural- and cultural resource data from a wide variety of sources. The quality of the data sets themselves, while not compromising the utility of the IEDROK as an initial screening tool, dictate its limitations. IEDROK should not be seen as a final product, but as an initial baseline data set upon which newer and more complete data layers can be added as they become available.

Notes

1. Since this paper was presented, NIMA has been renamed and is now the National Geospatial-Intelligence Agency (NGA).
2. ArcView is a commercial GIS software application, produced by the Environmental Systems Research Institute (ESRI), that was used in the development of IEDROK and as the framework for the IEDROK system interface. IEDROK is compatible with ArcView versions 3.1 to 3.3.
3. Shapefiles are a nonproprietary GIS data format introduced by ESRI that is widely used for GIS data storage and exchange.
4. Arc/Info is a commercial GIS software application produced by ESRI that was used in the development of IEDROK. A coverage is a GIS layer in a proprietary Arc/Info data format. This format was used only for data development, and completed data are included in IEDROK as shapefiles.
5. ESRI now produces another product, ArcView 8.x, that has the same name and general capabilities as ArcView 3.x, but has different software architecture. The IEDROK GIS database is fully compatible with this software, but the IEDROK interface tools are not. ArcView 3.x continues to be sold and supported by ESRI and has a large user base.

References

Choi, C-j., Korea's indigenous geomancy, in *Korean Cultural Heritage*, Vol. 4, *Traditional Lifestyles*, Korea Foundation, Seoul, 1997, pp. 72–77.

Choi, K-C., Jeon, S-R., Kim, I.-S., and Son, Y-M., *Coloured Illustrations of the Freshwater Fishes of Korea*, Hyang Moon Press, Seoul, 1990.

DoD, Tri-Service Legacy Project Help, U.S. Department of Defense; available on-line at http://Memphis.lmi.org/int/dienemann/legwelcome.nsf/about?readfo, accessed 2 February 2001.

Feuchtwang, S.D.R., *An Anthropological Analysis of Chinese Geomancy,* Vithgana, Vientiane, Laos, 1974.

KACN, *Rare and Endangered Wildlife in Gang-weon Do, Korea,* prepared for Gang-weon Do Provincial Government, Chunchon, Korean Association for Conservation of Nature, Seoul, 1980.

KACN, Current Condition of Korea's Rare and Endangered Fauna and Flora, Bulletin 10, Korean Association for Conservation of Nature, Seoul, 1990.

KACN, Endangered Wild Species of Korea, prepared for USFK, Seoul, Korean Association for Conservation of Nature, Seoul, October 2000a.

KACN, Reserved Wild Species in Korea, prepared for USFK, Seoul, Korean Association for Conservation of Nature, Seoul, October 2000b.

KCPA, *Management Atlas of State-Designated Cultural Properties,* Vols. 1–21, Korean Cultural Properties Administration, Taejon, 1997.

NIMA, Performance Specification Vector Product Interim Terrain Data (VITD), MIL-PRF-89040A, National Imagery and Mapping Agency, Reston, VA, 8 May 1996.

NIMA, Military Specification Vector Map (VMap) Level 1, MIL-PRF-89033, National Imagery and Mapping Agency, Reston, VA, 1 June 1995, amended 27 May 1998.

NRICP, *The Compendium of Cultural Relics,* Vols. 1–3, National Research Institute for Cultural Properties, Seoul, 1995–1997.

USFK, Environmental Governing Standards, USFK Pam 200-1, Headquarters, United States Forces, Korea, Seoul, 15 July 1997.

12

Appropriateness and Applicability of GIS and Predictive Models with Regard to Regulatory and Nonregulatory Archaeology

Kira E. Kaufmann

CONTENTS

12.1 Introduction

Regulatory and nonregulatory archaeological venues approach the use of new technology such as geographic information systems (GIS) and predictive modeling in different ways. There are important concerns for cultural resource managers and land planners because archaeological projects that are monitored by federal or state laws involve different factors that need to be taken into consideration to comply with the legislation. The Advisory Council on Historic Preservation has minimal guidelines for the implementation and application of GIS and predictive modeling for archaeological research that is accomplished to comply with Section 106 of the National Historic Preservation Act. This chapter presents current guidelines for the implementation of predictive models and describes how several states, U.S. territories, and tribes are interpreting them. Recommendations are also provided, from a regulatory archaeology perspective, for cultural resource managers and land managers who desire to use this type of analysis in archaeological research.

12.1.1   Nonregulatory Archaeological Venues

Nonregulatory archaeology involves individuals, organizations, or institutions that initiate archaeological investigations that are not a result of a need to fulfill federal, state, or local laws, regulations, permits, licenses, or ordinances. This type of archaeology is usually conducted under the auspices of independent or academic research.

12.1.2   Regulatory Archaeology Venues

Regulatory archaeology involves fulfilling requirements according to federal, state, or local preservation laws. This type of archaeology often entails various aspects of identification, evaluation, and treatment of historic properties as a result of fulfilling one or more federal, state, or local laws, regulations, permits, licenses, or ordinances. In regulatory archaeology, the authoritative agency or group sponsoring a project requires that certain aspects of cultural resource investigations be accounted for.

It is important to stress that different federal and state agencies will have different needs related to archaeology and compliance issues. Specific types of information are required by different agencies responsible for identifying or treating historic properties. It is recommended that an agency sponsoring a project consult with the State Historic Preservation Office (SHPO) concerning the appropriateness of the proposed methodology for a project prior to initiating an alternative method of investigation.

Each federal agency is responsible for ensuring that its programs comply with mandates set forth in the American Antiquities Act of 1906, the Archaeological Resources Protection Act of the 1979, and in Section 106 of the National Historic Preservation Act. Some suggestions are given in this chapter to provide guidance to project sponsors in fulfilling their legal responsibilities. Note that different state or municipal agencies may be affected by supplementary regulations in addition to federal regulations.

12.2 Theoretical Issues

GIS is more than geographic information systems; frequently it is referred to now as geographic information science. It is moving from a technological tool to form its own discipline. There are several culturally and socially embedded aspects of GIS. For example, concepts of space with regard to GIS not only vary between cultures, but sometimes between different users as well. Different users employ different terms to talk about space, and this discrepancy can affect our interpretation of the data. We also maintain different concepts of time. Some users think of GIS data as providing the most up-to-date information. However, GIS and predictive modeling provide the user a picture of just one moment in time. As we know, GIS is also evolving technology. It is moving from the realm of the specialist to many other applications. What the specialist sees as important, the general user may not, and vice versa.

12.3 Technical Issues

12.3.1 GIS

There are several technical issues pertaining to using electronic data and tools that affect the accuracy and reliability of the data for archaeological investigations and interpretation. First, the most visible technical issue is a matter of access compared with no access: Which people have access compared with those who do not have access, and how does this affect the data, results, and interpretation? Although many individuals in the business world, cultural resource management (CRM) firms, urban-planning facilities, and so forth have access to cutting-edge technology, many SHPOs do not.

Second, the type of software that a researcher uses affects many aspects of the data output. Is a raster-based or vector-based software being employed? For example, there are differences between the Environmental Systems Research Institute's (ESRI) ArcInfo, ArcView, and ARCGIS software.

Third, the sample size used for the data in the GIS system affects data output and interpretation. Previously, a larger sample size was considered more important. However, as many GIS users know, sometimes too large of a sample size becomes counterproductive when it locks up the system.

Fourth, the type of datum used affects the GIS information. For example, which maps were used? Which North American datum (NAD) was the map based on? Counties frequently use Department of Transportation (DOT) maps with the NAD 1927, whereas the USGS topographic maps and U.S. Census data use the NAD 1983. The different location of these two datums can skew GIS data exponentially.

Fifth, the data set used can affect the accuracy of the results. For example, primary data are much more reliable then secondary or third-generation data. However, primary data may not be available. Was the data entered by an automated or manual method? Each method contains inherent errors (for example, digitizing errors from manual entry).

Sixth, regarding data entry, what criteria were used for the mode of the data input (e.g., raster or vector — raster should be left as raster)? What is the lineage of the metadata? How many conversions were there from or since the primary data form? What stage or level of transformation is the data used are using: second, third, fourth, etc.? The higher the level of transformation, the greater will be the percentage of error.

Seventh, often overlooked is the issue of who owns the copyright of the data. Will royalties be expected for the use of the data? Which type of data set database was used? Census Tiger files, for example, have their own set of data-entry problems, misspelled streets, incorrect address locations, etc. Was the data set proofread and corrected for these types of errors?

Eighth, geographic concerns such as how we measure space (topographic, ellipsoid, geoid) also affect the way we enter data and interpret the results. How we use scale affects our data. The data were obtained from which scale-sized map? The effects of scale on data are threefold: we cannot use what is not on the data set; we can only query at the level applicable to the map generalizations; and the features on a specified map will change with the scale of the map. Map projections affect the accuracy of the data (cylindrical, secant, or any other). All of these issues affect levels of resolution, the accuracy of the results produced, and the compatibility of the data produced.

12.3.2 Predictive Modeling

Many of the above issues — sample size, type of datum, lineage of the metadata, and geographic concerns — also affect the reliability or strength of a predictive model. The sample size used for the GIS system affects data output and interpretation. The manner in which the model is established (e.g., nearest neighbor, standard values, etc.) will affect data results. The underlying assumptions and the data the model is based on will determine results. Also, specific models used will result in different output. For example, using different models results in different three-dimensional surfaces.

12.4 Linguistic Issues

Many GIS users have realized that the "techies" speak a very different language than they. Developer-oriented discussions concerning GIS and predictive modeling are very different from user-oriented discussions. This not only creates a gap in the effective transmission of information from the developer to the user, but also creates anxiety on the part of the user, to the point that many users may even abandon implementing GIS or predictive modeling.

Translation problems occur when the end-user language is different from that of the technical user or developer. Therefore, there may be an increase in interpretation problems when the developer and user are operating from different schemata. Translation problems also occur across cultures using this technology.

12.5 Policy Issues

The most important consideration for the project planner or sponsor is to determine whether or not GIS and predictive modeling are appropriate for the project. Although most agencies do not have a written policy concerning GIS or predictive modeling, they do have issues with these methods used for projects that are completed to comply with state or federal laws.

Regulatory archaeology has approached this technology of GIS and predictive modeling in two different ways. State Historic Preservation Offices (SHPOs), the regulatory review arm of the federal and state government, are truly interested in developing GIS as a tool for their states. However, they are more interested in GIS as a management tool that will increase their efficiency as an office, rather than using the relational database aspects of GIS to conduct research. Limited funding support has severely hampered their efforts to create, employ, and continue to support GIS applications for archaeology and other cultural resources.

Many states, U.S. territories, and tribes are in the process of developing and using GIS, but few have specific guidelines on the application of this technology to archaeological investigations. The Advisory Council on Historic Preservation has no specific standards or guidelines regarding the use of GIS. The National Park Service (1983) has some general recommendations regarding GIS included with the "Secretary of the Interior's Guidelines" for conducting archaeological investigations.

12.6 A Specific Example: Recommendations from the State of Iowa

The State Historic Preservation Office of Iowa developed recommendations for using GIS after many months of research, review, and consultation in the process of updating their guidelines for conducting archaeological investigations in Iowa. The Iowa guidelines (1999) recommend that when conducting archival or field research, the researcher should attempt to access all available sources via any new technology or methodology that may be appropriate. New technology sources such as e-mail, GIS, or the Internet should be credited and cited accordingly. When conducting field research or analysis, the use of new technology or methodology should be verified as consistent, reliable, and accurate. Some examples of new technology that may be applicable to field research are GPS units, remote-sensing equipment, or new survey technology. When conducting analysis, an attempt should be made to create data sets that are compatible between the sponsor of the project and the consultant. It may be necessary to use compatible computer programs to accomplish this. New technology or methodology should also be consistent and accurate when applied to analysis and curatorship.

Some considerations regarding GIS again concern the accuracy of the results produced. Individuals submitting maps or other data produced by using GIS should be aware that the information produced is only as accurate as the information that had been entered into the program. This is particularly important for maps that have been digitized or electronically scanned in. The Iowa State GIS laboratory addressed some specific points pertaining to GIS data entry. When digitizing data, control points should be carefully recorded and monitored. When submitting hand-drawn maps that will be digitized, rulers and thin lines should be used, as it is difficult to digitize thick and crooked lines.

Care should also be taken when choosing a GIS program or software. There are specific differences among the variety of GIS programs. There can be incompatibility problems between results from two different programs if the data are entered differently. For example, if certain data are entered as line segments in one program and a polygon in the other program, there may be discrepancies in the results. Further training in GIS technology is recommended prior to using GIS technology.

12.7 Predictive Modeling

A predictive model is a set of cumulative data that have been obtained from a thorough overview of background information and previous research pertinent to a specific area, landform, watershed, etc. The end product of a predictive model is a series of explicit and testable statements concerning the location and general characteristics (size, depth, age, cultural affiliation, integrity, etc.) of specific historic properties. The Advisory Council on Historic Preservation (1983) defines predictive modeling as "the generation of models of the likely nature and distribution of historic properties in an area that has not been subjected to intensive, complete survey." The Advisory Council on Historic Preservation (1983) recommends ten important overview data components to consider when developing predictive models:

1. Propositions derived from history, ethnography, ethnohistory, anthropological, sociological, and geographical theory, and other disciplines about the kinds of settlement patterns, subsistence practices, and social organizations that might have characterized the area under investigation, or the kinds of social groups that occupied it in prehistoric or more recent times

2. Background historical and ethnographic data on the actual locations of places, structures, and areas for different purposes (e.g., residence, industry, religion, and transportation) in the past

3. The known or postulated distribution of historic properties of different kinds in the region within which the study area lies, or within similar areas

4. Data on changes in landform, vegetation, and other environmental characteristics that may serve to obscure or reveal historic properties

5. Research questions of concern in the area, and/or in the general disciplines involved, which may serve as bases for evaluating archaeological significance

6. The cultural concerns and practices of local communities, American Indian groups, and others who may ascribe significance to historic properties, as a basis for evaluating associative significance

7. The architectural or artistic traditions of the region, as a basis for evaluating architectural and art-historical significance

8. Historical events and people of the region, as a basis for evaluating historical significance

9. The physical characteristics of different kinds of historic properties in the area, to serve as a basis for generating expectations about what to look for in the field

10. Where available, data on historic property types and distribution in the area based on sample survey

The resulting predictive models should be capable of projecting the distribution of different types of historic properties. The predictive model should contain a high level of resolution that infers that substantial data contributed to the formation and applicability of the model. The more information incorporated into a predictive model, the higher the predictive model resolution will be and the more useful the model may be. If very little data are available or incorporated into the predictive model, the resolution or reliability of the model will be low. If the model resolution is low, a few areas of historic properties may still be identified, but most areas will be identified as having unknown potential, and the use of the model may be very restricted.

It is important that testing substantiate predictive models. Predictive models should be tested in the real world by techniques and methods that can be duplicated by others. Testing creates confidence in the applicability and accuracy of the predictive model. A testing program should be designed that obtains a representative sample of the predictive model area and involves on-site inspection. Appropriate statistical techniques should be applied to the predictive model and authenticated in a manner that does not bias the model. It is recommended that those predictive models with a faulty information base or sample biases should not be used for projects. The resulting predictive model should be accompanied with detailed text explaining why predicted areas contain or do not contain historic properties of a certain type.

Sampling strategies were specifically addressed in the *Guidelines for Archaeological Investigations in Iowa* (1999) because certain consultants were attempting to use inappropriate sampling strategies to develop predictive models (e.g., windshield surveys to identify prehistoric Native American sites, some of which may have had the potential to be buried). Sampling strategies can be defined as different methodologies employed to obtain a representation of historic properties. Sampling strategies typically investigate a portion of the area that is the subject of investigation. Sampling strategies may constitute a separate investigation or may be supportive of a predictive model. Sampling strategies attempt to extrapolate or generalize about the distribution and nature of historic properties in an area. These strategies are often employed to acquire information about cultural resources within a broad area with minimal effort.

Generating a predictive model is an attempt to predict what kinds of historic properties will be found in a previously unsurveyed area. The resulting model may predict the likely locations where historic properties may be found and may permit an attempt to predict what kinds of significance such properties may have. A predictive model that conforms to one part of the state for certain resources may not be applicable to other dissimilar areas or cultural resources. It is important to make sure that the predictive model has been tested in the real world to ensure that it accurately predicts the locations and general characteristics of the cultural resources being modeled. This testing should encompass a process of continual adjustment to verify the continued validity of the predictive model. After a predictive model has been developed, the model should be revisited and reassessed for accuracy and appropriateness.

Predictive models and sampling strategies should be developed and employed after consultation with the SHPO for their appropriateness and applicability for a particular project. Most SHPOs work with predictive models on a case-by-case basis or are beginning to "use modeling as a tool to specify survey areas for 106 compliance" (respondent from 2003 survey).

Predictive models are most applicable to surveys for planning purposes or general land-management considerations. Sampling strategies are helpful when investigating extremely large areas of land. Sampling strategies are also very useful for testing predictive models. The greatest concern among SHPOs is that predictive models will be inappropriately applied to avoid proper investigation and evaluation of cultural resources.

The Advisory Council's position with regard to predictive modeling states that "the use of predictive modeling in projects carried out pursuant to Section 106 of the National Historic Preservation Act was addressed by the Council in 1983." This position is detailed as follows: "The usefulness of a model in Section 106 review is directly linked to its quality, its comprehensiveness, and its reliability." The Advisory Council cautions against the use of coarse-grained models, models that have not been tested, or models that fail to take into account certain classes of historic properties. The Advisory Council has stated in recent seminars that a predictive model should not be used by an agency to make a determination of "no historic properties affected" (Advisory Council on Historic Preservation 1999). It is strongly recommended to consult with the SHPO and receive written SHPO concurrence about predictive modeling strategies prior to the implementation of any predictive modeling strategy for any regulatory project.

12.8 Results of Survey

In 2001, the author sent a brief letter to the states, U.S. territories, and tribes within the U.S. asking what, if any, guidelines and recommendations they were using for GIS, predictive modeling, and electrical remote sensing as investigative tools when conducting archaeological research. In August 2003, a follow-up e-mail letter was successfully sent to 55 states, U.S. territories, and tribes requesting additional information regarding their archaeological guidelines that addressed GIS and predictive modeling use with archaeological projects. E-mail addresses were obtained from the National Council of Historic Preservation Officer's list of SHPOs. Four territories or tribes did not have e-mail addresses, and five e-mails were returned unopened. In addition, a Web-site search was attempted for those states that did not respond to determine if they offered Web-site access for their archaeological GIS database if they maintained one.

Nineteen states, U.S. territories, and tribes (hereinafter referred to as "states") responded to the first survey in 2001, and 14 states responded to the follow-up survey in 2003. In 2001, two states had guidelines for using GIS, and one state had draft guidelines for using GIS as an investigative tool when conducting archaeological research. Sixteen states did not have any guidelines or recommendations for this type of methodology. In 2001, eight states had GIS coordinators or were using GIS to manage their cultural resources. Eleven states did not have GIS coordinators or use GIS to account and manage the cultural resources within their states.

In 2003, three states had guidelines that address GIS use with archaeological investigations, and no states had draft guidelines. As of 2003, ten states did not have standard guidelines but made recommendations on using GIS, usually on a case-by-case basis. Twelve states had no guidelines regarding GIS use in archaeological investigations.

In 2003, eight states had full-time GIS coordinators, three states had part-time GIS coordinators, and five states were using GIS but hired outside contractors to develop or manage the GIS data (Figure 12.1). Currently, nine states do not have GIS coordinators or use GIS to account for cultural resources at the state level. Several states do use GIS at universities and research institutions to assist with cataloging and documenting cultural resources within the state. Eight states have Web-site access to their GIS database for archaeological resources and 18 do not (Figure 12.2).

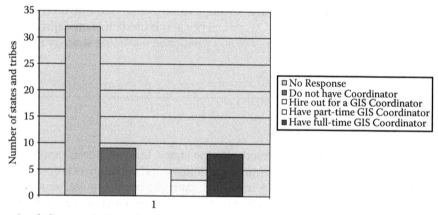

Level of states and tribes with GIS coordinators for archaeology

FIGURE 12.1
States and tribes with GIS coordinates.

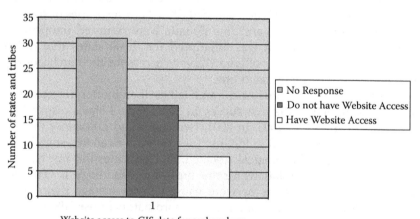

Website access to GIS data for archaeology

FIGURE 12.2
States and tribes with website access to GIS data for archaeology.

Most states are still working to develop GIS as a managerial tool but have not addressed standards with regard to the use of GIS or predictive modeling. However, many states polled shared the sentiments expressed by one respondent: "As far as the importance of GIS, I can't stress enough the advantages that it will have once completed, especially for predictive modeling. Our agency implements the Section 106 process for archaeological and architectural historic sites. Once the system is operational, not only will it assist our staff in determining the effects of requested projects to our cultural resources, but it will also give consultants the ability to avoid areas of potential impact in the planning stages of their projects. If cultural resources can be avoided in the planning phase of projects such as highways, sewer and water projects, cell towers, etc., then that would save the company time and money. It also saves our state's cultural resources."

Another respondent commented that "there is very little use of the GIS by ... state or federal agency staff to conduct research ... but at this time the 'agency' GIS and mapping capabilities are basically a very expensive toy used to make checks during reviews a little more efficient and to add pizzazz to written recommendations. It is not surprising that there is limited use for research given that research takes a long time to develop and our GIS is very new."

Predictive modeling is much less represented than GIS as a method for investigating archaeological resources. In 2001, two states, Iowa and Minnesota, had specific guidelines or recommendations for using and applying predictive modeling to archaeological resources. In 2003, three states (Iowa, Minnesota, and Mississippi) have specific guidelines for the use and application of GIS and predictive modeling. Five states do not have specific guidelines but make recommendations on a case-by-case basis. Sixteen states that responded do not have any specific recommendations pertaining to predictive modeling and cultural resources.

Although the Advisory Council on Historic Preservation has some recommendations pertaining to predictive modeling, these are 19 years old now and need to be readdressed as well as formalized. The National Park Service (1983) has no specific detailed guidelines for these methods as of yet. Although most states do not have guidelines concerning predictive modeling, respondents either did not comment on the use of predictive models or remarked that predictive modeling has not been encountered frequently as an approach to investigating or managing archaeological resources. As one respondent reported, "Predictive modeling's head has not yet been raised and will again be looked at on an individual basis, if the time comes." One state interjected that they do not have guidelines for predictive models but they "only ask that users not disclose site locations."

Another perspective on predictive modeling was aptly expressed by a respondent who observed "that most agency folks and a lot of archaeologists tend to see predictive modeling as a map for treasure hunting, 'Where can I go to dig a hole that is guaranteed to get me lots of neat artifacts' kind of thinking. I view predictive modeling as providing interpretations of spatial distributions that inform decision making on what level of survey intensity is needed. I have pretty much given up on trying to get agency archaeologists to develop informed uses of predictive modeling."

12.9 Conclusion

There is obviously a lack of standards or guidelines with regard to the use of GIS and predictive modeling in archaeological research from the federal and state level. Is there a need? I would argue that there is a need for some standards not only to maintain consistency, but also to continue the "use-life," so to speak, of the data produced. Also, there is a need for some standardization to maintain a professional approach to the methodology of GIS and predictive modeling.

I would encourage organizations or groups that conduct GIS and predictive modeling research to develop guidelines or standards with regard to using these approaches in archaeological research to create consistency in investigating, recording, and documenting historic properties. However, standards should not be developed to regiment or oversee the research conducted by academic archaeologists or other professional scholars.

These standards should specify the recommended amount of technical information necessary to carry out historic-properties investigations that will contribute to survey and inventory data and will lead to more informed decision making. The Society for American Archaeology (SAA) provides seminars on the use of GIS. However, I am unaware of a task force, such as there is for public archaeology, for GIS or predictive modeling. I would suggest that this is the next step in developing these methods as viable alternatives in conducting archaeological research.

12.10 Addendum

In conjunction with this research, an additional survey of how states, U.S. territories, and tribes are using geophysical remote sensing (GRS) for archaeological research was conducted. States were queried as to what guidelines exist or what recommendations they provide for these types of archaeological applications. In 2003, three states had guidelines that address the archaeological use of geophysical remote sensing, and one state had draft guidelines (Figure 12.3). Ten states did not have guidelines but made recommendations on using geophysical remote sensing, and 14 states had no guidelines regarding archaeological applications of geophysical remote sensing.

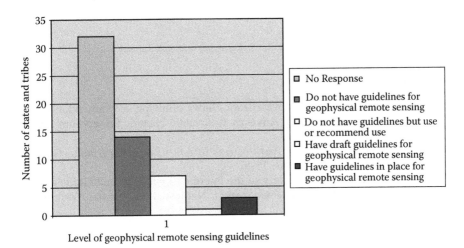

Level of geophysical remote sensing guidelines

FIGURE 12.3
States and tribes with guidelines for using geophysical remote sensing with archaeology.

One respondent aptly stated, "We try to encourage the use of these tools, but for Section 106 reviews this is very difficult. These strategies are often very expensive. Often the applicant or the agency will go with the low bid and we will get crummy archaeology. But for Section 106 reviews, even if it is crummy, it might be enough. We cannot demand the most expensive archaeology. We risk losing our credibility and integrity if we ask for too much. Mostly in Section 106 projects the archaeological site will be destroyed, so you are going to get to excavations for important sites regardless of whether you get there by adding a geophysical testing step or not. Geophysical testing is increasingly being recommended by our office and increasingly being used in situations where the efforts are directed towards preservation in place. For these situations, we generally try to find the right tool for the situation."

Again, this survey demonstrates the dilemma that archaeological applications of technology present for cultural resource managers that might aspire to use technology, such as GIS, GRS, and predictive modeling, in archaeological investigations. The dilemma is that, too frequently, these types of technological applications are avoided or not employed to assist with state or federally monitored archaeological projects. Partially, this is a result of a lack of funds available to cultural resource managers. Also, this dilemma is a result of preconceived notions about the validity and appropriateness of using these technologies to identify, document, and assess archaeological resources. However, the frustration and avoidance of using GIS, GRS, or predictive modeling for nonacademic-oriented archaeological investigations also results from minimal to no standardized guidance in applying these strategies successfully in archaeological applications. Most importantly, developing standardized guidelines for the archaeological utilization of these technologies will greatly increase their acceptability, application, and productivity in the future.

References

Advisory Council on Historic Preservation, letter outlining recommendations for using predictive modeling, Advisory Council on Historic Preservation, Washington, DC, 1983.

Kaufmann, K., Ed., *Guidelines for Archaeological Investigations in Iowa*, State Historical Society of Iowa, Des Moines, 1999.

National Park Service, The Secretary of the Interior's Standards and Guidelines for Archaeology and Historic Preservation, National Park Service, Washington, DC, 1983.

13

Archaeological GIS in Environmental Impact Assessment and Planning

Linda S. Naunapper

CONTENTS

ABSTRACT Preliminary assessment of the impacts of development on natural and cultural resources in federal projects is often relegated to the domain of engineering and environmental contractors in what is known as an environmental impact assessment (EIA). As such, budget and time constraints often undermine the depth to which background investigations can occur. In the case of cultural resources, environmental scientists are generally not aware of issues specific to historic and prehistoric resources. In this case study, an examination of the impact that a state-developed, archaeological geographic information system (GIS) (predictive model) had upon the accuracy of assessment in a federal project is explored. In sum, through consultation with the site file administrator during the assessment phase, the GIS served as a valuable tool in increasing the efficiency of the EIA process and made archaeological field survey possible at a number of project areas. Further, access to site file information in the earliest phase of the process substantially reduced the number of areas considered for development by the client, thereby reducing the amount of candidates submitted to the State Historic Preservation Office (SHPO) for review as part of the National Historic Preservation Act (NHPA) Section 106 process.

13.1 Introduction

This chapter is the written and expanded version of a poster session presented at the "GIS in Archaeology Conference" held at Argonne National Laboratory in 2001 and the "Society for American Archaeology Conference" at Denver, CO, in 2002. The original intention of developing the poster session was to present information gathered from the author's personal experience as an environmental scientist/assessor involved in a federally regulated project that was concerned with protection of natural and cultural resources. Given my professional experience in archaeology and cultural resource management (CRM) as well, the environmental firm looked to me as a specialist who they believed could provide additional skills and knowledge in this area to benefit the overall project. It was within the experiences of this project that the extent of the gap between management of natural versus cultural resources in the environmental industry became clear.

While my experiences were limited to a number of states located in the midwestern U.S., my discussions with colleagues during the course of these poster sessions suggested that similar circumstances have occurred throughout the country. A number of concerned archaeologists working within the compliance industry agreed with the observations made in my poster and questioned why such a difference exists between the use of GIS in management of cultural versus natural resources. These poster sessions afforded a unique opportunity to discuss these issues with colleagues in an informal manner, and the experience demonstrated that, as archaeologists, we need to discuss these issues more openly in the future.

As discussed with colleagues, what was most striking to me as an environmental assessor was the differential in availability of, and access to, data regarding natural versus cultural resources. Natural resources data are managed and maintained by numerous agencies (such as U.S. Fish and Wildlife Service, Department of Natural Resources, Federal Emergency Management Agency) and were readily accessible through consultation, most often received well within the strict time constraints of the project. Cultural resources data, on the other hand, as managed by State Historic Preservation Offices, were not as readily available, making our ability as assessors to meet project and budget deadlines difficult at best. Further, this threatened to compromise our ability to make complete and accurate assessments of potential impacts of development upon cultural resources.

What is the reason for this differential in access to cultural resources data? In all fairness, it is due to a variety of factors, including lack of funding and staff at the agency level, but also it is due to the hesitancy on the part of cultural resource managers to use GIS and predictive modeling as management tools. The negative effect of this has been the relative difference in the resulting data management strategies.

As will be discussed in the following case study, agencies managing natural resources were making use of GIS technology to maintain and update their data, thereby making it readily accessible and aiding in the assessment process. At the time of the case study, only one agency managing cultural resources data was making use of GIS technology and predictive modeling to manage their data. The result of that sole implementation was far more effective assessment of resources on the part of assessors and performance of field survey in that state. While much progress has been made by SHPOs managing cultural resources data since my project work in 1996–1997, much still needs to be accomplished in regard to providing access to data by professionals outside of CRM. This is where GIS and predictive modeling provide the greatest amount of potential.

13.2 The Environmental Impact Assessment (EIA) Process and Role of the Screening Phase

13.2.1 The National Environmental Policy Act (NEPA) vs. the National Historic Preservation Act (NHPA)

As described by Neumann and Sanford (2001: 44), the NEPA and NHPA are two separate and distinct pieces of federal legislation, although there is some confusion that the NHPA and Section 106 process are triggered into effect by NEPA. Under NEPA, all actions of a federal agency considered as an "undertaking" must assess the potential effects of that undertaking on all natural and cultural resources. This process of consideration involves research and consultation in what is called the "environmental impact assessment" process or the "NEPA" process (Canter 1996: 2) (Figure 13.1).

While the NEPA or EIA process involves consideration of impacts upon cultural resources, the scope of investigation is limited in comparison with review in the NHPA and Section 106 process. NEPA requires identification of known cultural resources, while Section 106 requires identification of potential cultural resources and eligibility for nomination to the National Register of Historic Places, a domain of research restricted to professional archaeologists, architectural historians, and historians. Although recent Section 106 revisions allow SHPOs to give some environmental scientists a greater amount of involvement in cultural resources assessment during the EIA process, this is not meant to undermine the necessity for the more

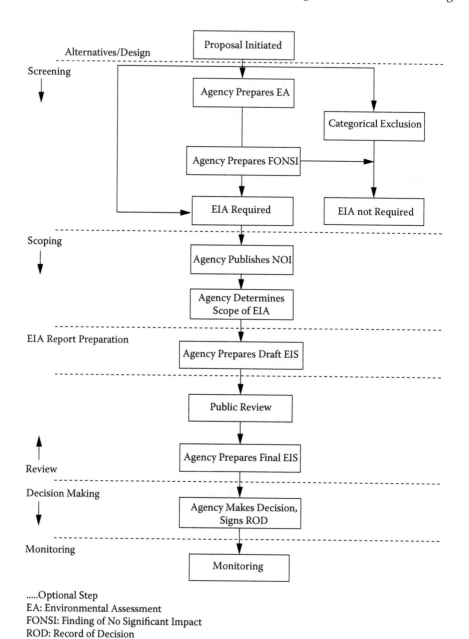

FIGURE 13.1
(From Wood, C., *Environmental Impact Assessment: A Comparative Review*, John Wiley and Sons, New York, 1995. With permission.)

rigorous review provided under Section 106 (Neumann and Sanford 2001: 46).

As mentioned previously, under NEPA, the EIA process is a very comprehensive, broad method of investigation and assessment involving evaluation of a wide variety of environmental parameters. The EIA process can be considered as a management strategy with the intent to identify, predict, interpret, and communicate impacts upon all aspects of the human environment, including air quality, surface water, soil and groundwater, noise, biological environment/habitat, cultural environment (including archaeological, architectural, and historical), visual impacts, and socioeconomic environment (Hussain 1996; Canter 1996: x–xiv). All aspects of the human environment must be given equal weight and consideration in the process; therefore, consultation and data availability are important factors involved in the proper assessment of resources and concerns.

To help in managing the bulk of information and data gathered through the process most effectively, the EIA consists of phases of investigation: screening, scoping, report preparation, review, decision making, and monitoring (Wood 1995: 22) (see Figure 13.1). These phases proceed in succession from most general to most case-specific, the screening phase being a good-faith effort on the part of assessors to compile the most accurate and complete picture of the potential effects of development upon all affected natural and cultural resources. As such, the ability of assessors to accomplish this goal is a function of their access to, and availability of, data regarding these resources.

Various resource managers maintain their data better than others. Management of data ranges from hard-copy versions of maps and resource lists to highly developed database systems and GIS. Some of the more common tools used by assessors during the screening phase include checklists and databases, matrices, network analyses, maps and map overlays, GIS, and computer modeling/photomontage (Figure 13.2).

While GIS and predictive models are used by many natural resource managers and planners as data-management tools, their use by archaeologists in the compliance industry is only recently becoming widespread. Funding and staffing issues at SHPOs have been suggested in the past as contributing to this situation. However, other factors have also contributed to the reluctance of archaeologists to develop and use GIS and predictive models. Most arguments against the use of GIS and predictive models revolve around factors that make cultural resources and the archaeological record unique. The idea is that cultural resources cannot be modeled based on static, two-dimensional data alone because factors such as vertical depth and time are not accounted for. Another argument is that the data universe used for developing a GIS is a biased sample, constrained by surveys generating data within a compliance environment.

Although these arguments are valid, they are not unique to the development of GIS and predictive models in other disciplines. A GIS as a "system" is, by definition, an interconnected series of variables and their relationships;

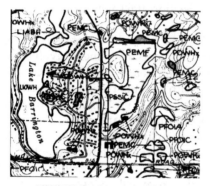

USDI Wetland Inventory Map

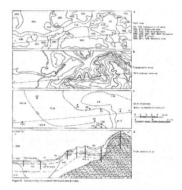

ISGS Map Overlay

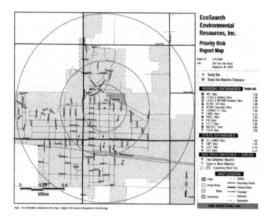

Environmental Risk Assessment Report

Fig. 17. Photomontages of the Avonmouth Clinical Waste Incinerator Unit site (a) before construction of the Unit; (b) after its construction (courtesy of Oldfield King Planning)

Photomontage/Visualization

FIGURE 13.2
(From Berg, R.C. et al., Potential for Contamination of Shallow Aquifers in Illinois, Circular 532, ISGS, Urbana, IL, 1984 and Singleton, R. et al., *Environmental Assessment*, Thomas Telford Publishing, London, 1997. With permission.)

therefore, initial assumptions must be made regarding the choice of variables. Further, a GIS is a "model" or a representation of some aspect of reality and is *not* a direct reflection of reality. No GIS is a direct reflection of reality because it is limited by the variables chosen for the model. However, a unique aspect of a GIS is that it is flexible enough that variables can be added or deleted from the system as necessary. A GIS is a "tool" that helps us to better understand and manage the part of reality that we intend to model. This is the reason that GIS and predictive models are of the greatest importance to the effectiveness of assessment in the EIA process.

13.2.2 The Role of Archaeological GIS (Predictive Models) in the Screening Phase of EIA: A Case Study

In 1996, ATC/ATEC Environmental Consultants (1997) entered into a contract to provide environmental services for a commercial personal-communications provider under guidelines of the Federal Communications Commission (FCC). Environmental scientists conducted all phases of environmental impact assessment (EIA), and the screening phase was performed according to the special-interest items outlined in 47 CFR, Section 1.1307, of the FCC regulations for risk-assessment of impacts upon natural and cultural resources (Table 13.1).

As mentioned previously, NEPA assessment differs from NHPA and Section 106 review in that it only requires assessment of impacts upon previously known and recorded resources (Neumann and Sanford 2001: 45; Scovill et al. 1977). Thus, as noted in Table 13.1, the various resources were investigated within a proposed project area using databases, checklists, and maps made available by resource managers. In the case of most State Department of Natural Resources and U.S. Fish and Wildlife Service offices managing natural resources, this involved consultation with a GIS or database manager. In terms of known historic sites, a state listing of National and State Register Sites was made available for review from SHPOs.

Most interesting to this case study, however, is the assessment process and capability regarding potential impact upon archaeological resources. Keeping in mind that NEPA compliance is concerned with assessment of known resources, only one state with which our project was concerned managed its archaeological data in a GIS and predictive model. In effect, this resulted in the difference between the process of consultation lasting a few days (correspondence with a database manager) to consultation lasting for months at a time (staff manually reviewing hard-copy files). Because of this, assessors

TABLE 13.1

Data Sources for Screening Phase of EIA

Resource	Data Source
Wilderness areas	USFWS and State DNR Database
Wildlife preserve	USFWS and State DNR Database
Threatened/endangered species	USFWS and State DNR Database
Designated critical habitats	USFWS and State DNR Database
Historic places	National Register listing by state
Archaeological	Consultation with SHPO (Sect. 106)
in Illinois	Consultation with GIS site file manager
Flood plain (50–100 years)	FEMA FIRM Maps
Wetlands	USDI Wetland Inventory Maps
Indian religious sites	Tribal consultation (case by case)
High-intensity lighting	Not applicable
Radio-frequency radiation	Not applicable

were more capable of providing timely assessments for effects upon archaeological resources in the one state, versus those that did not maintain their data in a GIS.

The fact that assessors had access to archaeological data early in the screening phase allowed the EIA process to continue to the next phase, environmental assessment (EA), as was the case with assessment of other resources (such as wetland areas). USDI (1984) wetland inventory maps delineate locations of known, protected wetland areas: the archaeological predictive model maps delineate areas of high, medium, and low probability for location of archaeological sites. These maps provide assessors with the data necessary to argue for the need for field survey and the performance of an EA. In this case study, the predictive model map provided the necessary information to subcontract with professional consulting archaeologists to conduct a field survey. The consultant's report was then sent to the SHPO office for review and was eventually included within the EA report submitted to the FCC.

The steps involved in the screening and EA phases of an archaeological resources review include:

Step 1: Determine project site location(s)

Step 2: Consult with archaeological GIS database manager regarding project site locations

Step 3: Determination if project site is

Near a known site

In or near a probability area (high, medium, low)

Step 4: Determination to proceed with development

Proceed — EA required

Not proceed — no further action required

Step 5: Subcontract with professional contractor (archaeologist/architectural historian) for Phase 1 field survey as part of the EA investigation

Step 6: Submit contractor's report of findings to SHPO for Section 106 review

Step 7: Incorporate contractor's report and SHPO review into final EA report to be submitted to FCC

In sum, access to predictive model maps at an early stage in the EIA process places assessors involved in NEPA review in an important position. Data regarding cultural resources at this stage of review can substantially reduce the number of project areas considered for development by a client, thereby reducing the bulk of proposed candidates sent to the SHPO office for NHPA Section 106 review. In other words, if a project area is located in proximity to a known site or within a high-, medium-, or low-probability area of a GIS predictive model, it is more likely that the project area will not be considered for development by the client.

13.3 Conclusions

Following are some important conclusions to consider regarding the previous case study and the use of GIS and predictive models in the EIA process:

1. This case study was performed in 1996–1997, and Federal Communications Commission (FCC) regulations regarding NEPA and NHPA Section 106 review have been revised since that time. Nevertheless, the access to archaeological resources data during the screening phase (through consultation with the archaeological GIS predictive model site file manager) proved to be the determining factor in regard to enabling the EIA process. The predictive model data were the only information available to assessors that could be used to argue in favor of conducting archaeological field survey and performing an EA.

2. Archaeological GIS predictive models are an important and effective tool in the management of cultural resources data (as is true with most natural resources), providing the potential to make data more readily available for use by assessors in the EIA process.

3. A certain amount of hesitancy continues to exist regarding the use of state-based archaeological GIS and predictive models in compliance. Reasons for this vary, but include claims that models are biased in sampling strategy, modeling parameters are inadequate, etc. These reasons are not unique to archaeological predictive models, however, and the most important aspects of a GIS are its flexibility and its potential for revision. This is why research regarding system design, revision, and maintenance of archaeological predictive models is so important in ensuring their future use and stability.

4. Concerns have also been raised regarding the confidentiality of data contained within predictive models. Should data only be accessible by those persons who meet the Secretary of the Interior's standards as a professional archaeologist? Should the data only be available to those professionals working for SHPOs? In the author's case as an environmental scientist/assessor, the confidentiality of data was assured by the professional agreement created between the assessor and the GIS site file administrator.

This case study has been presented with the hope of shedding light on a discrepancy that was encountered during the course of a federal project regarding resource data management and the environmental review process. Most natural resource data are currently readily available for review by environmental scientists/assessors, making the assessment of impacts on these resources a relatively smooth process. The availability of cultural resource information is a different matter entirely, especially when it comes to the screening phase of the environmental review process, the most crucial of all phases. Perhaps the development of better data-management techniques for cultural resources would make this data more readily accessible to environmental assessors as well, and I hope to have demonstrated that this can best be accomplished through the use of archaeological GIS and predictive models.

References

ATC Associates, Inc., Screening and Environmental Assessment Reports for Mathews Landing and Lake Forest High School, Projects 25105.0827 and 25105.0807, ATC Associates, Downers Grove, IL, 1997.

Berg, R.C. et al., Potential for Contamination of Shallow Aquifers in Illinois, Circular 532, ISGS, Urbana, IL, 1984.

Canter, L.S., *Environmental Impact Assessment*, Irwin McGraw Hill, Boston, 1996.

EcoSearch, Inc., Environmental Risk Assessment Report (sample map), EcoSearch, Indianapolis, IN, 2001.

Hussain, S.M., *Environmental Impact Assessment*, Carleton University Press, Canada , 1996.

Neumann, T.W. and Sanford, R.M., *Cultural Resources Archaeology: An Introduction*, Altamira Press, Walnut Creek, CA, 2001.

Public Service Archaeology Program, Archaeological Survey Short Report for Proposed Cellular Tower near Ringwood, Illinois, University of Illinois-Urbana/Champaign, 1996.

Scovill, D.H, Gordon, G.J., and Anderson, K.M., Guidelines for the preparation of statements of environmental impact on archaeological resources, in *Conservation Archaeology: A Guide for Cultural Resource Management Studies*, Schiffer, M.B. and Gumerman, G.J., Eds., Studies in Archaeology Series, Academic Press, New York, 1977.

Singleton, R. et al., *Environmental Assessment*, Thomas Telford Publishing, London, 1997.

USDI, Wetland Inventory Map, Barrington, Illinois, Quadrangle, U.S. Department of the Interior, Washington, DC, 1984.

Wood, C., *Environmental Impact Assessment: A Comparative Review*, John Wiley and Sons, New York, 1995.

Section 6:

Modeling Applications in Progress

14

Understanding Lines in the Roman Landscape: A Study of Ancient Roads and Field Systems Based on GIS Technology

Frank Vermeulen

CONTENTS

ABSTRACT   This chapter presents a summary of the basic methodology and some results of a recent (1997–1999) geoarchaeological project concerning the ancient landscapes of parts of northwestern France and Belgium. The primary goal of this project was to detect, reconstruct, and interpret the Roman road system and land organization in a well-delineated area of northern Roman Gaul, as well as to study other suspected pre-Medieval linear features in this region. To gain a better understanding of the structural organization in this poorly preserved ancient landscape, a full battery of techniques has been deployed: active aerial photography, systematic field survey, selective excavation, soil analysis, palynological research, regressive examination of cartographic material, etc. All spatially linked data obtained in this way were vectored and integrated in a unique GIS especially developed for the project. This work allowed the spatial analysis of all cartographic and archaeological information and was very helpful in tracing, mapping, and explaining prehistoric and Roman roads and field patterns. Important methodological problems that needed to be solved included (a) the use of oblique aerial photographs and cadastral map data when studying ancient patterns in a geographic information system (GIS) environment and (b) the separation of relevant ancient lines in the landscape from more recent linear features.

14.1 Aims, Area, and Field Methodology

Between 1997 and 1999, a small-scale geoarchaeological project was undertaken at Ghent University (Belgium).[1] This study was mainly concerned with the reconstruction of the landscapes in western Belgium and during the Roman occupation, essentially the first four centuries of our era. But as the Roman landscape cannot be isolated from earlier evolutions, it is the pre-Medieval landscape as a whole that was looked at.

The primary aim of this project, which combined a whole set of geographical and archaeological methods, was to reconstruct and interpret the Roman road system and land organization as well as other pre-Medieval linear features in this area. Results should ultimately lead us to an improved understanding of human behavior behind this aspect of material structure, especially in the light of the acculturation processes at work in Gaul during the Roman period. Local answers to political, economic, technological, and ideological stimuli should then be measured and the Roman influence on autochthonous land organization evaluated.[2]

Another main priority was to explore the potential of new methods of paleolandscape analysis in a geographical area of northern Gaul where, up until recently, very little was known about the late prehistoric ("protohistoric") and Roman landscape.[3]

The precise area chosen for this research is the Civitas Menapiorum. This former administrative unit of the Roman Empire covers the coastal plain, the flat sandy area, and the loamy and hilly lowlands of interior Flanders up to the River Scheldt (Figure 14.1). It stretches from northwestern France to the southeastern Netherlands. Even more interesting than its variety of natural landscape types is an unusually clear division during Roman times into a more romanized villa landscape with a more Roman-looking urban center, Cassel, in the south and a more indigenous, less-exploited rural area in the north (Vermeulen 1992).

The chosen region appeared at first sight not to have known any pre-Medieval landscape structures. On the one hand, it is situated just south of the regions where large protohistoric field systems of the Celtic field type have been discovered. On the other hand, our area is a bit too far to the north to expect any profound Roman influence on the landscape. The well-structured and essentially military Roman road network only stretched as far as the southern fringes of this region, and only vague traces indicate that the possible existence of an orthogonal Roman land division ("centuriation") should not be ruled out a priori.

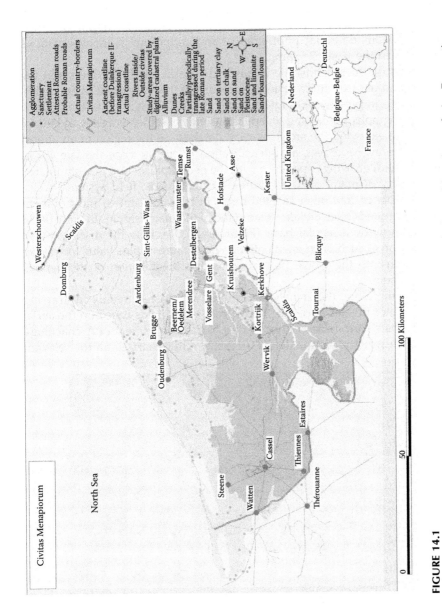

FIGURE 14.1

Localization of the case studies and the main natural landscape types in the Civitas Menapiorum during Roman times.

With the exception of the area around the region's Roman capital, most investigations were concentrated in the northern part of the Civitas Menapiorum, in so-called Sandy Flanders. This is the generally flat, sandy, and lightly sandy-loamy region situated between the River Scheldt and the coastal plain. The main reason for choosing this region is the high level of objectivity and quality of the available archaeological information obtained through systematic aerial photography and numerous field observations carried out in the past 15 years.

Although the rural landscape is much cut up in Sandy Flanders and the agrarian land use extremely varied, the results of this systematic oblique photography are impressive (Bourgeois et al. 1992). The now more than 50,000 aerial photographs, containing several thousands of settlement sites and burial places, have extended the archaeological database of the region in an almost "alarming" way. Moreover, an intensification of flights since the late 1980s, combined with growing experience with the geographical unit, has stimulated the discovery of many off-site features. Ancient roads, field systems, wells, pools, cult places, and other isolated traces in the landscape — most of them obviously of pre-Medieval date — are some of the main features that are being registered now on a regular basis (Figure 14.2). Precisely the discovery and interpretation of such phenomena offer an important surplus value to systematic research into the whole human landscape organization of the periods concerned here.

Initially, therefore, the archaeological study consisted mainly in drawing up a computer inventory of the "fossil" roads and fields based on the aerial photographs stored in the computer files of the Department of Archaeology of Ghent University (Vermeulen et al. 2002). After charting all the relevant

FIGURE 14.2
Ancient land divisions and Bronze Age burial monuments at the archaeological site of Aartrijke-Gemene Veld (prov. of West-Flanders). (Aerial photography: J. Semey, Ghent University. With permission.)

FIGURE 14.3
Ancient field system at Koekelare "Bergbeek." (Aerial photography: J. Semey, Ghent University. With permission.)

data, we had to study the central issue of the relationship of the detected linear patterns with archaeologically datable elements (Figure 14.3). This meant that these fossil structures, more than 2000 probable ancient roads and field boundaries, had to be separated from the more recent and modern elements in the landscape and that a number of the presumed protohistoric and Roman fields and roads had to be interpreted and dated correctly. This, in turn, required the introduction of historical-geographical and primary archaeological research on a detailed scale. While the former involved classic regressive examination of cartographic and vertical photographic material, the latter comprised a certain amount of fieldwork (Vermeulen and Antrop 2001).

This fieldwork included first a campaign of field walking and augering on and near several hundred selected ancient road traces and field systems. Borings with a simple Edelman auger enabled us to make a detailed study of a selection of roadlike features and a few supposed field ditches. Some 30 double-line traces that were clearly visible on aerial photographs were verified systematically and on different locations by means of augering. This technique allowed us to draw some general conclusions regarding the nature of the traces, and more than once they confirmed our assumptions that these traces were indeed fossil roads. In general, it was fairly easy to retrace these ditches, to determine their location, and to specify their dimensions (width, depth) and backfill. Sometimes a thin hardened stratum of soil constituting the original earthen road surface between the draining ditches was found. These borings could provide only sporadic dating evidence.

However, only by excavating the traces is it possible to determine their precise nature and date. Within the framework of this research, two working methods were used: (1) a systematic and focused study of some carefully selected traces by means of narrow trial trenches, and (2) the large-scale excavation of entire systems in an open area, often as part of current rescue archaeology in the region.

The first method involved the digging of a series of limited trial trenches cutting across a number of fossil roads and field boundaries selected from the database of aerial photographs. This type of fieldwork confirmed the existence of pre-Medieval, mostly Roman, tracks and of some fields that were more or less simultaneously parceled out. They are reassuring evidence of the fact that a considerable but as yet indeterminable part of the archaeological database consists of road and land-division structures that were used in the Roman period and sometimes even before.

The second method, the one of open-area excavations, was only applied within the scope of a number of large-scale rescue digs within the study area. These four excavations were the direct result of indications on aerial photographs, which were used to guide archaeological excavations to these areas threatened with destruction. Such case studies made a very important contribution to our research. They made it possible to excavate several hundred meters of Roman roads, and in one particular case (at Sint-Gillis-Waas), research could be done on the excavated parcel boundaries (Figure 14.4).

Integrated in this fieldwork, colleagues from Ghent University were able to do quite a number of paleobiological and pedological analyses. These pollen analyses, studies of macroplant remains, and detailed pedological observations allowed us to reveal the environmental context of the line patterns detected by means of aerial photography as well as the function and evolution of the various field structures.

14.2 Spatial Analysis Based on GIS

To be able to develop efficient methods of spatial analysis of all gathered data, an original geographic information system (GIS), based on ArcInfo and ArcView software, was built (Vermeulen 2000; Johnson and MacLaren 1997; Jones 1997). In this study, ArcView GIS 3.1 from Environmental Systems Research Institute (ESRI) formed the core for the integration of data from different sources, and the spatial analysis was mainly performed with ArcView's extensions Spatial Analyst 1.1 and 3D Analyst 1.0. For extraction and exploring geometrical data procedures, ArcInfo 3.5.1 and AutoCad LT were used as well, although Microsoft Excel 97 offered interesting possibilities. Data management was achieved with Microsoft Access 97 and statistical analysis with Microsoft Excel 97 and Statistica.

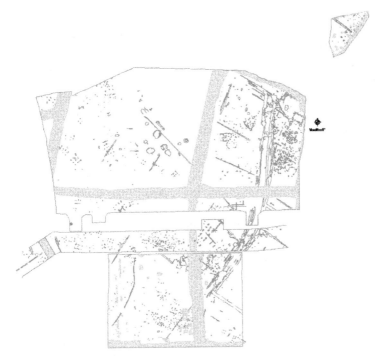

FIGURE 14.4
General plan of the Roman traces (red) of roads, field ditches, and farms found during the 1996–1997 excavation campaign at Sint-Gillis-Waas.

First of all, as many as possible relevant primary sources, such as maps and vertical aerial photographs, were stored in this GIS in digital or scanned versions. For the region considered, this means that geographical information had to be derived from very different data sources provided by three countries: France, Belgium, and the Netherlands. The integration of these data was not a simple task, as we had to deal with a multitude of sources, scales, and especially projections. The following map layers were integrated in the GIS: scanned topographic maps, digital orthophotographs, maps of waterways, maps of the municipality borders, a geological map, soil maps, a rastered version of an 18th-century military map of the Low Countries, and various vectored maps from archaeological literature. For specific study areas, we also vectored and georeferenced some 19th-century cadastral plans. As the three-dimensional landform information is fundamental in all spatial analysis, we also extracted and transformed contour and height information from the scanned topographical map into a digital elevation model (DEM) of the same study microareas.

Secondly, the GIS comprised a database of all relevant archaeological information on the protohistoric and Roman periods in the Civitas Menapiorum. This not only includes the Roman settlement sites and other ancient finds known from earlier publications or recent fieldwork, but also all vectored excavation plans of relevant archaeological fieldwork in the area. Furthermore, the archaeological database includes all processed oblique aerial photographs, of which a majority is now scanned and digitally available on CD-ROM. The main purpose is to store this archaeological information in a suitable way so as to examine vertical and horizontal relations between different archaeological sites and other elements of the landscape, like soil type, terrain, rivers, etc. in time and space.

For the selected linear traces observed on aerial photographs, this means that we had to use and further develop methods for georeferencing oblique aerial photographs. The retrieval system that our team has developed works as follows. Each photograph is described in a relational database. The description contains, among others, the shape of the marks and the coordinates of one point, which is visible on the slide and can easily be located on the topographic base map. For each point on this map, a "hot link" can be defined to the corresponding images, which makes it possible to bring the photograph on the screen simply by clicking on the map. By comparing the situation of the photograph with other themes, such as known Roman settlements, roads, and excavation plans, photographs can be selected by map queries. Using imaging software such as Corel PhotoPaint, the most promising images are treated by methods such as edge enhancement to raise the visibility of the marks on the picture. By so-called warping, the image can be fitted on topographic maps or orthophotoplans. Locations that can be easily found and precisely located are usually chosen as control points. By using the corner-points of the field (which contains linear marks) as tie points, the corresponding real-world coordinates can be identified on the actual orthographic photoplans.

14.3 Tracing Ancient Roads[4]

An attempt has been made to examine the common characteristics of the limited number of well-attested and archaeologically known major Roman roads within the Civitas Menapiorum. The purpose is to derive a model to reconstruct possible connections between large Roman settlements where no exact road is known so far. The characteristics of archaeologically known Roman roads can be evaluated by map-overlay analysis and statistical tests.

After the detection of correlations between Roman roads and thematic maps, we tried to develop a so-called cost surface, which represents areas suitable for Roman roads by low costs and areas less suitable by high costs. By applying a cost-path algorithm, we then tried theoretically to detect the "cheapest" connection between known Roman settlements of central importance whose interconnection in Roman times is at least very plausible. The results of this type of analysis can stimulate and help further research in areas where the major Roman roads are still hard to trace in the field.

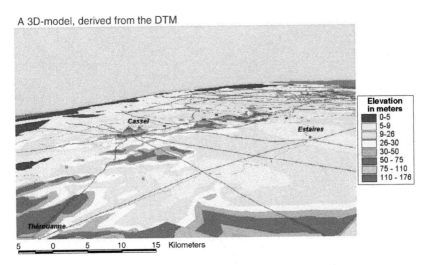

FIGURE 14.5
Three-dimensional visualization looking from the SW to the NE, with known and supposed Roman roads and settlements overlaid on the area around Mount Cassel.

This cost-path analysis was first applied on the Roman capital site of Cassel and its connection with the neighboring major town of Thérouanne, south of our region. Roman Cassel was situated on a hilltop and became the core of a whole network of well-identified Roman roads (Figure 14.5). The connection between both cities follows the interfluvium between the alluvial lowlands of the River Aa basin in the northwest and the River Lys in the southeast. The present road reflects in its major part an almost straight connection between these two *civitas* capitals, and there is a lot of evidence that it corresponds partially to the original Roman road trace.

If we compare the Roman road with the direct connection, we see that there are several areas where the difference between the road and the direct connection is more than 500 m (Figure 14.6). One of them is situated near Bavinchove, where the reason for the choice for this particular trajectory might have been a small isolated hill. Visibility analysis shows that visibility is much better from here than from the direct connection.

The resulting cost-path, calculated from Cassel as source to Thérouanne as target, fits quite well with the Roman road. The model seems to indicate that topographic depressions and the later developed municipality borders, often remnants of ancient roads in this region, are the main factors that have an influence on the differences between direct and real connections.

A comparable model, based on similar parameters, was then applied to the possible reconstruction of major Roman roads between the central settlements of the northern part of our region, where archaeological proof is still lacking. In some instances there seems to be a real connection between the hypothetical road trajectory and the location of recently discovered Roman sites and of fossil roads detected on aerial photographs (Figure 14.7).

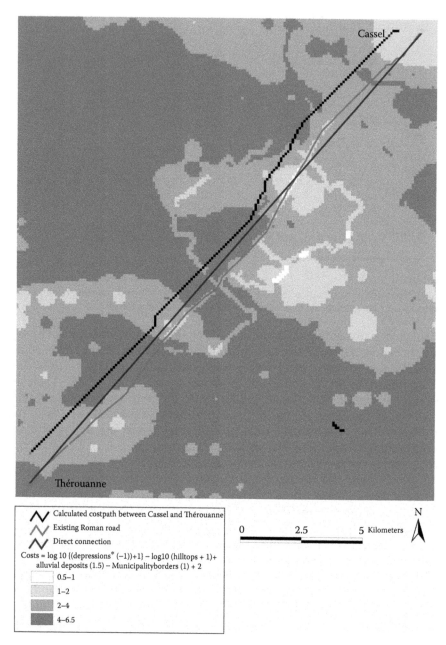

FIGURE 14.6
Cost-path analysis of the Roman road between Thérouanne and Cassel based upon the digital elevation model, the geological map, and the viewshed analysis.

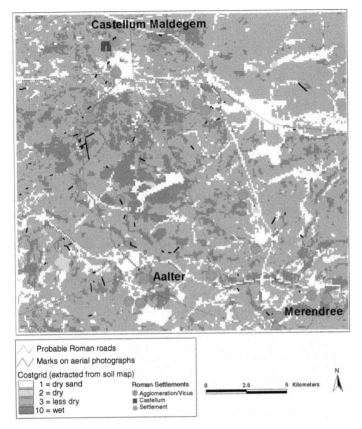

FIGURE 14.7
Cost-path analysis between the Roman camp at Maldegem and the stone quarry area of Aalter in the north of Sandy Flanders.

It looks as if this GIS analysis, if further developed, can be extremely helpful in detecting corridors of high probability for major Roman road trajectories. It could prove to be a valuable tool for defining microregions in the search for possible Roman roads in areas where archaeological and historical proof is still lacking.

Such simple models evidently cannot be applied for the localization of the large number of secondary and local roads. The irregularity of these roads, interconnecting minor settlements or leading from Roman or protohistoric farms toward their fields, is notorious. It seems, for the time being, that only a combination of systematic aerial photography and selected fieldwork can help to locate these.

Our excavation work in Sandy Flanders has shown that, although only earthen roads and tracks seem to occur here before the 18th century, these minor connections can be investigated quite well. So far, these fieldwork campaigns indicate that a large part of the double-linear features visible on aerial photographs can be dated in the Roman period (Figure 14.8). Yet dating

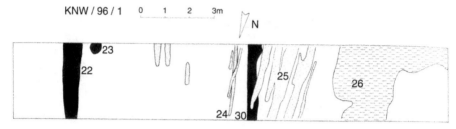

FIGURE 14.8
Knesselare: plan of the excavated pre-Medieval road with ditches (22–30) and cart tracks (24) and of the cart tracks of a medieval road (25).

them precisely is not an easy task, as the ditch fillings seldom contain numerous artifacts. The excavated traces of these Roman roads may vary:

Sometimes only the drainage ditches are preserved, and the road surface has been completely eroded.

When this road surface survives somehow, then a bundle or a couple of cart tracks can be distinguished. Pedological processes such as iron precipitation often stress their presence.

In other cases, the Roman road is succeeded by a medieval one, and a slight shift in location or orientation can be distinguished.

When the Roman road is excavated in an open area, then its connection with adjoining and contemporary field ditches can become apparent.

It is clear, however, that our selection of traces checked in the field by way of excavations is somewhat biased. Most of them already displayed surface indications, such as scatters of pottery or the nearness of a known Roman settlement, pointing toward a possible Roman date. Therefore, it is also much too early to draw statistically meaningful conclusions from these field checks.

There can be no doubt that many of the still-unchecked road traces have a medieval or even a post-Medieval date. A part of these can be isolated by using techniques that analyze the relation of the old road traces seen from the air with cultural elements in the landscape. A buffering of municipality borders, which often coincide with medieval roads, is one of these techniques. In Figure 14.9 we see that the abandoned roads at the northern rim of the municipality of Aalter are very much related to the municipality border between Aalter and Ursel. A buffering zone of 200 m around this border allows a selection of those linear traces that are situated very close by. Several fragments of single and double lines with a mainly E-W orientation appear over a distance of some 2 km. The ditches are probably the remains of an early medieval connection between Aalter and Ghent and are situated mainly on the somewhat higher parts in the landscape.

But not all traces of abandoned roads seen from the air are Roman or medieval. The use of different GIS analyses seems to indicate that some of

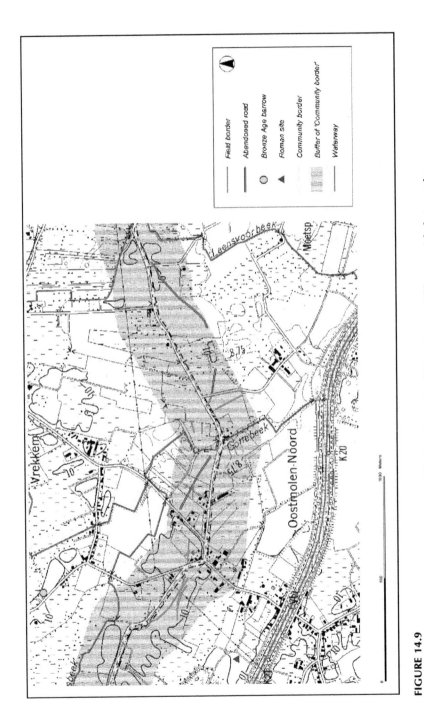

FIGURE 14.9
The municipality borders between Aalter and Ursel and associated linear traces visible on aerial photographs.

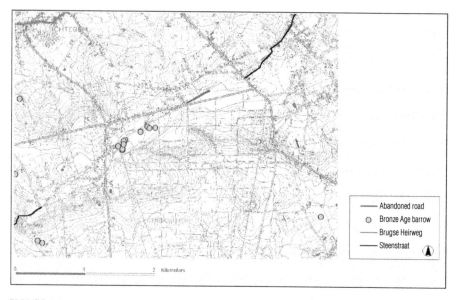

FIGURE 14.10
Fragment of an ancient road connecting two parts of the "Roman" Steenstraat and other ancient features visible on aerial photographs.

the ancient roads and tracks have a pre- or protohistoric date. A simple example from the northern edge of Sandy Flanders demonstrates this well. A clearly visible lining of some 30 Bronze Age barrows has been discovered near the municipality border of Maldegem and Oedelem (Figure 14.10). The barrows are spread over an area of more than 25 Ha, and they all are situated on dry to moderately wet sand. In the immediate vicinity of the barrows, a large number of field borders and a few fragments of ancient roads appear. The field borders do not show one main orientation, which may indicate their different chronology, but the orientation of the longest ancient road segment might suggest contemporaneity with the Bronze Age cemetery.

To conclude this section on the ancient roads, it must be clear by now that the implementation of the results of aerial photography in the GIS environment opens up a whole set of possibilities for studying so-called fossil roads, routes, and tracks. The overlay of the archaeological traces with different kinds of maps and geographical information and their analysis in relation to physical and cultural aspects of the landscape offer great opportunities. Although confirmation in the field remains essential, this approach facilitates the construction of firm hypotheses about the date and interrelation of a wide range of road traces.

14.4 Unraveling Field Systems

The study of grids and networks of field systems is a more intricate matter that needs a different approach on a well-chosen scale. Most of the detailed research concerning field systems was carried out in smaller microregions on which we had recent and reliable information. Six of these test areas lie within the sandy region, one in the south around the Roman urban center of Cassel. Different types of analyses were carried out, of which we will summarize only a selection here.

Around the town of Cassel, we used GIS techniques to check the possible existence of a Roman normative land division (Vermeulen and Antrop 2001; Antrop et al. 2002). Several authors suggested recently that the rural landscape in a wide area to the south/southwest of Cassel and possibly even in other directions shows clear marks of Roman centuriation (Jacques 1987; Roumegoux and Termote 1993; Malvache and Pouchain 1994). In particular, the hypothesis put forward by the French archaeologist François Jacques, concerning "classic" 710 × 710-m module land divisions with orientations ranging from 31° NW (north of Cassel) to 35° NW (south of Cassel), has many followers. This work, mainly based on the visual and manual analysis of 19th- and 20th-century topographic maps, has never been tested in the field.

As systematic field research within our project is restricted to Sandy Flanders (northwest Belgium), we have not yet taken the opportunity to really test this attractive hypothesis in the field. We did, however, check the centuriation idea by using digital overlays of different maps and then filtering the major orientations in the area. Figure 14.11 shows the classification of the field-boundary orientation overlaid upon the presumed centuriation grid for the area around Cassel. The most significant concentrations of line segments are situated here in the intervals −25° to −35° and 55° to 65°, confirming a dominant NW orientation of 31° ± 5°, as proposed by Jacques. Interesting to note is that some orientations agree rather well with the direction of the Roman roads, such as the ones toward the salt-producing sectors on the edge of the former coastal area. We must, however, also keep in mind that some orientations are also parallel to the drainage flow of the nearby Peene Stream, which could well have influenced the direction of field systems in Roman times.

In certain areas, a more detailed observation of the cadastral field boundaries filtered for directions between 26° and 36° shows much more "centuriation-oriented" field edges than published by F. Jacques in 1987. This illustrates well the potential of GIS-based filtering using digitized ancient cadastral documents. It procures not only more detailed and much faster approaches toward the testing of centuriation hypotheses, but it also delivers basic maps for precise testing in the field.

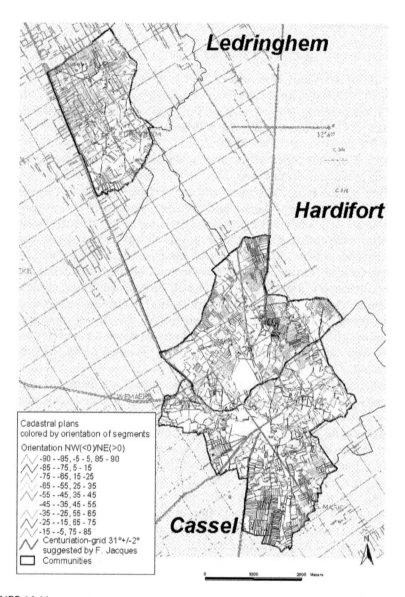

FIGURE 14.11
Presumed centuriation grids (Jacques 1987) overlaid by cadastral field boundaries of the zones
Ledringhem, Hardifort, and Cassel. Filtering according to main orientations.

After an inventory had been made of all ancient field systems on oblique
aerial photographs in the region of Sandy Flanders, and after the relevant
traces were charted, the problem of interpreting and dating the structures
became crucial. The surveyed microregion, Sandy Flanders, is almost completely covered with sand, a soil type that facilitates agriculture, requiring
lower manual labor input in easily penetrated topsoil. This soil factor

increases the effect of multiple viewing, which means that fields from totally different periods can be seen as one complex. One of the most problematic aspects of interpreting aerial photographs is this cumulative view that results from the superposition of marks deriving from different periods.

Tracing the fields involves an extra problem: as these ditches are often not very deep, they can only be revealed in the best climatographic conditions or on the soils that show features best. As these soils often provide the best basis for agriculture, there is a good chance they have been used for longer periods in history. The perceptible ditches will most probably be a combination of multiperiodical land use. As such, it is impossible to attach a chronological label to these features without having excavated them.

It is even very hard to establish a typology of the traces,[5] and although an attempt has been made (Figure 14.12), it is almost impossible to attach any chronological meaning to the individual types of field systems. The typology was based on the purely formal aspects of the line structures.

Beside the purely formal approach of linear ditches and field borders, it is clear that in many cases the relationship of fields with the surrounding elements is more significant (Chouquer 1997). These elements can consist of environmental features, such as a river or a difference in altitude, or cultural features caused by human intervention, such as a road or other ancient structures seen on photographs or dated in the field. Both can, in some cases, offer good possibilities to collect some information on the chronology of the field structures. Two examples might illustrate this.

In the close vicinity of Ghent, we selected several entities with a surface area between 15 and 70 Ha where the correlation of the traces with the environmental properties, in this case slightly higher and drier sandy soils, cannot be neglected. The close relationship of Bronze Age burial mounds and field ditches could form an indication of the nearness of the graveyards and the supposed settlement sites and their surrounding fields. The environmental entity shown in Figure 14.13 is cut in two by the community border and consists of land sections that form a well-preserved half-circle, truncated by a brook. The total zone has a surface of 70 Ha, providing more than enough space for living and for burying the dead.

The relation between the attested field borders and parallel traces or roads has also been examined. In this example, situated in the municipality of Zingem, a regular arrangement of a field system has been detected (Figure 14.14). The field borders, spreading over an area of about 13 Ha, are situated in the immediate vicinity of a recently found Roman settlement, cemetery, and coin hoard. The ensemble consists of a framework of regular lines with a parallel or perpendicular orientation toward the presumed protohistoric and Roman road between the Rivers Lys and Scheldt. The intermediate distance between the ditches is rather small; the estimated plot size

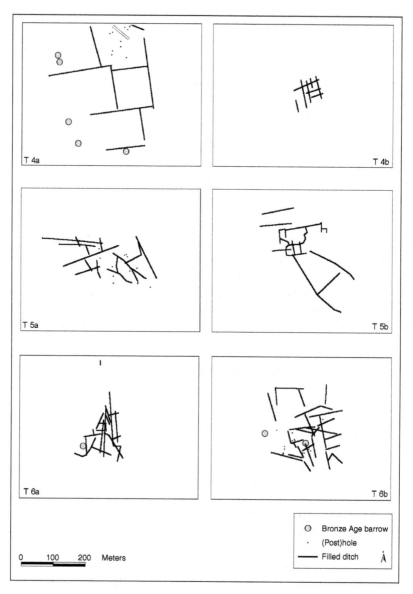

FIGURE 14.12
A selection of types of ancient field systems on aerial photographs in Sandy Flanders.

varies between 6000 m^2 and 1 ha. The field system is clearly the result of an organized construction, but the supposed Roman character can only be proved by means of excavation.

This technique of excavation has been used in a number of cases such as with the field system discovered in Kruishoutem, in the more romanized landscape of the south of Sandy Flanders. Low-altitude aerial photographs

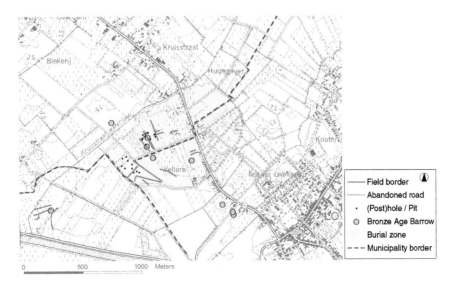

FIGURE 14.13
A possible burial zone in the communities of Zomergem and Lovendegem.

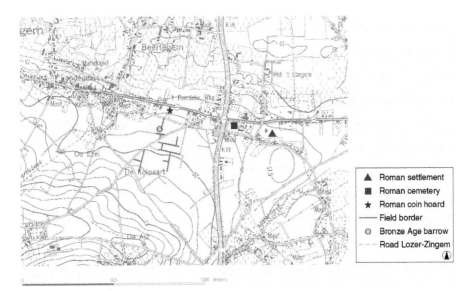

FIGURE 14.14
A possible framework of Roman field boundaries near Kruishoutem.

and our excavations revealed here the existence of several field borders and a possible local road lying perpendicular to the Roman road between two regional centers (Figure 14.15). Like the major road, they were dated to the 1st and 2nd centuries A.D. Pedological and paleobotanical evidence suggest that the fields adjacent to the road system were mainly in use as meadows.

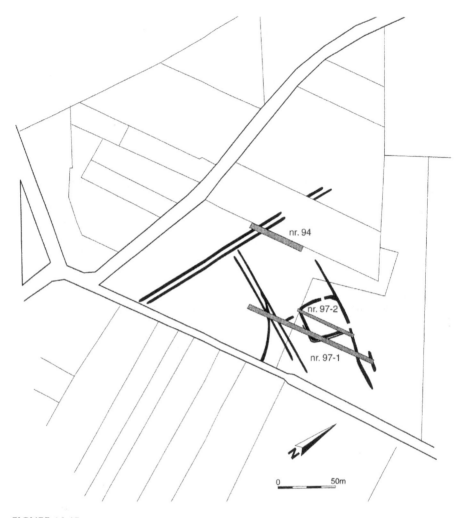

FIGURE 14.15

Nokere Kruishoutem: interpretation of the traces visible on aerial photographs and indication of the trial trenches of 1994 and 1997.

The discovery and precise dating of these systems gave us the opportunity to do some further GIS analysis on surrounding areas of these ancient features. As we have vectored the cadastral situation of the beginning of the 19th century, we can easily compare the orientations of the dated Roman fields with the land division of the whole municipality, which reflects the land organization of older periods (Figure 14.16).

On this orientations map we see four different areas, with concentrations of one dominant orientation, corresponding to four different land facets with different directions of slope, drainage, and different soil types:

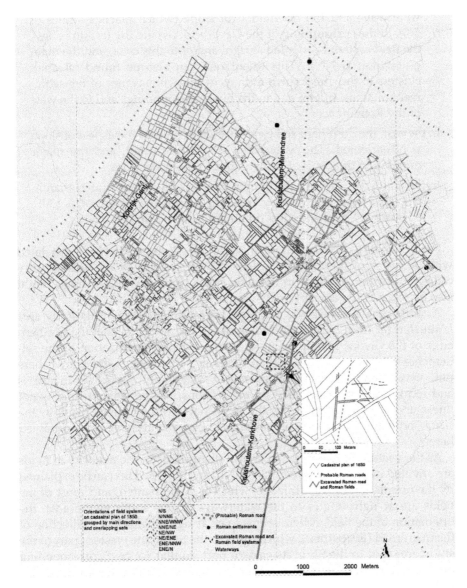

FIGURE 14.16

Kruishoutem: cadastral plan of 1850 and excavation plan of a Roman road and possible Roman fields. GIS-filtering of the field boundaries according to the major orientations.

The center is situated on the highest points of this study area; the terrain is flat; and sandy soils are dominant. The main orientation is NE, and the fields are quite large. In 1770 the landscape was character- ized by open fields.

In the northwest, on the slope to the valley of the Lys, we find the same dominant NE orientation, corresponding to the orientation of the waterways and the orientation of slope (which reaches values of 5%). In the surrounding of the Gaverbeek (where the terrain is flat), the field system is oriented at right angles to this river, and the main orientation is NNE. This orientation can also be found in small clusters in the direct surroundings of the actual center of the settlement of Kruishoutem and in the direct surroundings and to the west of the Roman *vicus*.

In the east, the dominant soil type is sandy loam, and the aspect of slope is NE-oriented. The waterways are ENE-oriented, and the major orientation of arcs is at right angles to the waterways.

In the south, the dispersion of orientations is less concentrated than in the other parts of the study area. The dominant orientation is ENE, "disturbed" by the second dominant orientation: NE.

Further analysis shows that the marks of our Roman cluster are correlated with a cluster of NNE/WNW-oriented field systems of the 19th-century cadastre. Further research in the field must confirm whether these might also be relics of the ancient field systems.

A comparable test case was developed in Sint-Gillis-Waas. This study-area is situated far from important Roman roads and settlements at the northern edge of the Civitas Menapiorum in an area where romanization of the settlements is rather low. No Roman roads or settlements were known here until very recently, when two Roman field systems and some Roman houses and local roads were revealed by excavations. As the Roman road is not linear and is situated far from the main network of Roman roads in the *civitas*, only the orientations of the ancient field systems have been examined here.

As the vectored cadastral map shows (Figure 14.17), the majority of fields are oriented in a range between –30° (ENE) and 10° (N). This can be explained by the orientation of the drainage, which is W-E-oriented, and most of the fields are at right angles to the waterways. In the south of the area, the orientation of the field system seems to be at right angles to the direction of the main road (Reepstraat), which turns to the SW in the western part of the study area and to the SE in its eastern part, so that we find corresponding orientations of the field system ranging from –40° (NE) in the west and 10° (N) in the east.

In the part of the study area that is situated in the polders of the River Scheldt, there is a concentration of ENE-oriented field systems. In the center and in the north, where sandy soils of the Land van Waas dominate, the dominant orientation is NE. Here, where the study area was covered by woods in 1770, the fields are much bigger than in the rest of the study area.

The most complex cluster of marks is concordant with the 19th-century cadastral divisions and completely parallel to the excavation plan of Roman fields near the Reepstraat (Figure 14.18). Because the marks complement the Roman fields very well, we may suppose that the excavated Roman field

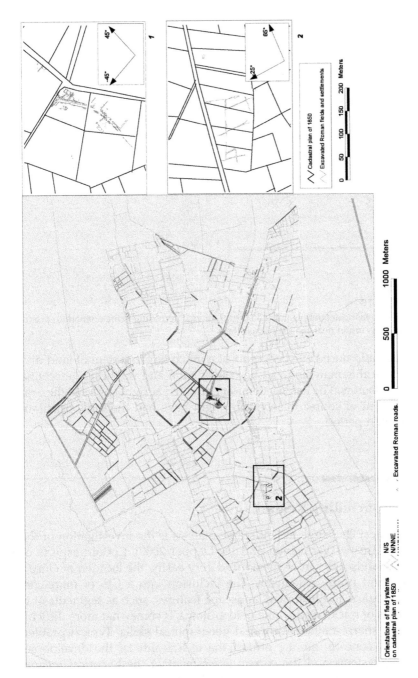

FIGURE 14.17
Sint-Gillis-Waas: GIS-filtered cadastral plan of 1850 and excavation plans of Roman field systems.

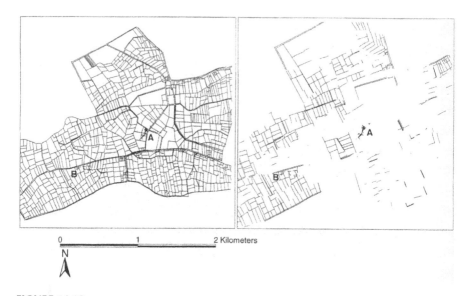

FIGURE 14.18
Sint-Gillis-Waas: cadastral plan of 1850 (right), filtered according to the orientation of excavated Roman fields found near the Reepstraat (left).

system could once have been part of a much larger system of land division. Of course this matching can be coincidental and might be determined by physical factors. The field systems are oriented at right angles to the drainage system. But of course this system might have had the same orientation in the Roman period.

14.5 Concluding Remarks

The greatest challenge in landscape analysis is the investigation of the cultural landscape (Vermeulen and De Dapper 2000). Certain aspects of this cultural landscape can be introduced very easily: the location of cities, cemeteries, and rural settlements. The inclusion into a GIS of more-complex archaeological and historical landscape features, such as segments of roads, networks of tracks, and whole field systems, is somewhat more difficult and involves sharp mathematical and geographical skills. Typical problems in this field concern, among others, the technicality of the development of filtering systems and the correct assessment of the historical meaning of ancient and present-day patterns as observed on aerial photographs and maps.

However, now that this project on the Civitas Menapiorum has come to an end, we are convinced that some of the GIS technology briefly presented here could prove useful for an efficient processing and study of ancient land organization. The advantages are many. We will mention only three here:

Information about the physical environment (topography, landform, soils) can be well integrated in spatial analysis for explaining Roman and other ancient roads and field patterns.

A fuller use of oblique aerial photographs when studying such patterns is now within reach.

The cadastral map data, which are so important because of their historical dimension, can be more efficiently used for the spatial analysis of ancient field systems.

The examples shown here demonstrate, however, that fine archaeological surveys and especially stratigraphic excavations will still be necessary to support even the best hypotheses. This is especially true when confronted with a type of archaeological features, in this case just some lines in the landscape, that present a whole range of problems concerning their historical interpretation.

Notes

1. The research directed by the author was undertaken in close collaboration with M. Antrop, B. Hageman, and T. Wiedemann, all at Ghent University. The main archaeological results, as well as specific methodological contributions, were published in Vermeulen and Antrop (2001).
2. This lies in the line of former research of the author into romanization processes in northern Gaul (e.g., see Vermeulen [1992, 1995]).
3. This aspect of the work is well integrated in recent developments in Europe concerning the geo-archaeological approach of the landscapes of Classical Antiquity (Vermeulen and De Dapper 2000).
4. As space is limited here we will present only a brief overview of some of the most important approaches concerning our road analysis. For the full publication of the results, see Vermeulen and Antrop (2001).
5. Compare with the excellent work in Brittany (Boissinot and Brochier 1997).

References

Antrop, M., Vermeulen, F., and Wiedemann, T., Une organisation cadastrale dans la partie méridionale de la Civitas Menapiorum (Gallia Belgica), in *Atlas Historique des Cadastres d'Europe II*, Clavel-Lévêque, M. and Orejos, A., Eds., Editions Errance, Paris, 2002.

Boissinot, Ph. and Brochier, J.-E., Pour une archéologie du champ, in *Les formes des paysages, 3: L'analyse des systèmes spatiaux*, Chouquer, G., Ed., Editions Errance, Paris, 1997, pp. 35–65.

Bourgeois, J., Crombé, Ph., and Semey, J., L'archéologie aérienne en Flandre sablonneuse: Initiatives pluridisciplinaires de l'Université de Gand, in *Actes du Colloque international d'Archéologie Aérienne*, Amiens, 15–18 October 1992.

Chouquer, G., Ed., *Les formes des paysages, 3: L'analyse des systèmes spatiaux*, Editions Errance, Paris, 1997.

Jacques, F., Témoins de cadastres romains dans la région de Cassel, *Revue du Nord*, 69 (272), 101–108, 1987.

Johnson, I. and MacLaren, N., Eds., *Archaeological Applications of GIS*, Proceedings of Colloquium II, UISPP XIIIth Congress, Forli, Italy, Sept. 1996, Sydney University Archaeological Methods, Series 5 (CD-ROM), 1997.

Jones, C.B., *Geographical Information Systems and Computer Cartography*, Longman, London, 1997.

Malvache, E. and Pouchain, G., La recherche des cadastres antiques dans la région Nord-Pas-de-Calais (état provisoire de la recherche), *Revue du Nord, Archéologie*, 76 (308), 83–98, 1994.

Roumegoux, Y. and Termote, J., Op de rand van een imperium de Romeinen in de Westhoek, *Westvlaamse Archaeologica*, 9 (2), 61–80, 1993.

Vermeulen, F., Moderate acculturation in the fringe area of the Roman Empire: some archaeological indications from the Civitas Menapiorum, *Bulletin de l'Institut Historique Belge à Rome*, 62, 5–41, 1992.

Vermeulen, F., The role of local centres in the romanisation of northern Belgica, in *Integration in the Early Roman West: The Role of Culture and Ideology*, Metzler, J., Millet, M., Roymans, N., and Slofstra, J., Eds., Luxembourg, Museé National d'Histoire et d'Art, OOPEC, Luxembourg, 1995, pp. 183–198.

Vermeulen, F., The potential of GIS in landscape archaeology, in *On the Good Use of GIS in Ancient Landscape Studies*, Slapszak, B. and Stančič, Z., Eds., proceedings of international workshop, Ljubljana, Dec. 1998, OOPEC, Luxembourg, 2000.

Vermeulen, F. and Antrop, M., Eds., *Ancient Lines in the Landscape: A Geo-Archaeological Study of Protohistoric and Roman Roads and Field Systems in Northwestern Gaul*, Babesch Supplement 7, Peeters Publishers, Leuven, Belgium, 2001.

Vermeulen, F. and De Dapper, M., Eds., *Geoarchaeology of the Landscapes of Classical Antiquity*, proceedings of the congress in Gent, Oct. 1998, Babesch Supplement 5, Peeters Publishers, Leuven, Belgium, 2000.

Vermeulen, F., Hageman, B., and Wiedemann, T., Photo-interprétation et cartographie des systèmes spatiaux anciennes: l'archéologie des routes et des parcellaires en Belgique, in *Atlas Historique des Cadastres d'Europe*, Vol. II, Clavel-Lévêque, M. and Orejas, A., Eds., OOPEC, Luxembourg, 2002.

15

A GIS-Based Archaeological Predictive Model and Decision Support System for the North Carolina Department of Transportation

Scott Madry, Matthew Cole, Steve Gould, Ben Resnick, Scott Seibel, and Matt Wilkerson

CONTENTS

ABSTRACT The North Carolina Department of Transportation (NCDOT) has identified advanced technologies, including Geographic Information Systems (GIS) and GIS-based archaeological predictive modeling, as having potential for improving the planning of new multi-lane highways. These models, and the infrastructure of GIS data, easy user interfaces, and updated cultural resource databases integrated into a Decision Support

System (DSS), will create a tool to analyze and rank proposed highway corridor alternatives. The long range project tasks include: 1) development of digitized environmental and cultural information for the State of North Carolina, including computerizing all existing archaeological site files and site maps at the North Carolina Office of State Archaeology (OSA); 2) development of GIS archaeological predictive models using the digitized information for North Carolina and verifying their accuracy; 3) creation of an internet-based DSS and graphical user interface (GUI) for use by NCDOT and OSA staff; 4) application of the GIS archaeological predictive models to multiple transportation projects to aid in the selection of a preferred corridor/alternative; 5) field testing of the models in actual NCDOT projects using intense archaeological survey and GPS mapping to quantitatively test and refine the models.

15.1 Introduction

From October 2002 through September 2005, Environmental Services, Inc. (ESI) and GAI Consultants, Inc. (GAI) conducted a pilot study for a state-wide Geographic Information System (GIS)-based Archaeological Predictive Model on behalf of the North Carolina Department of Transportation (NCDOT). The team was led by Dr. Scott Madry of ESI and the University of North Carolina at Chapel Hill as Principal Investigator. Dr. Kenneth Kvamme of the University of Arkansas acted as project consultant for spatial statistics and analysis. This work was conducted at the request of NCDOT and funded by the Federal Highway Administration (FHwA) with the intention of using the model in the planning of multi-lane highways in new locations throughout the state. The final model integrates available environmental and archaeological site data in order to rank proposed highway corridors and alternatives as reflecting High, Medium, or Low probability for containing prehistoric archaeological sites. A Decision Support System approach was used to allow managers access to computerized data in a variety of formats using easy-to-use web-based tools and databases. Predictive models were built by comparing the distribution of archaeological sites on the landscape with digital environmental data. Models were verified by several independent methods, and the final model is being field-verified. This verification is in the form of an archaeological survey of a new transportation corridor around Asheboro, NC as part of NCDOT's Transportation Improvement Program (TIP).

The ultimate goal is to combine all environmental and cultural resources analysis into a single, integrated Decision Support System that will permit the interactive analysis of all issues relating to the National Environmental Policy Act (NEPA)/ Section 106 of the National Historic Preservation Act (NHPA) process. This project is a first step in that direction.

15.2 Project Benefits

Preparation of GIS-based archaeological predictive models will benefit NCDOT by expediting the selection of preferred highway alternatives, thereby decreasing costs of archaeological investigations and reducing project schedules, and should serve as a cost effective management tool when planning NEPA/ Section 106 projects. The merged NEPA and United States Army Corps of Engineers Section 404 permit process allows for preliminary design and environmental data to be gathered and analyzed for selection of the best alternative for construction of a given project, designated as an Environmental Assessment (EA) or Environmental Impact Statement (EIS) level undertaking. This increase of environmental consideration at the front end of the NEPA process is designed to foster better decision-making regarding selection of a given EA or EIS build alternative. Of all of the environmental constraints, only archaeological resources are not carried forward to the same level of identification and evaluation as community impacts, wetland delineation, air/ noise issues, etc.

The use of GIS technology for predicting and quantifying potential impacts to archaeological sites is a tool that can significantly streamline the identification of prehistoric archaeological resources by NCDOT early in the NEPA process. This approach will allow mapping of highway corridor constraints and summary tables quantifying a project's probable archaeological impact to be available early in project planning. Just as important will be the dynamic nature of the information produced. The GIS approach will allow for ready adaptability to changes that occur throughout the life of a given project, including modifications to existing study alternatives or the addition of new corridor alternatives. By using GIS, a clear understanding of the archaeological potential of a new or revised alternative can be generated quickly without the need to conduct additional fieldwork or create/revise addenda to an existing report. Avoiding areas of high archaeological potential is expected to preserve many important sites, while significantly reducing costs for NCDOT. By incorporating a DSS interface to access all relevant data, managers and decision-makers can do a better job of planning and responding to changes.

Another key benefit of this approach involves coordination with state and federal agencies responsible for compliance with NEPA and Section 106. In order to make the GIS approach work, the massive amount of archaeological site data at OSA must be available for review and analysis in digital format. These data were maintained primarily on paper and microfiche. This project created framework to maintain a modern digital site database to use in probability model development and other uses. By using project funding to accomplish this task, the regulatory agency obtains digital archaeological site information to use for future planning and research purposes with a minimum of expense. Ultimately, OSA will be able to significantly streamline their site review process and quickly produce information currently generated by those using the archaeological site files. OSA will also be able to integrate new site information in a more timely and efficient manner. Other benefits include making the archeological site probability information available to other planning organizations via a web-based environment. Once the initial information is created in a digital format and refined through field efforts, appropriate local and state agencies and regional planning organizations will be able t0 better plan their undertakings at a reduced cost in both time and money.

15.3 The Example of Mn/Model

The Minnesota DOT is the first such agency in the nation to use a GIS-based archaeological predictive model, known as Mn/Model (Hudak and Hobbs, n.d.), to better predict the potential for encountering archeological sites. This model predicts that about 85.5% of pre-1837 cultural resources in Minnesota are located in 23% of the land, with significant areas being 'unknown' due to lack of data in those areas. The total cost of the project was approximately $5 million, and savings over the last four years have been documented at $3 million per year since the model has been used in planning new projects. The total cost of the project was recouped in two years.

In addition to these savings, the Mn/Model project:

- Allows Mn/DOT's Cultural Resources staff to clear approximately 35% more projects per year.
- Reduces the number of Memorandum of Agreements (MOA) required by nearly 60%.
- Reduces project turnaround time by 30%. Some projects have saved one or two construction seasons in survey time alone.
- Rreduces schedule and budget uncertainty by minimizing "surprises."
- Reduces disturbance to cultural resources.
- Supports coordination among governmental organizations.

Mn/DOT is providing the Mn/Model and training in its use to Minnesota's State Historic Preservation Office (SHPO), the Minnesota OSA, and the Tribal Historic Preservation Offices (THPO).

15.4 Project Structure

The overall project was divided into two major tasks. Task 1 involved the collection of archaeological and environmental data covering the seven counties of the initial project area (Cabarrus, Chatham, Forsyth, Granville, Guilford, Randolph, and Wake) in the Piedmont region of North Carolina, and the integration of the data into a project database that was used for modeling during Task 2. Specific activities included updating and quality review of the existing archaeological database at OSA for use in the modeling process, collecting and converting all site data for the study area to digital form compatible with GIS analysis, and collecting all relevant environmental GIS data for use in the modeling process.

Task 2 included the statistical manipulation of the digital archaeological site data with the environmental data, creation and validation of the predictive models, and creation of the easy-to-use, web-based graphical user interface DSS to access and query the digital data. Recommendations were made regarding operationalization of the system for continuing NCDOT and OSA use. Future tasks will continue the digitizing and modeling process across the rest of the state.

15.4.1 Initial Project Area

The initial project area was identified by NCDOT based upon the locations of anticipated NCDOT highway projects where multiple alternatives are present. This area includes seven Piedmont counties (Figure 15.1). The Piedmont makes up 38% of the state, and will see a significant number of new NCDOT projects in the near future. The initial project area provides a good cross section of the Piedmont, ranging from the coastal plain boundary in the East to near mountainous areas in the West, and from the Virginia border to near South Carolina.

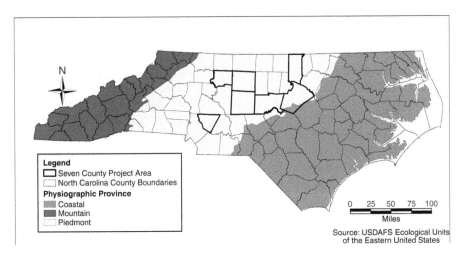

FIGURE 15.1
The Mountain (left), Piedmont, and Coastal Plain regions of North Carolina, showing the 100 counties and 7 county project areas at center.

15.4.2 Task 1 Work

Task 1 work was conducted between October, 2002 and September, 2003. The purpose was to collect the data that was required for the development of the archaeological predictive models for the project area in Task 2.

15.5 Existing NC Archaeology Data

The Office of State Archaeology (OSA) is the state agency responsible for maintaining all North Carolina archaeological records. There are more than 35,000 archaeological sites that have been recorded in the state. Of these, some 21,000 had been entered into a UniVerse database designed in the 1980s. This system was not built to current standards, was difficult to maintain, and provided only limited search and query of data records. It was not a relational database, and stored all information in flat files. All of these factors made the UniVerse system problematic for future use. Between 50 and 450 new archaeological sites are recorded each month at OSA, and much of the recently recorded site information was not available in a timely manner for planning purposes due to delays in processing data into the system.

Surveyed areas, archaeological site locations, and related information such as bibliographic numbers and reports are currently recorded onto a master set of USGS 1:24,000 quadrangle maps that are archived at the OSA office in Raleigh, North Carolina. Data access was limited to physical visits to the OSA. The Survey and Planning Branch (S&P) of the Division of Historical Records (DHR) separately maintains additional information on historic architectural properties and historic districts. More than 50,000 historic properties in North Carolina are currently recorded. None of these records are computerized at this time, with the exception of historic districts, which are available in digital format. Architectural data are not included in the current phase of the project, but there is significant interest in including this information in the future. Addition of these data should lead to a better understanding of the location of potential Section 4(f) resources.

In order to accomplish the goal of providing NCDOT with a comprehensive set of GIS and database tools that utilize a DSS approach to improve the integration of archaeological data into the process of road planning, the current North Carolina archaeological site file database had to be updated. This included scanning and digitizing archaeological site maps from OSA and historic maps from the N.C. Office of Archives and History. Since many of the archaeological site forms for the project area were only available on paper or microfiche, 1,122 site forms were manually entered into the Uni-Verse database. In order to make the digital archaeological site data GIS compatible, a new Microsoft Access database was designed and populated with site data from UniVerse (Figure 15.2).

In order to make the digital site data GIS compatible, a new MS Access archaeological site database was designed, replacing the existing OSA UniVerse system. This provides OSA with a true relational database system with a variety of standardized forms, queries, and report capabilities. Over 26,000 records are now in the database, with a total of 30 parent tables and 52 lookup tables. A complex data migration plan had to be created and followed, to allow OSA to continue to function in all aspects while the conversion project was underway. In addition, 1,122 archaeological site forms not yet in the UniVerse database were manually entered into the system, thus computerizing virtually all archaeological site data for the state, including the initial project area. This MS Access database has become the foundation for all future OSA activities, permitting the retirement of the UniVerse system. A through quality control review process was conducted by OSA staff to ensure that the database and master quad maps contained correct information within the seven county study area. A new version of the OSA archaeological site form was also created in consultation with OSA that will be more compatible with the new MS Access database, and that is more compatible with GIS and future capabilities, such as recording GPS coordinates of sites.

The second major activity in Task 1 was converting the existing archaeological site data and surveyed areas currently stored on USGS quad maps into GIS

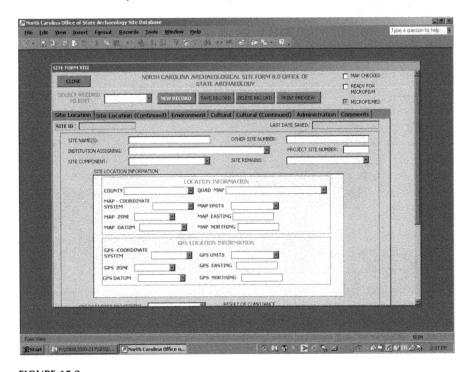

FIGURE 15.2
North Carolina Office of State Archaeology Microsoft Access site database interface that was developed for this project.

data format. A total of 142 master archaeological site maps from OSA covering the seven county project area were scanned using a high resolution color scanner. These maps were scanned, quality inspected, georeferenced, and had archaeological features and surveyed areas extracted into the GIS (Figure 15.3). All quads containing any portion of the seven counties were scanned, so the actual study area included a significant buffer zone around the county boundaries. A total of 14,446 features were extracted using ArcGIS. These include 7,103 polygons, 5,179 points, and 2,184 lines representing the archaeological sites, surveyed areas, and other relevant features. (Figure 15.3). The data from these maps were manually verified, and issues such as duplicate site numbers were resolved with the help of OSA staff. This process required a significant investment of time and effort. In addition to the master OSA quad maps, more than 440 original historic maps of different dates, scales, and extents were also scanned for future project use. Most of these original maps came from the North Carolina Division of Archives and History, and they represent the history of cartography in the state dating back to the early 1700s. Of these, a total of 44 maps, some covering the entire state, others historical county maps, were georeferenced with important cultural features extracted into the GIS.

15.5.1 GIS Data

Existing GIS environmental data to be used in the predictive model development were collected during Task 1. This includes all available data on county

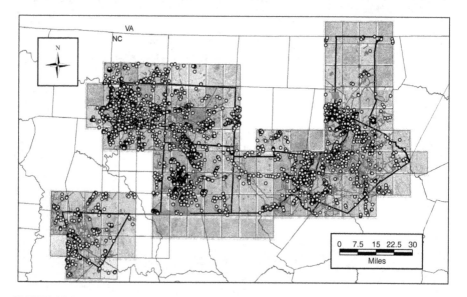

FIGURE 15.3
The seven county study area, showing the georeferenced master OSA quad maps and extracted GIS data representing archaeological sites and surveyed areas. Note that the data includes significant 'edge' data outside the seven county boundaries.

soils, relevant drainage system information, geological base mapping, USGS 1:24,000 and 1:62,500 quadrangles, Digital Elevation Models (DEMs), geomorphological data, digital ortho-quarter quads, additional aerial photography and remote sensing data, and other available relevant information. North Carolina has an excellent state GIS database at the North Carolina Center for Geographic Information and Analysis (NCGIA). The NCDOT GIS Unit was able to provide all of this data, as well as other data useful for this project.

A total of 372 point, vector, and raster GIS data layers covering the project area and the immediate surroundings were acquired, reviewed, and integrated with the new archaeological GIS data created for this project. All major sources of GIS data were searched, and data were reformatted and reprojected as needed to create a database covering the seven county study area. While creating the GIS database, a minimum 2 Km buffer extended beyond the county boundaries to ensure that statistical analysis of data (such as distance to hydrology) was not misrepresented at the edges of the database.

The data collected during Task 1 represent a 20% sample of the Piedmont physiographic province. This area contains 16% of the recorded archaeological sites in the state. This provided a sufficient data sample covering a wide range of environmental variability within the Piedmont region for Task 2, where the Team created the archaeological predictive models for the project area.

15.5.2 TASK 2 Work

Task 2 work was conducted between July 2004 and September 2005. The three main objectives were to create, test, and validate a workable archaeological site predictive model for the test area, to develop a web-based DSS, and to conduct further additions and alterations to the new OSA site form database developed in Task 1.

The DSS development was conducted using the ESRI ArcIMS Internet Mapping System (Figure 15.4). A DSS can be defined as an interactive computer-based system that is used to help managers make decisions. The goal is to improve the ability to access, summarize, and analyze relevant data from multiple sources. In this context, the goal was to provide NCDOT and OSA with a systematized and improved ability to process, access, and integrate archaeological site information and GIS data using a combination of tools that would be delivered through a web-based graphical user interface for a variety of uses. The GIS predictive models and GIS data including the scanned OSA master maps, vector site and surveyed area locations, orthophotos, and georeferenced historic maps are available for viewing using appropriate password protection. Analysis can be conducted, including the generation of buffers, and queries can be conducted on site information. Various reports and maps can be printed. This is done over the internet using a java web browser and high speed internet connection, removing the need for expensive and difficult to use software, broadening

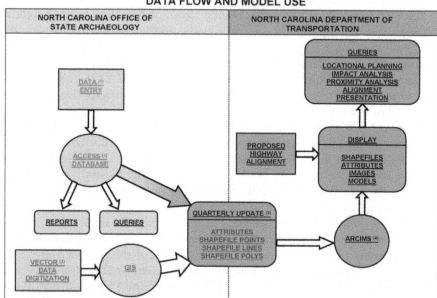

FIGURE 15.4
Project Decision Support System (DSS) Architecture Flow Diagram.

the general utility of the system. Security of archaeological data is an important consideration. As part of Task 2 work, a data sharing mechanism was created (Figure 15.4). This mechanism of data flow was designed to facilitate the combination of archaeological data captured and stored at OSA with the predictive models, GIS data, and highway alignment information housed at NCDOT. Under this design the site data are captured at OSA, shared with NCDOT, and NCDOT serves the data and analyzes it with their road corridors.

15.6 Project Predictive Modeling

The expanding use of GIS in archaeological research in general and site location modeling in particular lies in its ability to capture and manipulate large data sets for analysis and display (Kvamme and Kohler 1988) The generally accepted method of archaeological site predictive modeling in use today is an inductive modeling procedure using a logistic regression analysis technique (Kvamme 1999). In this, the location of a sufficiently large sample of known archaeological sites is statistically analyzed with various other data in the environmental data at the same location (usually, but not always, environmental variables such as elevation, soils, and distance to water). Relationships between archaeological and environmental data are statistically determined, and areas with similar patterns of occurrence are then "predicted" to be of higher likelihood to contain similar archaeological remains. Dr. Ken Kvamme, who was involved in the Mn/Model project and is a recognized leader in the field of archaeological predictive modeling and spatial statistics served as the primary statistical and spatial analysis consultant for this work. All aspects of the spatial analysis and modeling process were reviewed by him over the course of the project.

ArcGIS 9.0 (ESRI) and StatView 5.0.1 (SAS) software was used for the modeling and analysis. The modeling process and results utilized a 30-meter grid cell resolution in order to coincide with the existing 30-meter grid of the National Elevation Dataset (NED). This has sufficient spatial resolution to allow for quantitative field testing of user-defined areas, while being manageable computationally and over the internet.

An initial literature review was conducted of GIS-based archaeological site predictive models to define which environmental variables to use for the modeling. This review included all known modeling projects in the southeastern U.S., as well as major projects such as Mn/Model. The following environmental variables were identified:

- Elevation
- Slope
- Aspect
- Aspect (north/south axis)
- Aspect (east/west axis)
- Solar radiation
- Distance to water
- Vertical distance to water
- Cost distance to water
- Distance to water confluences
- Topographic variation
- Soils
- Landuse
- Distance to historic trading paths

Not all GIS data collected for use in modeling were useful and much of it required manipulation. Elevation data collected from multiple sources had various problems and issues, including: edge effects, biased values near water, etc. Hydrography for the state included many man-made features such as reservoirs and ponds. These all had to be removed and replaced with historic hydrography by digitizing historic maps and stream beds available from USGS quads. STATSGO soil data were available for the entire state, but were far too generalized and spatially homogeneous to be useful in this project. Detailed SSURGO soils data were missing for Chatham county, and so could not be used in this project. There were no existing data for stream confluences or vertical distance to water, so these had to be created using the ArcHydro data model for use with ArcGIS. The data for prehistoric trading paths were determined to be too generalized to be useful for modeling at this scale.

The environmental variables initially created for modeling included: distance to water, cost-distance to water, distance to stream confluence, cost-distance to stream confluence, slope, aspect, solar radiation, and topographic variation. The distance to water and stream confluence were calculated using a Euclidean distance algorithm. The cost-distance to water and confluences were calculated by combining the slope and Euclidean distance. It should be noted that multiple versions of the distance and cost-distance variables were made based on streams created with different thresholds for starting to map a stream. Although slope and aspect are traditionally used for archaeological predictive modeling, we divided aspect into 2 variables. Aspect east to west and aspect north to south variables were created to capture directional trends in the data. Related to the aspect variables, solar radiation was calculated to represent the southwest to northeast exposure of the landscape. Topographic variation was calculated by analyzing the diversity of elevation values in a neighborhood analysis.

The archaeology data that had been generated in Task 1 by digitizing the location of all archaeological resources also had to be further manipulated. Smaller sites were represented as points, but larger sites were digitized as polygons. We standardized these data for the statistical analysis by using the site centroids of the polygon data. The impact of this decision on the statistical analysis is uncertain, as a site may cover a large area with different environmental variables, but since we were dealing with a total of 4,838 archaeological sites in the study area, it was necessary to have a single location for each site to be compared with the environmental data.

Once all of the potential environmental variables were created, the Pearson's Correlation statistic was used to eliminate variables that contained redundant information. Redundant variables can falsely increase the amount of variation explained by a logistic regression model. The Kilmorogov-Smirnov (K-S) test was used to identify variables that showed a relationship with the archaeological data. These data were identified as being useful or relevant to the modeling process.

A total of 11 GIS data layers emerged out of this process:

1. Aspect East/West
2. Aspect North/South

3. Cost Distance to Water 100*

4. Cost Distance to Water 1000

5. Cost Distance to Water 10000

6. Cost Distance to Confluence 1000

7. Cost Distance to Confluence 10000

8. Distance to Stream Confluence 500

9. Elevation

10. Slope

11. Topographic Variation

These were the environmental GIS data used in the statistical analysis of the distribution of the archaeological site data. Logistic regression was used as the statistical technique because of its' ability to be normalized and mapped as probability values (Chou 1997).

15.7 Modeling Research Design

A major concern in this project was the issue of appropriate scope and scale of the predictive model. North Carolina is generally thought to consist of three physiographic provinces, but where does the piedmont end and the mountains begin? How many models are needed to cover all of North Carolina? One single model might be more consistent, but it would certainly be less robust in predictive power, especially at the scale of an individual highway corridor 100 meters wide. The various cultural components are also very different across the state. One hundred county models may be more robust, but we would have significant edge effects between counties that would make this difficult for use in highway planning. Several counties have had very few archaeological surveys, thus providing a limited or biased sample. Models developed at the quad map level might be the most robust in predictive power, but you again have the edge effect problem and quads with either no sites or an insufficient or biased sample. How we balance these issues for this project became a major topic of discussion in project design. In the end, we decided to construct a variety of models, using the same logistic regression technique and environmental data, to quantitatively test these questions. A total of eight different predictive models were created; two at the level of the entire seven county project area, two at the county level (for Randolph and Guilford counties), one at the quad level (for the Asheboro quad within Randolph county), and three using a new technique

---

* Data layers with a number represent calculations based on different stream thresholds.

developed for this project (f0r the Asheboro, Pleasant Garden, and Randelman quads). The results of these different models were evaluated to provide one functional model that covers the seven county study area and recommendations on future modeling efforts by NCDOT.

A new technique was developed by Madry and refined by Madry, Cole, and Kvamme for this project. The goal was to find a new way to balance these competing constraints of model scale and precision. In this approach, a new model is developed for each individual quad that is based on the environmental and site data for not only that quad but the eight surrounding quads as well. This 'roving window' approach was adapted from raster filtering techniques used in image processing. The idea is that the surrounding eight quads of data will create a more locally appropriate model for that area in terms of both the environment and cultural resources, but with a sufficient sample size and with a reduced edge effect.

15.8 Model Validation

One traditional problem with archaeological predictive modeling has been the lack of appropriate validation procedures. Many such models have been created and never validated at all, or have been tested only against a small 'withheld' sample. As the results of this work will be used in a very practical manner by NCDOT, it was our intention to test the models as rigorously as possible, within the time and cost constraints. A total of 4,838 archaeological sites recorded by OSA are located in the study area. A 10% random sample (484 sites) was withheld from the statistical analysis for use as an internal 'jackknife' sample for model verification, leaving 4,354 use in the regression analysis. A second validation sample was acquired from the Research laboratories of Archaeology at the University of North Carolina at Chapel Hill. This is a state-wide database of lithic diagnostics that is separate from the OSA database, which contained a total of 2,242 sites in our study area and includes a significant number of sites not included in the OSA database. It is different in that it only contains the location of diagnostic lithics (stone points). A third validation sample is the archaeological sites located in the 'edge' areas of the study area. All site information was extracted from the quads that border the seven county area, but only the sites within the counties were used in the analysis.

A final validation process consists of a detailed field sample for the NCDOT bypass being planned around the town of Asheboro. A field crew from ESI, with no knowledge of the results of the GIS analysis, conducted a survey of the entire proposed corridor using the standard technique of 30-meter shovel tests. All positive shovel tests were located using differential GPS for comparison with the predictive model for that area. These allowed us to accurately compare and measure model effectiveness.

15.8.1 Results

This work is ongoing, but preliminary results are available at this time. Upon initial evaluation, the county models used different environmental inputs, and the resulting models were very different and the edge effects between counties were pronounced. The roving window procedure worked very well in two cases, but poorly in a third, perhaps due to a limited and biased site sample in that area. The edge effect was significantly reduced. One of the two models of the entire area was the best predictor of sites overall.

This final, seven county model was developed using the following variables: Aspect NS, Cost-Distance to Water_100, Cost Distance to Confluence 1000 and 10000 (major and minor confluences), Slope, and Topographic Variation. This model was then iteratively processed into high, medium, and low probability zones. Figure 15.5 shows the site distribution of the preliminary final model in terms of area and compared with site distribution in the 10% withheld sample, the 90% sample used to create the model, and the UNC lithic database. Also shown are the sites per square mile located in each of these. This model contains 20% of the project area and accounts for 70.8% of the UNC diagnostic lithic sites.

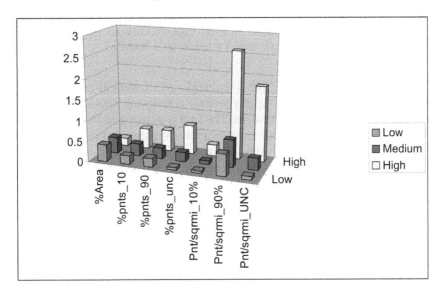

	%Area	%Pnts 10	%Pnts 90	%Pnts unc	Pnt/sqrmi 10%	Pnt/sqrmi 90%	Pnt/sqrmi UNC
Low	0.399447136	0.202479339	0.21221865	0.062890277	0.056711995	0.534713099	0.08159583
Medium	0.398881244	0.29338843	0.263665595	0.228367529	0.082291105	0.665283016	0.296711589
High	0.20167162	0.504132231	0.524115756	0.708742194	0.279674715	2.615646307	1.821324269

FIGURE 15.5
Final seven County Archaeological Predictive Model. Variables Include: Aspect N/S, Cost-Distance to Water, Cost Distance to Water Confluences, Slope, and Topographic Variation. Major river drainages are visible as high probability areas.

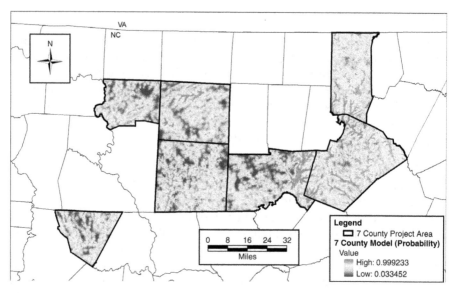

FIGURE 15.6

Results of the preliminary predictive model. At bottom is the low probability zone, middle is medium, and top is high. This is shown for the 10% withheld sample, the 90% sample used for the model, and the UNC diagnostic lithic database. The high zone in the model contains 20% of the area, and 70.8% of the sites in the UNC diagnostic database. On the right are the archaeological sites per square mile for each of the same three datasets.

Task 2 results consisted of both graphical and numerical representations of the probability of a given 30-meter area for containing archaeological sites. Probability is expressed in two ways. The first is as areas of High, Moderate, or Low potential. These were defined as areas that are most likely to contain cultural resources, areas that may contain such resources, and areas least likely to contain them. Some cultural resources will be located in all three categories.

A second product shows the range of probability for site location in a range from 0.0-1.0 (Figure 15.6). This gives the analyst a wider range of information regarding the patterns of probability across the varying landscape.

Refinements to these basic categories are expected as the system is used over time. Graphic representations (Figure 15.7) will include GIS layers delineating a transportation project area's potential for containing archaeological sites. Together, these products should lead to a streamlined road planning and design process, expediting selection of preferred alternatives and project schedules. This project should lead to a more comprehensive understanding of archaeological potential over a given project area and improved decision making.

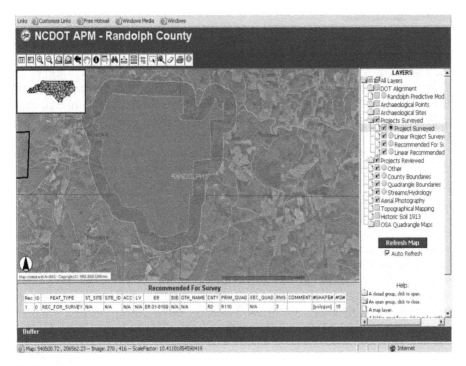

FIGURE 15.7
An example of the ArcIMS web-based Decision Support System interface developed for the
project. This shows an orthophoto and surveyed area, with a buffer zone built around it. Data
for an archaeological site within the area are displayed below. At right are options for viewing
different data. At top are the various display and analysis functions.

15.9 Conclusions

The North Carolina DOT has developed a GIS-based archaeological predictive model for the piedmont region of North Carolina for use in the
merged NEPA/404 permit process. Final model development is underway.
The modeling process we have used shows great potential for predicting
archaeological sites. Existing archaeological site files and relevant other data
(such as historic map information) have been digitized. Also created was a
user friendly, web-accessible DSS that should provide benefit to the NCDOT,
OSA, and other users throughout the state, as well as a user friendly, web-
accessible DSS interface that should provide benefit to the NCDOT, OSA,
and other users throughout the state.

15.10 Future Directions and Recommendations

We have significant additional work to complete on this project. We still have to scan the remaining 827 OSA master quads and digitize the site data from these. We also have to extrapolate our Piedmont model to the rest of that area and test its effectiveness. We must then develop predictive models for the mountains and coastal plain areas of the state.

Detailed SSURGO soils were not used in this analysis because one county did not have them available. Based on our previous experience, this could significantly improve model effectiveness in addition to having general utility for NCDOT.

All sites were considered equal in this analysis, which is arguable, but since the NCDOT is interested in only the presence or absence of cultural resources, no attempt was made at analysis by culture period or site significance, and there is certainly great potential for more detailed analysis of site types and cultural periods in the future.

Traditionally, GIS applications in archaeological research have focused on modeling prehistoric site locations based on the distribution of a suite of environmental variables similar to those mentioned above. To date, the use of GIS in historical research has focused more on the display of cartographic data, reference to site function, and locational distribution. We collected and processed historic archaeological data for this analysis, but we were not able to conduct the historic analysis due to time and funding constraints. We hope to do this in the future, as the NEPA/404 process concerns all cultural resources.

Beyond the immediate project goals we see benefits in improved access to integrated data, ongoing updating and processing of state archaeological records, and improved teaching and research potential.

References

Chou, Y-H., Exploring Spatial Analysis in Geographic Information Systems, Onword Press, 1997, pp. 474.

Kvamme, K., Recent directions and developments in geographical information systems, *Journal of Archaeological Research*, 7(2):153-201, 1999.

Kvamme, K. L. and Kohler, T. A., Geographic Information Systems: Technical Aids for Data Collection, Analysis, and Display. In Quantifying the Present and Predicting the Past: Theory, Method, and Application of Archaeological Predictive Modeling, W. J. Judge and L. Sebastian, Eds., U.S. Government Printing Office, Washington, D.C., 1988, pp. 493-547.

Hudak, G.J. and Hobbs, E., Mn/Model: A Statewide. Archaeological Predictive Model for Minnesota, Minnesota Department of Transportation, St. Paul, MN, nd.

16

Multicriteria/Multiobjective Predictive Modeling: A Tool for Simulating Hunter-Gatherer Decision Making and Behavior

Frank J. Krist, Jr.

CONTENTS

16.1 Introduction

Geographic information system (GIS)-based multicriteria/multiobjective predictive models have been successfully used to predict the location of archaeological sites. However, such models often lack explanatory power and are unable to identify the range of behaviors occurring at the archaeological sites they locate. The research presented here — making use of the multicriteria/multiobjective decision-support tools found within the GIS environment and with guidance from "satisficer" approaches and decision theory — proposes a model that simulates behaviors resulting from the decisions made by prehistoric hunter-gatherers. The proposed model, based on a hypothesized hunter-gatherer adaptive strategy, can be used to simulate the location and nature of activities relating to resource use and settlement within the landscape and, thus, to predict the types of sites or activity areas that would be expected within a region.

This chapter describes the development of a model that simulates the behavior of hunter-gatherers based on their decisions or adaptive strategies in various real-world settings. Unlike other approaches (Binford 1980; Keene 1979, 1981; Belovsky 1987; Kelly 1995), the research design provides an efficient means for interpreting the settlement systems of hunter-gatherers without relying on direct comparisons between specific ethnographic and archaeological data sets. Following the lead of Jochim (1976, 1981, 1998), this model considers hunter-gatherer subsistence and settlement behaviors as the result of an adaptive strategy with a series of embedded decisions chosen for the ability to resolve problems while satisfying various objectives/goals (Figure 16.1). This approach, rooted in Jochim's satisfier theory, provides a means to simulate the adaptive responses (behaviors) of hunter-gather groups that result from the utilization of a particular strategy in a particular real-world setting. Bettinger (1980) argues that traditional hunter-gatherer models have been unsuccessful in this area.

This research is applicable to general hunter-gatherer studies in several ways:

- The approach provides an alternative to the more traditional models of hunter-gatherer adaptive strategies, which depend on archaeological or ethnographic data for construction (Bettinger 1991). Models constructed solely from archaeological data contain biases or inaccuracies as a consequence of regional and local taphonomic processes, site-formation processes, and data-recovery methods (Keene 1981; Raab and Goodyear 1984). According to Wobst (1978), the use of ethnographic data to decipher the past places a perceived ethnographic reality or construct onto the interpretation of past hunter-gatherer lifeways, leading to biases in the interpretation of the archaeological record. The proposed model does not rely on either ethnographic analogies or a direct comparison between the ethnographic and archaeological records, thus allowing archaeologists to test hypotheses about past human behaviors in regions with a paucity of archaeological data or ethnographic analogs. This capacity also allows researchers to develop simulations (scenarios) or to predict past adaptive strategies independent of preconceived notions about hunter-gatherer behaviors. However, the model can also accommodate available ethnographic and archaeological data to provide valuable insights into the construction of alternative hypotheses that can be used as a point of departure for further model development.

- The approach is dynamic, with the ability to simulate hunter-gatherer behaviors for any given time and at any spatial resolution.

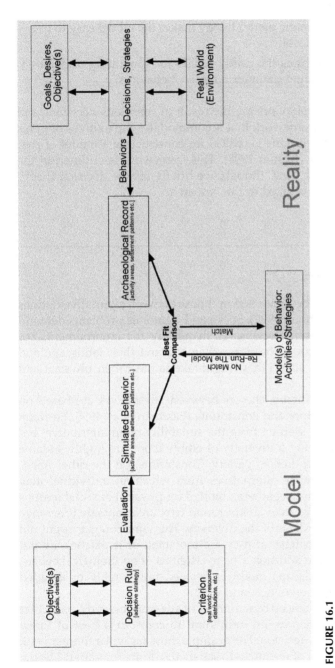

FIGURE 16.1
The approach advocated in this research, which simulates the behavior of hunter/gatherers.

338 *GIS and Archaeological Site Location Modeling*

- The model is flexible, accommodating a wide range of hunter-gatherer adaptive strategies, goals, and objectives, which can be social or economic in nature.

- The proposed systematic methodology makes the model easy to use and the results replicable.

- The study demonstrates the utility of GIS for generating decision-based models of hunter-gatherer adaptive behaviors.

Because most human behaviors are the result of conscious decisions and traditions, a theoretical framework that accommodates the motives or goals guiding (constraining) decisions is critical for constructing a model of past hunter-gatherer behavior (Jochim 1998). This framework is outlined in the remainder of this chapter. First, though, we briefly review decision theory as it relates to the model utilized in this research.

16.2 Decision Theory

Decision theory outlines the logic used to choose between alternative options (Eastman et al. 1995; Eastman 1999). Social and economic problems demand a response, and the chosen response is shaped by the environment. For example, imagine the view of hunter-gatherers about their landscape and the alternative courses of action they might take to meet their physical and social desires.

According to decision theory, choices between alternatives are based on two forms of criteria: factors and constraints (Eastman et al. 1995, Eastman 1999). Factors enhance or detract from the suitability of an alternative. For example, patches containing a diversity of edible foods are highly suitable for an individual seeking dietary variety. Constraints, on the other hand, limit the number of possible alternatives from which an individual may choose. In the case of a hunter-gatherer, limited for physical or social reasons to foraging trips no greater than 20 km round trip, areas outside this range would be excluded, regardless of the diversity. The same forager might not utilize diverse resource patches if they are too small or if patches are not adjacent to one another. In addition, due to religious or mythological beliefs, some regions of the landscape might be viewed unfavorably or avoided altogether by hunter-gatherers (Cleland 1992).

Many decisions are influenced by a number of alternatives, and a particular real-world setting may require an individual to consider a host of fuzzy criteria prior to making a decision. For example, how far will a forager walk to reach a particular patch of resources to satisfy the desire for variety? Using this example, the forager contemplates a complex set of factors and constraints, such as walking distances between and within resource patches, patch size, patch diversity, terrain roughness, social and political boundaries, and possibly mythological beliefs about the landscape. Therefore, the researcher considers the degree to which varying criteria trade off to affect the final decision-making process.

Objectives help determine the real-world features that become criteria for an individual or a band of hunter-gatherers. Although objectives are based on social perspective and motives that may vary between bands or individuals, a limited set of universal goals guide the choices that hunter-gatherers make about such issues as resource use and settlement placement (Eastman et al. 1995; Jochim 1976, 1998). This limited set of goals is reinforced as a population begins to meet its needs, and the objectives are either consciously or subconsciously reinforced, resulting in patterned and redundant decision making (Egan 1993; Jochim 1998).

The final selection and evaluation of criteria is aided by a strategy, or set of strategies, known as a decision rule (Eastman et al. 1995; Eastman 1999). The decision rule, formulated in the context of a given set of objectives, also contains the means by which selected alternatives are acted upon. For example, hunter-gatherers, utilizing a forager strategy (Binford 1980) in which settlement-location decisions and resource patches (criteria) are guided by a desire to move the entire group to resources, would find locations that provide direct access to resources more favorable.

16.3 A Predictive Model of Hunter-Gatherer Decision Making

Predictive modeling in archaeology really began with Jochim's progressive 1976 work *Hunter-Gatherer Subsistence and Settlement: A Predictive Model* (Bettinger 1998). Jochim demonstrated that predictive models of subsistence, settlement, and population size could be generated from the presumed goals of a population and the resource characteristics of a region that a society occupied. The shortcoming of Jochim's predictive model was that it relied heavily on quantitative measures as input (Bettinger 1998; Jochim 1998). Precise quantitative measures are difficult to obtain, considering the unknowns about past ecosystems and the beliefs and strategies that prehistoric peoples adopted. Despite these limitations, archaeologists continued to develop predictive models relying heavily on quantitative measures for model inputs (Reidhead 1979, 1980; Keene 1981; Mithen 1990). Undoubtedly, these models contributed greatly to our understanding of the behaviors of past societies; however, such models are often difficult to utilize because of their quantitative requirements (Jochim 1998). An alternative to building predictive models on precise mathematical measurements is to utilize general relationships among factors. However, simplistic models are more likely to be unrealistic. Egan (1993) has worked to resolve this problem through the use of detailed information, producing rankings that depict the general relationships between resources or factors. To address these issues, this research takes a unique approach by standardizing information about factors affecting the behavior of hunter-gatherers through the production of ratings within a comparison matrix. Ratings can be based on either general relationships among factors, i.e., resource A tastes better than resource B, or quantitative measures that specifically assess the ability of a criterion to satisfy an objective, i.e., resource A has ten more grams of sugar than resource B. The ability to utilize both precise and nonprecise measures when assigning ratings to every criterion enables the model to take advantage of all available information about the past while allowing the model to grow as the archaeological knowledge base increases.

Over the last 20 years, the development of geographic information systems (GIS) has stimulated the use of predictive models in archaeology (Kvamme 1995). These models are useful in identifying regions with a high archaeological site potential. Such information is particularly useful for cultural resource management studies and the preservation/avoidance of regions likely to contain archaeological resources (Brooks et al. 1996). In general, the ability of GIS to efficiently manipulate and analyze large amounts of spatial and tabular data makes it a particularly useful tool for the construction of multivariate predictive models (Kvamme 1999).

Traditionally, predictive models constructed within GIS are limited to confirming the existence of spatial correlations between past activity areas and physiographic features (Ebert 2000). Recently, archaeologists have developed GIS-based predictive models that consider how humans make choices (Dalla Bona 2000). However, these models fail to consider the objectives of past hunter-gatherers or how those goals affected decision making. Because such models are unable to explain why hunter-gatherers conducted activities in the locations they did, the applicability of GIS-based predictive modeling for archaeological problem solving is in question. The research presented here demonstrates how a predictive-modeling approach rooted in both anthropological and decision theory can be efficiently implemented within a GIS system and used to simulate the hunter-gatherer decision-making process and thus archaeological problem solving.

The primary assumption of the behavioral model outlined here is that the conscious decisions made by hunter-gatherers are directed toward achieving or satisfying a set of specific goals or objectives, such as those outlined by Jochim (1976, 1998). Another assumption is that hunter-gatherers make rational decisions based on their ability to satisfy their objectives. Therefore, goals and objectives are the guiding force behind the development of a hunter-gatherer adaptive strategy.

The approach offered here is designed to simulate the outcome of hunter-gatherer problem solving, and although it is easily implemented within a raster GIS system, such a system is not required to run the model, and evaluations can be calculated by hand. The model comprises three modules in which both hypothesized and observed information about hunter-gatherer behavior and the environment, related to a particular problem such as settlement placement, can be input: (1) objectives, (2) decision rule, and (3) criteria (Figure 16.1). The first module requires the researcher to determine the objectives of the hunter-gatherer group under study and the relative importance of these objectives in solving a particular problem. For the second module, the researcher establishes the adaptive strategy utilized by the hunter-gatherer group to meet their objectives. For example, a group of hunter-gatherers might determine that, in order to meet an objective of minimizing energy expenditure while obtaining enough food resources for survival, a forager strategy dependent on large game is the best solution. The decision to pursue large game and the objectives surrounding this choice help to formulate a worldview by determining which features on the landscape become the factors and constraints that are entered into the model using the criterion module. Factors and constraints are not limited to features on the landscape. They may be culturally derived, for example, in which groups or individuals avoid areas of the landscape due to social or religious beliefs. Once data are collected for each module, suitability values are calculated (using a weighted linear multicriterion/multiobjective evaluation) for comparison with the archaeological record to predict the likelihood that human behavior(s) occurred at a particular location(s).

Although GIS software is not required to run the modules outlined below, it is particularly useful when examining an entire region or evaluating a large number of objectives and criteria. The raster GIS software IDRISI is recommended because it possesses a comprehensive set of tools for conducting multicriterion/multiobjective evaluations (Eastman 1999; Malczewski 1999).

16.3.1 Module One: Objectives

Once a set of objectives has been defined, the relative importance of each objective is entered into a pairwise comparison matrix. This matrix is used to generate a set of weights representing the relative importance of each objective in affecting hunter-gatherer behaviors. Prior to entering values into the pairwise matrix, comparisons must be made between each objective using a continuous rating scale. Ratings represent the relative importance of each objective compared with the others (Table 16.1). For example, if direct access to resources is significantly more important in determining settlement placement, this objective receives a value of 5. Table 16.2 depicts a sample pairwise comparison when every possible pairing of objectives is entered into the matrix.

After entering the ratings into the comparison matrix, the principle eigenvector is calculated and then used to produce a best-fit set of weights from the criteria ratings (Table 16.3). Within IDRISI, the Weight command is used to calculate these weights, which sum to one. The Weight module also determines the degree of uniformity that was utilized in generating the criteria ratings by producing a consistency ratio every time a set of weights

TABLE 16.1

Continuous Rating Scale

Description	Comparison Rating
Most important	1
	1/2
Moderately less	1/3
	1/4
Strongly less	1/5
	1/6
Very strongly	1/7
	1/8
Extremely less	1/9
	1/10
Unsuitable	N/A

Source: After Eastman, J.R., *IDRISI 32: Guide to GIS and Image Processing*, Vol. 2, Clark Labs, Clark University, Worcester, MA, 1999.

TABLE 16.2

Sample Pairwise Comparison Matrix for the Objective of Settlement Placement

	Resource Proximity	Shelter	View
Resource Proximity	1	5	7
Shelter	1/5	1	2
View	1/7	1/2	1

TABLE 16.3

Sample Weights Resulting from Calculation of Principal Eigenvector of the Pairwise Comparison Matrix for Settlement Placement

Criteria	Weight
Resource proximity	0.7396
Shelter	0.1666
View	0.0938

is calculated. A ratio of less than 0.10 indicates that the pairwise matrix values were not generated at random. When values greater than 0.10 are obtained, the matrix ratings require reevaluation. An approximation of the weights generated from the principal eigenvector can be calculated by filling out the entire matrix and summing each column to get the column marginal total. Each rating in the matrix is then divided by the marginal total of its column. Finally, the weights across the rows are averaged.

16.3.2 Module Two: Decision Rule

The decision rule requires the researcher to determine what strategy a group of hunter-gatherers might have used to meet their objectives and is generally derived from a hypothesis, or set of expectations, of how peoples behaved in the past. Information about the decision rule is not entered directly into the model *per se*; rather, these data determine the way in which objectives and criteria interact to produce a final outcome. Once a group of hunter-gatherers select a strategy to meet their objectives, that strategy determines how resources, settlement locations, other bands, etc. are considered within the decision-making process (Jochim 1998).

Two broad, economically focused examples of a decision rule are Binford's (1980) forager and collector settlement strategies. Food resources play a different role in the decision-making process for each of these strategies. Foragers choose residential locations based on the proximity to food resources including water, while collectors select residential settlement locations for their proximity to key nonfood resources required for immediate survival, such as water (Kelly 1995). Collectors use logistical forays as a primary means of bringing edible resources to settlements. When simulating the activities of foragers, immediate access to food resources is the dominant criterion affecting decisions regarding forager settlement patterns. Each resource acts as a criterion, since foragers move people from resource to resource. For collectors, adjacency to key nonfood resources in areas that allow remote access to a wide range of edible resources is the main criterion affecting decisions about collector settlement patterns. Thus, direct access to food resources plays only a minimal role in collector settlement selection.

16.3.3 Module Three: Criteria

Criteria, or factors, are features both physical and cultural that enhance or detract from the suitability of a location toward meeting an objective (Eastman 1999). As discussed earlier, determining which criteria play a role in hunter-gatherer decision making is based on the objectives and adaptive strategy (decision rule) found within a group of hunter-gatherers. Depending on the number of criteria the researcher identifies for each objective, the model will be relatively simple or quite complex. Factors can also be changed and the model recalculated to test different hypotheses. The criteria themselves are often generated from, but not limited to, known facts about the landscape, such as the availability of plant or animal resources. The objective of obtaining shelter from the elements, for example, may consist of several factors, such as forest cover, slope, and aspect (Table 16.4). As a result, this type of model places the driving force behind the criterion rankings in the pairwise matrix on the desires of the hunter-gatherers under study. The model is a vehicle by which hypotheses about the desires of hunter-gatherers can be tested and the outcome simulated.

TABLE 16.4

Sample Pairwise Comparison Matrix for Shelter Factors

	Forest Cover	Slope	Aspect
Forest Cover	1	—	—
Slope	2	1	—
Aspect	2	1	1

Note: Only the lower half of the table needs to be filled out.

Each objective is assigned a group of criteria for which a comparison matrix is constructed (Figure 16.2a). After calculating factor weights for each objective (Table 16.5), a list of constraints, if any exist, must be identified. Constraints are not ranked, but represent areas to avoid regardless of the factors that may be present. Regions may be inaccessible due to territorial boundaries or because of a physical desire to walk no more than a specific distance.

To represent each factor within a GIS, raster cells are assigned numeric values based on the potential to satisfy a particular criterion. For the factor of resource proximity as it relates to the objective of logistical settlement placement, areas that are adjacent to a suitable patch are assigned a value of 10 (most suitable), while a location greater than a specified distance receives a value of 0 (unsuitable). In this example, factors receive a numeric value ranging from 0 to 10 to ensure standardization and allow the comparison of criteria that are based on different techniques or values, such as slope and tree cover. Other common standardized scales include 0 to 1 and 0 to 255. Standardization is achieved by performing a linear scaling between the minimum and maximum factor values as follows (Eastman 1999):

$$x_i = (R_i - R_{min})/(R_{max} - R_{min})^*standard\_range$$

where R = raw score

Depending on the adaptive strategy, objectives may be satisfied by only a single criterion. For example, resource use at logistical camps is often intensely focused on a single resource patch or activity (Binford 1980; Jackson 1998). To identify which resource a group of foragers might focus on at a logistical site, the suitability of each resource is evaluated individually in the context of the group's objectives.

When a single criterion exists for each objective, a comparison matrix is not necessary because factors are not being compared with one another to determine how well a group of criteria satisfies each objective (Figure 16.2b). Instead of assigning a weight to a criterion using a continuous rating scale, an individual criterion is ranked (by percent) based on how well it satisfies each objective (Table 16.6). Percents range from zero to 100, with 100% fully satisfying an objective. The percent rank adjusts the original factor value according to its ability to satisfy an objective. To adjust the value, each factor rank is multiplied by the standardized criterion value, thus lowering the

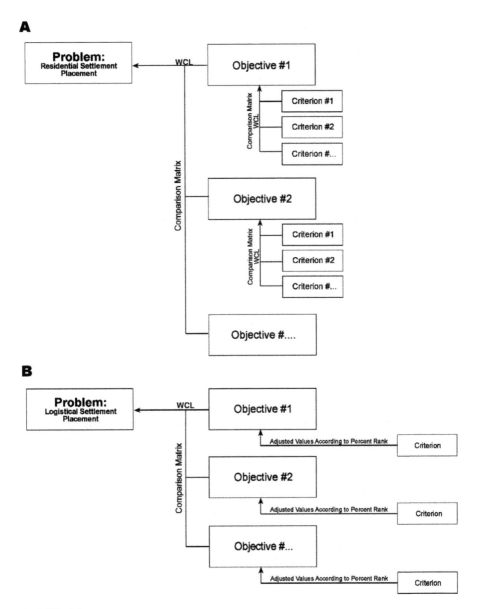

FIGURE 16.2
Method of combining criteria for objectives with multiple criteria (A) single criterion (B).

original value. For example, an unreliable and unpredictable resource would not satisfy an objective to minimize risk, and therefore it receives a low rank. If the resource has a suitability value of 8 at a particular location and the rank is 40%, the adjusted value is 3.2. On the other hand, this same resource

TABLE 16.5

Sample Weights Resulting from Calculation
of Principal Eigenvector of the Pairwise
Comparison Matrix for Shelter Factors

Criteria	Weight
Forest cover	0.2
Slope	0.4
Aspect	0.4

TABLE 16.6

Rank Percent Equivalents for Criteria Ratings

Percent Rank	Importance Ranking
100%	Most important (1)
90%	(1/2)
80%	Moderately less important (1/3)
70%	(1/4)
60%	Strongly less important (1/5)
50%	(1/6)
40%	Very strongly less important (1/7)
30%	(1/8)
20%	Extremely less important (1/9)
10%	(1/10)
0%	Not a factor (0)

might satisfy the population aggregation objective of a group and thus receive a high rank for this goal.

16.3.4 Multicriteria/Multiobjective Evaluations

Once the factor values and a set of weights are generated for each objective and the objectives themselves, the information is combined in the IDRISI MCE module using a series of weighted linear combinations (Figure 16.2). Within the first set of weighted linear combinations, the factors are combined by multiplying the weight of each criterion by its value followed by a summation of the results (Eastman 1999):

$$S = \text{suitability}$$

$$S = \sum w_i x_i^* \quad \text{where} \quad w_i = \text{weight of factor}$$

$$x_i = \text{criterion score of factor } i$$

In the locations containing constraints, the following equation would apply:

$$S = \sum w_i x_i^* \prod c_j \quad \text{where} \quad c_j = \text{criterion score of constraint } j$$

$$\prod = \text{product}$$

The output from the weighted linear combination (WLC) is typically a value ranging from 0 to 10 (depending on the common scale), with the highest values representing the most suitable locations for meeting a particular objective. The results of the weighted linear combinations for all the objective's criteria are combined using another WLC, with the weights generated from a pairwise comparison matrix created for the objectives. This produces a final suitability value depicting the likelihood that a location was utilized as a solution to a problem that hunter-gatherers encountered. The entire procedure for developing objective/factor values and weights and their combination is summarized in the flowchart in Figure 16.3.

If a set of objectives contains a single factor, such as the problem of logistical resource use, the adjusted values for this factor are also combined using the WLC (Figure 16.4). When considering logistical resource use, a WLC is calculated for each resource type based on its ability to satisfy each objective. In addition to these calculations, the IDRISI MDChoice (multidimensional choice) module can be used to produce a summary map depicting the regions in which logistical-resource-extraction activities are most likely to have taken place. The MDChoice command performs a multidimensional choice procedure identifying the highest suitable value occurring at each cell. MDChoice records the resource type with the highest suitability at any one location. A suitability threshold is set within MDChoice that represents the minimum value needed before any resource is recorded for a location.

The results from the weighted linear combinations identify the range of potential behaviors that may have occurred at a particular location based on hypothesized objectives of the hunter-gatherers under study. Correlations between simulated activity areas and the archaeological record suggest that the culture under study did indeed base their decisions on the simulated adaptive strategy.

Hunter-gatherer decision making is complex and multidimensional, and care must be taken not to oversimplify. Despite this complexity, some criteria and objectives likely play a more significant role in the decision-making process. For example, Jochim (1976) has identified a set of cross-cultural goals or motives related to resource use that often guide hunter-gatherer decision making:

- Attainment of a minimum amount of food and manufacturing materials
- Population aggregation
- Efficiency in the form of minimizing expenditures of energy and time
- Risk minimization

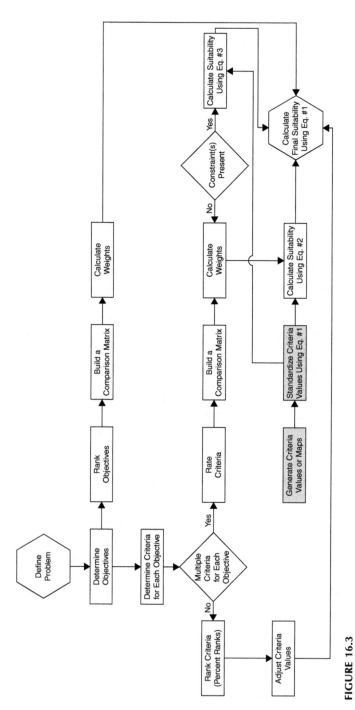

FIGURE 16.3
Flow chart for the behavioral model outlined in this chapter.

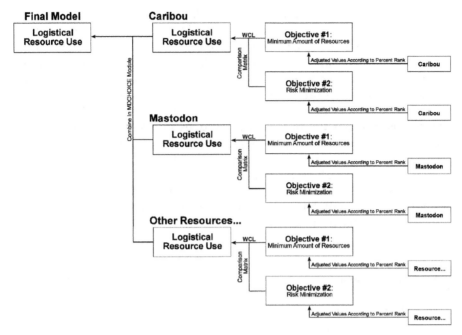

FIGURE 16.4
Method of combining criteria when only a single criterion exists for each objective.

- Attainment of good-tasting foods
- Attainment of a variety of foods
- Attainment of resources that carry prestige
- Maintenance of differentiation of sex roles

Jochim's work provides a point of departure for model building using the approach advocated in this chapter.

How do we account for risk taking that several researchers (Cashdan 1990; Halstead and O'Shea 1989; Jochim 1998) have identified as a critical part of hunter-gatherer decision making? Risk can be accounted for in this model in three ways. The simplest way to examine the possible effects of risk minimization is to use the highest suitability values in the final output, eliminating any regions that have the potential to be risk averse. The second means of modeling risk minimization is to incorporate it as an objective of the hunter-gatherers that are being studied, as Jochim (1976) did. Having risk as an objective allows control over the weight, or effect, that risk has on behavior. This method has the potential to address the question of the extent to which hunter-gatherers factored risk minimization into their decision making. The third method, which is the most complicated, can be used in conjunction with the second. An ordered weighted averaging (OWA) approach can be used to combine the factor values. The OWA method is similar to the WLC, but the OWA accommodates a second set of weights. This second set of weights control the degree to which weighted factors are aggregated (Eastman 1999). In the OWA approach, factors with low values receive extra weight in the outcome regardless of their original weight. Thus, highly weighted factors are prevented from dominating suitability scores in locations where all other factor scores are much lower. Therefore, orienting the second set of weights toward the minimum factor values ensures that the final suitability does not contain high values in locations based on any single factor. For example, weighting minimum-factor values to a higher value creates a low-risk suitability model in which higher suitability values represent areas containing the likelihood that several resources will be available in that area. If one resource fails, the hunter-gatherer will have something to fall back on. The latter two approaches are used in this chapter.

16.4 Discussion

The model presented in this chapter provides a framework for simulating the behaviors of hunter-gatherers. Recent research conducted by Krist (2001) has demonstrated the utility of this approach in the study of prehistoric peoples. With the aid of detailed digital cartographic representations of late Pleistocene floral and faunal distributions, Krist successfully simulated behaviors related to logistical and residential resource use and settlement for early paleo-Indian hunter-gatherers occupying Lower Michigan.

In addition to its applicability to prehistoric societies, the model can also aid research on existing hunter-gatherers. For example, within a modern culture, where the goals and objectives are not clearly understood, the model can be run and rerun to determine which results match the observed patterns found within a culture. Anthropologists can then objectively reevaluate cultural data recorded in the ethnographic record. Once a firm set of criteria and objectives are generated for a culture, the model can be used as a tool to evaluate the effects of potential environmental and cultural change on existing indigenous cultures. Such an approach could be particularly useful in regions where significant ecological and cultural changes are taking place.

The overarching methodology behind the behavioral model outlined within this work is rigorous for several reasons:

- The model is relatively easy to implement and is repeatable. The approach is also flexible, accommodating a wide range of information (criteria and objectives). This flexibility allows the model to be used at any scale, from the site to the regional level.

- Data with different units of measure, such as taste and soil dryness, can be incorporated into a single model. Upon review, or with the introduction of new evidence, criteria scores and rankings can be changed and the model recalculated to produce updated results. Quantitative approaches for archaeological predictive modeling can be used to provide insights into criteria weights and scores.

- The data being entered into the model need not be highly accurate. Therefore, the model can be run using different levels of accuracy.

- The model was designed to accommodate the hunter-gatherer decision-making process. In implementation, the design of the model enables the incorporation of the goals and desires of a group of hunter-gatherers being studied.

- Although relatively simple, the model can incorporate an unlimited amount of data layers. With the development of GIS, it has become easier to incorporate a large number of criteria, goals, and objectives into a model such as the one presented here.

- Once a set of criteria and objectives have been tested and entered into the behavioral model, the model can be used to predict the location of various activity areas on a landscape that has not undergone archaeological survey. This can aid archaeologists in the practical aspects of resource management by pinpointing regions that should be avoided during destructive land-altering activities.

There are two limitations to the model presented here. First, the accuracy of the results is, of course, limited by the least-accurate data set entered into the model. Therefore, great care must be taken to understand the limitations of data entered into the model. Second, the results generated within a multicriterion approach are subject to the effects of error propagation. When uncertainty exists within the data entered into the model, error will propagate throughout the analysis (Eastman 1999; Malczewski 1999). In addition, inaccuracies from different layers will combine, compounding the errors within the final output. This is a particularly acute problem for archaeologists who utilize data collected about the past, which frequently possess some degree of uncertainty (Jochim 1998). Further compounding this shortcoming, testing the output of such models is frequently difficult because of the lack of data for comparison. However, there are several techniques for quantifying these errors, such as the Monte Carlo simulation and the analytical error-propagation method (Malczewski 1999). A simpler way to minimize the effects of error propagation is to reclassify the output suitability values into generalized classes, such as the high, medium, and low categories, that can be used until the accuracy of input data can be determined.

Despite these limitations, the model presented in this chapter holds great promise of providing anthropologists with an objective means of simulating and understanding the behavior of hunter-gatherers. The model also demonstrates the importance of anthropological theory in the development of archaeological models constructed within a GIS environment, and it shows how the GIS environment can be used as a tool for archaeological problem solving.

References

Belovsky, G.E., Hunter-gatherer foraging: a linear programming approach, *Journal of Anthropological Archaeology*, 6, 29–76, 1987.

Bettinger, R.L., Explanatory/predictive models of hunter-gatherer adaptations, in *Advances in Archeological Method and Theory*, Vol. 3, Schiffer, M., Ed., Academic Press, New York, 1980, pp. 189–255.

Bettinger, R.L., *Hunter-Gatherers: Archaeological and Evolutionary Theory*, Plenum Press, New York, 1991.

Bettinger, R.L., foreword to *A Hunter-Gatherer Landscape: Southwest Germany in the Late Paleolithic and Mesolithic*, by Jochim, M.A., Plenum Press, New York, 1998.

Binford, L.R., Willow smoke and dog's tails: hunter-gatherer settlement systems and archaeological site formation, *American Antiquity*, 45, 4–20, 1980.

Brooks, A., Hudcak, G.J., Gibbon, G.E., and Hobbs, E., *A Predictive Model of Precontact Archaeological Site Location for the State of Minnesota*, report prepared for Minnesota Department of Transportation, Minneapolis, 1996.

Cashdan, E., *Risk and Uncertainty in Tribal and Peasant Economics*, Westview Press, Boulder, 1990.

Cleland, C.E., *Rites of Conquest: The History and Culture of Michigan's Native Americans*, University of Michigan Press, Ann Arbor, 1992.

Dalla Bona, L., Protecting cultural resources through forest management planning in Ontario using archaeological predictive modeling, in *Practical Applications of GIS for Archaeologists: A Predictive Modeling Kit*, Wescott, K.L. and Brandon, R.J., Eds., Taylor and Francis, London, 2000, pp. 73–99.

Eastman, J.R., *IDRISI 32: Guide to GIS and Image Processing*, Vol. 2, software manual, Clark Labs, Clark University, Worcester, MA, 1999.

Eastman, J.R., Jin, W., Kyem, P.A.K., and Toledano, J., Raster procedures for multi-criteria/multi-objective decisions, *Photogrammetric Engineering and Remote Sensing*, 61 (5), 539–547, 1995.

Egan, K.C., Hunter-Gatherer Subsistence Adaptation in the Saginaw Valley, Michigan, Ph.D. dissertation, Department of Anthropology, Michigan State University, East Lansing, 1993.

Halstead, P. and O'Shea, J., Introduction to *Bad Year Economics*, Halstead, P. and O'Shea, J., Eds., Cambridge University Press, Cambridge, 1989, pp. 1–7.

Jochim, M.A., *Hunter-Gatherer Subsistence and Settlement: A Predictive Model*, Academic Press, New York, 1976.

Jochim, M.A., *Strategies for Survival: Cultural Behavior in an Ecological Context*, Academic Press, New York, 1981.

Jochim, M.A., *A Hunter-Gatherer Landscape: Southwest Germany in the Late Paleolithic and Mesolithic*, Plenum Press, New York, 1998.

Keene, A.S., Economic optimization models and the study of hunter-gatherer subsistence settlement systems, in *Transformations: Mathematical Approaches to Culture Change*, Renfrew, C. and Cooke, K., Eds., Academic Press, New York, 1979, pp. 369–404.

Keene, A.S., *Prehistoric Foraging in a Temperate Forest*, Academic Press, New York, 1981.

Kelly, R.L., *The Foraging Spectrum: Diversity in Hunter-Gatherer Lifeways*, Smithsonian Institution Press, Washington, DC, 1995.

Krist, F.J., Jr., A Predictive Model of Paleo-Indian Subsistence and Settlement, Ph.D. dissertation, Department of Anthropology, Michigan State University, East Lansing, 2001.

Kvamme, K.L., A view from across the water: the North American experience in archaeological GIS, in *Archaeology and Geographical Information Systems: A European Perspective*, Lock, G. and Stančič, Z., Eds., Taylor and Francis, Bristol, PA, 1995, pp. 1–14.

Kvamme, K.L., Recent directions and developments in geographic information systems, *Journal of Archaeological Research*, 7 (2), 153–201, 1999.

Malczewski, J., *GIS and Multicriteria Decision Analysis*, John Wiley and Sons, New York, 1999.

Mithen, S., *Thoughtful Foragers*, Cambridge University Press, Cambridge, 1990.

Raab, L.M. and Goodyear, A.C., Middle-range theory in archaeology: a critical review of origins and applications, *American Antiquity*, 49, 255–268, 1984.

Reidhead, V.A., Linear programming models in anthropology, *Annual Review of Anthropology*, 8, 543–578, 1979.

Reidhead, V.A., The economics of subsistence change: a test of an optimization model, in *Modeling Change in Prehistoric Subsistence Economies*, Earle, T. and Christenson, A., Eds., Academic Press, New York, 1980, pp. 141–186.

Wobst, H.M., The archaeo-ethnology of hunter-gatherers, or the tyranny of the ethnographic record in archaeology, *American Antiquity*, 43, 303–309, 1978.

17

Predictive Modeling in a Homogeneous Environment: An Example from the Charleston Naval Weapons Station, South Carolina

Thomas G. Whitley

CONTENTS

ABSTRACT   Using what has typically been referred to as a "deductive" approach to creating probabilistic formulas seems to be more reliable than traditional "inductive" methods in the context of very homogeneous environments, where slope and distance to water are not discriminating factors. However, the terms "deductive" and "inductive" are incorrect in the context of probabilistic modeling, and the ultimate dichotomy must be considered one of probabilistic causality versus inferential determinism. Inferentially deterministic models are limited by the nature of inductive-statistical scientific explanation and their dependency on flawed data sets. The alternative model presented here, though limited by the reliability of research in the region as a whole and the nature of the environmental and social variables employed in the model, produces multiple "expert system"-based formulas that cast a much wider net over site-selection behavior. The result is a model that can address very homogeneous environments, very specific site types, and sites with poor representation in the archaeological record. The approach also forces us to think about the ways in which site placement is cognized, rather than providing just a single "lowest common denominator" type of predictive formula.

17.1 Introduction

The research presented in this chapter was part of a project to evaluate the potential for archaeological resources located within the boundaries of the Charleston Naval Weapons Station (NWS), Berkeley and Charleston Counties, South Carolina (Whitley 1999). Brockington and Associates, Inc., conducted a geographic information system (GIS) evaluation of the known archaeological sites and the available environmental information to understand (a) the distributions of potentially significant archaeological sites and (b) the areas where such sites might occur but are unknown at present. Identifying particularly sensitive areas allows foresight into alternative evaluations for future projects, potential avoidance or mitigation strategies, and the application of differential survey priorities.

Following a brief overview of archaeological probabilistic modeling, this chapter presents a discussion of modeling in a homogeneous region, addresses some of the factors that led to the model strategy, and presents modeled results. A detailed analysis of the environmental or cultural background of the region is not included, but a number of sources can be consulted for additional information (Adams 1987; Anderson and Logan 1981; Anderson et al. 1979; Brooks and Canouts 1984; Cable 1993, 1996; Drucker and Anthony 1979; Goodyear and Hanson 1989; Goodyear et al. 1979; Panamerican Consultants 1997; Rust 1997; Soil Systems 1982; Stephenson 1998; Stine 1991; Stine et al. 1993; Tidewater Atlantic Research 1995; Zierden et al. 1986).

17.2 Archaeological Probabilistic Modeling

The potential for any hypothesis to be true is a factor of its relationship to all other mutually exclusive possibilities. Although the mathematics for evaluating probability can become quite complex, even for simple parameters (see Pearl 2000: 2–6), in archaeological probabilistic modeling this usually translates into a simple declarative statement that any given land point has the potential to be either part of an archaeological site or not. Of course, this does not take into consideration the complex nature of defining sites by the presence of artifacts or features, the use of so-called nonsite areas, the nature of social and cognitive landscapes, the effects of postdepositional processes, and the methods and motivations of archaeologists looking for sites. All of these factors greatly influence the definition of "site" itself, and it might be argued that the concept is not a useful designation in any sense of the word (Dunnell 1992). The nature of regulatory archaeology, though, ensures that some categorical distinction be made. Therefore, in applications of probabilistic modeling, the definition of site is assumed to be an aggregation of artifacts or features either previously recorded as a site or that can be recognized in the field, based upon the experience of the researcher, as a focal point of human activity.

Archaeological probabilistic modeling relies on an understanding of the relationships between such "sites" and a series of independent variables that are assumed to be the conscious or subconscious causal factors in the intentional or unintentional placement of those artifacts or features by previous inhabitants of the area. Thus it makes several assumptions:

1. Humans make cognitive decisions about where to locate themselves in the available environment. Such decisions need not be conscious, but they are reflected in either explicit or implicit behavioral rules or tendencies.

2. Behavioral rules or tendencies are correlative with environmental parameters. The environment does not imply strictly ecological variables, but social and temporal ones as well. For instance, social territorial boundaries are every bit as real to site-placement decisions as distance to specific resources.

3. Reconstructed behavioral rules or tendencies can be applied to unsurveyed areas to reveal patterns of site potential or, in contrast, to causally explain patterns of settlement.

The two primary approaches to archaeological probabilistic modeling are traditionally referred to as inductive and deductive. The pluses and minuses of these two perceived categories have been addressed in numerous studies of probabilistic modeling (most notably Kohler and Parker 1986; Judge and Sebastian 1988; Dalla Bona 1994). Most succinctly, Ebert (2000) has pointed out the differences between the two methods in terms of explanatory capacity. In the end, almost all archaeological models have, in practice, been of the so-called inductive kind. Several factors, though, have greatly influenced the choice of modeling approaches. Foremost among these is the existing paradigm of scientific explanation currently in place in North American archaeology.

Although, the term "deductive" has been applied to models that rely on previously constructed formulas rather than formulas inductively derived from a statistical evaluation, the use of both terms is in fact a misnomer. The dichotomy between the two perceived approaches to probabilistic modeling has little to do with the origins of hypotheses and their subsequent testing (i.e., Hempel 1965). Rather it has to do with the nature of inductive inference, probabilistic causality, and determinism. The assumption has been that very strong empirically driven correlative models (those based on Hempel's [1965] inductive-statistical [I-S] model of scientific explanation) are required for accurate probabilistic forecasting (i.e., prediction, or in the case of archaeology, retrodiction). This argument mistakenly assumes that inductive models are producing viable statistical inferences. This has been severely called into question (Jeffrey 1971).

Regression, Bayesian, and many other approaches to archaeological probabilistic modeling use an existing set of archaeological data to build algorithmic "inferences" (Kohler and Parker 1986; Judge and Sebastian 1988; Kvamme 1990a, 1990b, 1995; Van Dalen 1999; Wescott and Brandon 2000). By default, that data is limited to identified sites located in previously surveyed areas. Even if we accept the representativeness and completeness of the archaeological data (and the inherent usefulness of the site concept), the statistical methods of correlative analysis are dependent on the notion that explanation is limited to evaluating the strong correlations between site location and the observed environmental or cultural factors. Any weak correlations are inherently not conducive to analysis and must be excluded from the examination. The result is a single formulaic statement that is essentially the "lowest common denominator" for site location prediction.

In any causal explanation of a phenomenon we would expect to generate a discussion of both how and why, not merely that an observation occurs (Jeffrey 1971: 21). Empiric-correlative type probabilistic models, in fact, demonstrate that correlations occur and summarize them in a mathematical relationship (the probabilistic formula). However, in an I-S model, the explanation of that formula is actually *post facto* applied as an interpretation (if it is applied at all) to the received mathematics and mistaken for logical inferential substantiation. Multiple nonlinear or logistic regression analysis, though, has no inherent explanatory capacity. It is merely a correlative evaluation that becomes tied to our observations of human behavior and used as statistical justification. This sets up a situation in which explanatory knowledge is relegated to an insistence on deterministic causality (cf. Salmon 1998: 38–42).

This is clearly illustrated in Jeffrey's (1971: 21–22) example of the weather glass (barometer) problem. When the barometer falls, the weather turns bad. A statistical analysis of numerous instances of observing both the weather and the barometer would produce a formula that can be expressed in terms of a very accurate prediction for the weather based on the height of the liquid in the barometer. But it does not clearly explain why or how the correlation occurs. That interpretation is dependent on known relationships between the expansion of liquids and atmospheric pressure plus those between atmospheric pressure and inclement weather. Unless they are specified, the explanation is not complete (it is merely an observation of the correlation). In fact, without the concurrent knowledge of the laws of chemistry and physics, we could logically infer that not only does bad weather cause the barometer to fall, but the falling barometer may equally cause the weather to turn bad.

In archaeological probabilistic models, I-S-based empiric-correlative analysis is but a quantification of the observed correlations. In fact, it is actually limited to a generalization of only statistically significant correlations strictly in the context of the known and measured environmental variables (not all of the potential factors). The nature of how or why these correlations occur is not specified in the statistical analysis, and therefore any *post facto* interpretation of how or why they occur is no more statistically inferred than any other hypothesis that could result in the same generalized correlations (cf. Jeffrey 1971: 26–27).

The perceived notion that inductive (that is to say I-S based) probabilistic models are somehow more accurate than deductive models (where how or why relationships are constructed before application of archaeological data) because they have been generated by explanatory statistical inference is patently absurd. In fact, several alternative approaches to I-S based probabilistic models have been presented in the literature that fit much more within the context of a statistical-relevance (S-R) model of explanation (cf. Salmon 1971, 1984, 1998), but that are generally acceptable to the peer group of probabilistic modelers because they have been couched in positivist terminology for what are essentially simplified forms of cognitive decision making (e.g., Boolean models, see Stančič and Kvamme 2000; Stančič et al. 2001).

We need to bear in mind that all probabilistic models are sterile, unproductive exercises if we do not explore the nature and relevance of the cause-and-effect relationships identified. The lack of large data sets, though, does not imply the absence of explanatory reasoning (as suggested by inferential I-S based modeling). We must recognize the inherent unpredictability of some cultural behavior that will never be amenable to deductive-nomological (D-N) or I-S type explanation. Complexity theory, fuzzy modeling, and cognitive studies are mature and detailed enough to strongly suggest that limiting archaeological explanation to predictable empirical correlations (and assuming a wide range of applicability for the correlative generalizations extracted) is much too simplistic of an approach. Instead, we need to address mechanisms of behavior (e.g., cognition) and to evaluate the ways within which empirically observed phenomena probabilistically reflect such causal mechanisms.

In the long run, we should not characterize probabilistic modeling with the simplified and incorrectly applied terms of "inductive" and "deductive." Similarly, I have replaced the term "predictive" with "probabilistic" to reflect a greater degree of indeterminacy. We need to focus on understanding that hypotheses of settlement tendencies or site placement "rules" should be based on our interpretations of the cognitive decision making employed by prehistoric or historic populations, and not on a simplistic mathematical device. If we choose to build models that are merely highly empirical-correlative and have no explanatory capacity, we have to realize that, in such cases, the goal of predictive modeling cannot be to understand prehistoric settlement; rather, it must be an application of modern cost-benefit analysis in the context of large regulatory projects. Such models may illuminate the best alternatives with which to avoid cultural resources, but they should be seen in the same light as barometric measures of site potential, and not as interpretations.

The salient differences between approaches to probabilistic modeling are, in essence, exemplified by either the acceptance or rejection of prior interpretation of settlement and whether it is appropriate to allow a statistical method to generate an "organic" predictive formula and then apply it as if correlation shows causality. These two approaches are, perhaps naively, described here as "data dependence" and "data independence."

17.2.1   Data-Dependent Models

Traditionally, the most common archaeological probabilistic models are derived from observing and statistically measuring the correlations between environmental characteristics and the locations of archaeological sites. They depend on observations of the archaeological record. As such, they tend to be limited to the ecological variables that can be easily measured and classified, and I have identified them here as data-dependent models.

There are, though, some major problems with data-dependent models:

1. They are reliant on the archaeological data as it is currently known. Whatever biases are inherent in the collection of the data are transferred to the model itself. For instance, any intuitive probabilistic model used by the surveyors will be inherent in the hypothetically objective formulas created by regression analysis or some other statistic. Unsurveyable areas also are going to be *de facto* excluded from proper consideration, since no known archaeological sites generally correlate with them.

2. Areas in which the most-recent human settlement has occurred affect inferential models in two ways. Sites in modern settlement areas are more likely to have been destroyed than those in the unsettled or

back-country areas, leading to a possible overemphasis on marginally placed sites than might be natural. Likewise, sites in areas developed since the implementation of the National Historic Preservation Act are more likely to have sites recorded for them, leading to possible overemphasis on those areas as well.

3. Most such models are based on the survey results compiled by many different archaeologists, sometimes even generations of archaeologists. The nature of the data is usually very inconsistent and unlikely to generate much confidence when sites are classified by temporal period or function. The vast majority of previously recorded sites (in the U.S.) are not classifiable, anyway, being simply nondiagnostic lithic scatters.

4. The definition of "site" is a fluid generalization made by the archaeologists doing the survey. Some states record sites on the basis of a single artifact; others require a minimum of ten (or three different kinds). Site boundaries can be as much as 10,000 m² for a site containing two lithic flakes, or they can be tightly drawn around clusters of high artifact density in a relatively small area. Frequently, they are bound to modern roads or rights-of-way and have no relationship to the prehistoric landscape.

5. Many states record the locations of archaeological sites but not the areas within which the archaeological surveys occurred. Thus any correlations between sites and environmental variables are often built without knowing the nature of how much was surveyed or where. This can create very misleading correlations.

6. Multiple, nonlinear, or logistic regressions are typically the statistical methods used for establishing site/environment correlations, yet these statistics make the assumption that the independent environmental variables used are the only ones required in the model. Additional environmental data added to the model at a later date requires a recalculation of the formula and can radically alter the outcome.

7. Perhaps most commonly encountered, the sample size of the archaeological sites must be sufficient to produce statistically significant results for each and every site type, temporal period, and function that would result in a different settlement pattern. This can only be accomplished in many regions by expanding the study area to such a large extent that it has likely captured sites across unknown territorial or cultural boundaries.

Some of these problems can be worked around. For instance, it is possible to use only data from an area that has been uniformly surveyed specifically for creation of the probabilistic model. The techniques used for the survey, though, must provide equal weight to all possible landforms (including unsurveyable ones) and derive enough archaeological sites to adequately develop regression formulas for each and every site type or temporal period that may have resulted in a different settlement pattern.

17.2.2   Data-Independent Models

If a model relies on one particular data set to estimate site/environmental correlations, it must be realized that those correlations can only represent a partial abstraction of the entire cultural decision-making system, due to the limitations of observable archaeological and historical data. Data-independent models are constructed by creating the model prior to the application of a data-based testing regime. Though they are not truly independent of the influence of archaeological data, because they are based on an "expert system" (the previous experience of regional archaeologists), it is convenient to refer to them as data-independent because they are not based on a formula statistically derived from the data set of known sites.

Data-independent models are much more like meteorological forecasting models and much less like barometers. The formulas for weather prediction are not based on a database of past statistical correlations between weather systems and environmental factors (i.e., barometrical observations). Rather, they are based on an understanding of the way that (what we call) weather systems react to current atmospheric conditions. Of course, that understanding is based on an interpretation of past weather systems (hence it is an expert system). For archaeology, such models face several challenges. Kohler and Parker (1986: 432–440) address several of these issues. A "deductive" model must:

1. Consider how humans make choices concerning location (a subset of the larger problem of how any decision is made), which requires considering:
   a. A mechanism for decision making
   b. An end for decision making (what is the goal?)
2. Specify the variables affecting location decisions for each significant chronological or functional subset of sites
3. Be capable of operationalization; it must propose a means for measuring each of the relevant variables and must allow for a set of predictions that can be compared with the archaeological data (Kohler and Parker 1986: 422)

In considering the first requirement postulated above, it is clear that a number of conditions might exist that greatly affect the process of making decisions on site placement. If too much information is available, a decision may not reflect rational assessment of important factors, but may be reliant upon "cultural precedent" or "simplified logics" (heuristics). Likewise, rational assessment may not occur in situations where not enough information is available (uncertainty of future conditions, or unfamiliarity with the area). The ability to model site locations probabilistically is always responsive to the context in which the site-placement decisions are made, and these limitations are inherent in inferential models as well.

Another problem arises in consideration of the decisions that are cost-benefit neutral. In other words, clear preferential alternatives may not be evident in decision-making processes that do not bear on the individual's (or group's) ability to extract resources from the environment efficiently. The cost-benefit strategy of decision making might not always apply for all important factors. An understanding of complex economic benefits may be beyond the grasp of the model if all constraints are not fully understood.

Third, some decisions may be based on circumstances that are inherently unpredictable. Even if the distributions of major food resources can be predicted in a given environment, it is just as likely that the fluctuations in patterns of warfare between two neighboring groups may be unpredictable, yet have a profound influence on site placement.

The benefits of a data-independent model, however, are:

1. A more complete comprehension of the cognitive decision-making processes that have led to the variety of settlement patterns in the region. In essence, the creation of a data-independent model is a simulation of the process by which prehistoric or historic inhabitants of a region thought about their environment and how it affected their settlement.

2. No dependence on the usually questionable data of previous archaeologists, their surveys, and their interpretations. To some degree, data-independent models are still reliant on the abilities and judgment of regional archaeologists, in that hypotheses about settlement are always subjective, but this need not be a limiting factor given the potential to test many multiple theories, even ones not previously envisioned for the region.

3. No reliance on having a large accurate sample size of previously recorded archaeological sites on which to base the formulas. A good sample size of sites is of course required to provide a sufficient test for the model, but it becomes possible to address issues of very specific site types or functions that may have a very small representation in the archaeological record.

4. Similarly, it becomes possible to develop formulas that address the use of specific areas or habitats in which it is unlikely that site remains would be identified. This would not only include areas that are currently unsurveyable or heavily disturbed, but also those that may have been used strictly for resource procurement.

The principal arguments against data-independent models have been their complexity and difficulty in implementation. It has always been easier to create a regression-based inferential statistical model (assuming the database of sites was large enough) and apply the resulting formulas to unsurveyed areas. In contrast, archaeologists have had little precedent and even less guidance in creating data-independent formulas. Where to begin? What variables to include? How to weight them appropriately?

In the last decade, GIS has facilitated the application of inferential models, and it can no longer be considered overly difficult to implement data-independent ones either. The statistics are not particularly difficult or obtuse (unlike regression-based inferential models), and I believe that data-independent models make more intuitive sense and thus can be more readily understood and used by nonspecialist archaeologists.

17.2.3   Modeling in a Homogeneous Environment

The real advantage of a data-independent model is the ability to make conclusions that are causally related to settlement systems and not merely reflective of site preservation or obvious generalities. The primary reason that inferential probabilistic models have been most frequently employed in archaeology is that they have the immediacy of a real-world data set, but very little need to understand or interpret the values produced by the regression statistics. The statistically generated formulas need only be reapplied to unsurveyed areas, and very often the model produces somewhat reliable results. The reliability of the results, though, is predicated upon the ubiquitous use of two very important environmental factors: terrain slope and distance to water.

In most regions where a useful semiaccurate archaeological probabilistic model has been applied, the major site-delimiting factors are almost invariably slope of the terrain and site distance to water. Not surprisingly, these regions are usually highly dissected and often arid, or the models refer only to large-scale village or agriculturally based sites. For areas that are very homogeneous and well-watered, there are few highly accurate and precise probabilistic models; this is likewise true for hunter-gatherer groups. Where water is limited and slope is quite variable, it makes sense that those factors should strongly correlate with archaeological sites. But how different is that from an intuitive model? Archaeologists have long used intuitive models for limiting survey strategies. Does it make sense to develop a statistically sophisticated inferential probabilistic model that merely supports an already well-understood intuitive model?

More difficulty is encountered when addressing site potential in a homogeneous environment. The primary fall-back environmental factor often used in probabilistic modeling, when slope and distance to water are not sufficient, is soil type. But is it legitimate to argue that certain soil types were prehistorically preferred when they are largely autocorrelative with modern wetland distributions (which are unsurveyable) or with highly plowed surfaces (where sites are more visible)? Clearly, the basis of prehistoric soil-type preference must lie in the nature of the ecological factors to which it is linked, in other words, to soil fertility, potential for wildlife habitats, and resource procurement (such as clay sources). Each of the factors associated with soil types needs to be understood on its own, not as a simple site-to-soil-type relationship.

Unfortunately, just as in every other type of environment, data-independent probabilistic modeling in a homogeneous region must rely on the accuracy and availability of the environmental data. Environmental data problems may include:

1. Inaccurate spatial data
2. Spatial autocorrelation of variables
3. Inability to model past environments
4. Incomplete data sets
5. Inappropriate scaling
6. Map-edge effects

In the case of a data-dependent model, though, the reliance is on the accuracy of both the environmental data and the archaeological data. The following sections describe in detail one attempt to produce a data-independent probabilistic model for a homogeneous environment.

17.3 Probabilistic Modeling at NWS

This study was not intended to identify areas of archaeological significance so that they could be entirely avoided. Rather, the identification of high-probability zones allowed management decisions to be made prior to the undertaking of any construction project. Alternative treatment options for different probability areas are a possibility, but being able to identify the likelihood for some areas to contain archaeological sites prior to survey allowed full consideration of project options and created a prioritization and appropriate survey strategies for the unsurveyed land units. In addition, examination of ongoing or projected adverse effects is likely to benefit from knowing which land units are probably the most sensitive.

The study area includes the four USGS 7.5-min quadrangles encompassing the Naval Weapons Station in Berkeley County, South Carolina: North Charleston; Ladson; Mount Holly; and Kittredge (Figure 17.1). Limiting the study to these quadrangles was based on the estimated computer analytical capacity available to process the chosen digital data. Larger study areas

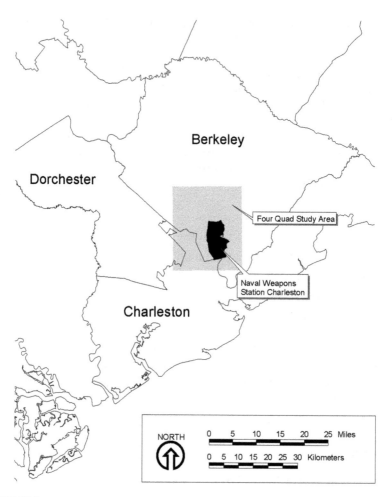

FIGURE 17.1
Study area location.

would require considerably more time to process, and smaller areas would
be less useful.

17.3.1   Previous Modeling in the Study Area

Berkeley County was included in the Charleston Harbor study (Cable 1996),
a comprehensive probabilistic model produced by New South Associates,
Inc. (New South). The Charleston Harbor study included assessment of soil,
spatial, and distance values to produce several regression equations that
were overlaid on prospective project areas within two distinct geophysical
districts (maritime and interior). A sample of archaeological sites from both
districts was included in the analysis. The model was developed without

the use of GIS data and was intended to be applicable without the use of GIS evaluation.

In general, Cable's (1996) results suggest that some correlations between soil classes, streams, site densities, and distances between sites may exist, which would allow an argument for restricted survey strategies in designated probability areas. Cable (1996) found that his modeled probability areas tended to show a weak positive correlation between where sites were predicted and where they actually occurred. Unfortunately, this is already the case for what archaeologists have routinely employed as intuitive probabilistic models (i.e., limiting fieldwork away from heavily sloped, disturbed, or saturated areas, as well as concentrating on streamside locales).

If the goal of probabilistic modeling is to understand prehistoric settlement strategies in the region, the Charleston Harbor study provided little new information. No causal explanation for site/soil or site/stream correlations is tested, so it cannot be assumed that prehistoric people were consciously seeking out specific soil types, soil-type interfaces, soil-drainage variability, or stream sizes. Locations of sites on particular soil types, near certain stream orders, or near specific soil interfaces may be causally related to some other ecological or social variables not considered.

If the goal of probabilistic modeling is to provide a tool for archaeologists and planners to reduce costs by making alternative decisions prior to investigation, the Charleston Harbor study also created little new potential. This is in large part due to the choice of variables, the tedious way in which they were measured (and would have to be measured in future applications), the marginal results, and the limited applicability to unsurveyed areas. Survey strategies are not likely to be developed based on an assessment of the distances to different soil-class interfaces or particular stream classes for every point in a potential project area. It may be possible to manually assess a small project area, but this could only be done with a GIS analysis for any large tract of land.

The contrast between these two possible goals is no small matter. The idea that we can save time and money by predicting site locations involves two very distinctive assumptions. One assumption is that we can understand and predict the behavior of long-dead people. The second assumption is that such an understanding will allow us to control or limit spatial types of archaeological investigations. For this project, the Cable (1996) model provided little usable data. Because it was not a GIS-based, easily replicated process, there was no potential to just apply the Cable (1996) model to the study area. Instead, it was determined that a new model would need to be created. For this new model, three major phases of analysis took place: data gathering and extraction, creation of probability surfaces, and formula evaluation.

The most appropriate means of assessing the regional patterning of probability areas was to acquire the necessary data for the four adjacent USGS 7.5-min quadrangle maps surrounding the study area. The digital data from these maps — downloaded from the South Carolina Department of Natural

TABLE 17.1

Original Data Sources

Data Source	Name	Type of Data
1	North Charleston 30-m digital elevation model	30-m grid data set
2	Ladson 30-m digital elevation model	30-m grid data set
3	Kittredge 30-m digital elevation model	30-m grid data set
4	Mount Holly 30-m digital elevation model	30-m grid data set
5	North Charleston 7.5-min USGS digital quadrangle	Georeferenced TIFF image
6	Ladson 7.5-min USGS digital quadrangle	Georeferenced TIFF image
7	Kittredge 7.5-min USGS digital quadrangle	Georeferenced TIFF image
8	Mount Holly 7.5-min USGS digital quadrangle	Georeferenced TIFF image
9	North Charleston hydrology	Polygon data set
10	Ladson hydrology	Polygon data set
11	Kittredge hydrology	Polygon data set
12	Mount Holly hydrology	Polygon data set
13	North Charleston USEPA wetlands	Polygon data set
14	Ladson USEPA wetlands	Polygon data set
15	Kittredge USEPA wetlands	Polygon data set
16	Mount Holly USEPA wetlands	Polygon data set
17	North Charleston soils	Polygon data set
18	Ladson soils	Polygon data set
19	Kittredge soils	Polygon data set
20	Mount Holly soils	Polygon data set
21	North Charleston digital orthophoto quarter quads	Georeferenced TIFF image
22	Ladson digital orthophoto quarter quads	Georeferenced TIFF image
23	Kittredge Digital orthophoto quarter quads	Georeferenced TIFF image
24	Mount Holly digital orthophoto quarter quads	Georeferenced TIFF image
25	North Charleston SCIAA archaeological site data	Polygon data set
26	Ladson SCIAA archaeological site data	Polygon data set
27	Kittredge SCIAA archaeological site data	Polygon data set
28	Mount Holly SCIAA archaeological site data	Polygon data set
29	1825 Mills's Atlas map of Charleston District	Georeferenced TIFF image
30	Naval Weapons Station boundaries	Polygon data set

Resources (SCDNR) Data Clearinghouse Web page, various USGS data clearinghouses, and through a data request from the South Carolina Institute of Archaeology and Anthropology (SCIAA) — provide the background from which the model was developed.

17.3.2 Data Gathering and Extraction

The original base map data sources used in the analysis are listed in Table 17.1. Data sources 1 through 20 were downloaded from the SCDNR GIS Data Clearinghouse (http://www.dnr.state.sc.us/gisdata/). The digital orthophoto quads (data sources 21 through 24) were ordered from the USGS through Microsoft's Terraserver Web page. The archaeological site boundaries (data sources 25 through 28) were requested from the SCIAA. The 1825 Mills's Atlas map (data source 29) was scanned at 300 dpi from a hard copy

TABLE 17.2

Derived Data Sources

Data Source	Name	Type of Data
31	Merged-grid 30-m digital elevation model	30-m grid data set
32	Merged hydrology polygons	Polygon data set
33	Merged USEPA wetlands polygons	Polygon data set
34	Merged-soils polygons	Polygon data set
35	Merged archaeological site data	Polygon data set
36	Degree of slope	30-m grid data set
37	Navigable waterways	Polygon data set
38	Rubber-sheeted 1825 Mills's Atlas map	30-m grid data set
39	1825 Mills's Atlas-based historic resources	Polygon data set
40	Cost surface: combined slope and wetlands travel costs	30-m grid data set
41	Cost distance to navigable waterways	30-m grid data set
42	Cost distance to historic resources	30-m grid data set
43	Cost distance to marsh ecotones	30-m grid data set
44	Size of nearest marsh ecotone	30-m grid data set
45	Soil codes: drainage rankings	30-m grid data set
46	Soil codes: seed/grain crop capacity values	30-m grid data set
47	Soil codes: wild plants capacity values	30-m grid data set
48	Soil codes: woodland wildlife capacity values	30-m grid data set
49	Soil codes: wetland wildlife capacity values	30-m grid data set
50	Soil codes: open land wildlife capacity values	30-m grid data set
51	Standardized soils codes: drainage (SDR)	30-m grid data set
52	Standardized soils codes: seed/grain crop (SSCC)	30-m grid data set
53	Standardized soils codes: wild plants (SWPC)	30-m grid data set
54	Standardized soils codes: woodland wildlife (SWOWC)	30-m grid data set
55	Standardized soils codes: wetland wildlife (SWTWC)	30-m grid data set
56	Standardized soils codes: open land wildlife (SOWC)	30-m grid data set
57	Standardized slope (DOS and DOS2)	30-m grid data set
58	Standardized cost distance to navigable water (CDNW and CDNW2)	30-m grid data set
59	Standardized cost distance to historic resources (CDHR and CDHR2)	30-m grid data set
60	Standardized cost distance to marsh ecotone (CDME and CDME2)	30-m grid data set
61	Standardized size of nearest marsh ecotone (SNME)	30-m grid data set

(Mills 1979) and georeferenced to overlay the quadrangle maps. The Naval Weapons Station Boundaries (data source 30) were digitized directly on top of the digital 7.5-min quads. All received data sources were projected into the UTM projection, North American 1927 datum.

A series of additional data layers was created using the 30 data sources listed in Table 17.1. These are referred to as derived data sources (Table 17.2). Derived data sources 31 through 35 were merely combined versions of the specific data layer for each of the four quads. Data source 36 (degree of slope) was calculated using ArcView's internal slope calculation routine applied to data source 31 (the merged DEM). Data source 37 (navigable waterways) was extracted from the combined hydrology layers (data source

32). Navigable waterways implies that the hydrologic area is not considered a wetland but is traversable by small watercraft (i.e., canoe, etc.). It does not imply modern shipping capability.

Using the digital image of the 1825 Mills's Atlas map, 12 landmarks were identified and then colocated on the digital quads. The index table identifying the UTM coordinates of each of the 12 points on both historic and modern maps was used to stretch and fit (rubber sheet) the Mills's Atlas to the modern quadrangles (data source 38). This aligned the known landmarks and created a starting point from which historic roadways and probable historic structure locations were digitized into data source 39.

Wetland areas were extracted from the hydrology features layer (data source 32) and combined with the degree of slope (data source 36) to create a cost-surface map (data source 40). This map was used in the cost-distance analysis for data sources 41, 42, and 43. "Cost surface" refers to the cost of traveling through any given land unit. When measuring distance variables, it is often realized that direct linear distance does not accurately take into consideration the effort that would be expended by anyone traveling between two points. Because we know that it takes a greater effort to traverse a steep hillside than flat terrain, and it is more difficult to walk through thick marshland than dry uplands, those factors need to be considered in evaluating distance variables. Therefore, cost distance from navigable waterways (data source 41) was calculated using degree of slope and presence/absence of wetlands as mitigating travel factors and navigable waterways (data source 37) as the origin. Similarly, cost distance to historic resources (data source 42) was calculated with the same travel costs and the Mills's Atlas historic resources (data source 39) as the origin.

Cost distance to marsh ecotones (data source 43) takes the wetland values from the hydrology data source (data source 32) as an origin source and uses data source 40 as travel cost. No distinction is made between salt and freshwater marshes (that identification is not possible for two of the four quads). Size of nearest marsh ecotone (data source 44) was calculated by a proximity analysis that assigns the acreage value of the nearest marsh ecotone to each grid point in the study area.

The soils polygons (data source 34) were used to assign drainage rankings, seed/grain crop, wild plants, woodland wildlife, wetland wildlife, and open land wildlife capacity values (data sources 45 through 50) based on the Berkeley and Dorchester County soil surveys (Eppinette 1990; Long 1980). The Charleston County soils data were not digitally available. Rankings were established by translating the qualitative values of good, fair, poor, and very poor into numeric classes of 1, 2, 4, and 5; soils not measured were given a value of 3.

The soil code rankings (data sources 45 through 50) were converted to standardized probabilistic indices (data sources 51 through 56) by changing the numeric categories (1 through 5, representing "good" through "very poor") to 0.75, 0.25, 0, –0.25, and –0.75, respectively. Thus a "good" ranking for any of the soil code variables earned a discrete 0.75 value, indicating a presumed positive relationship with archaeological sites. Discrete values were only assigned to the soil codes data layers, since they are categorical in nature.

Data sources 57 through 60 are standardized versions of each continuous variable, so that data values range between −1 and 1. This means that the values associated with each grid unit (from data sources 41 through 50) were translated proportionally to fit the range of −1 to 1, with −1 representing the furthest observed cost-distance value and 1 representing the nearest. The size of the nearest marsh ecotone variable (data source 43) was recalculated to create data source 61 by finding the average-size marsh value and translating it into 0. The range of −1 to 0 represents the values between 0 and the average marsh size. The range of 0 to 1 represents the values between the average marsh size and twice the average size. Any value more than twice the size of the average was recalculated as equal to 1. This implies that above-average-size marsh ecotones are preferred and that below-average ones are not.

When proximity or slope variables were translated into exponential values using data sources 57 through 60, a transformation routine was used to transform the data into a positive range of 0 to 2 prior to squaring the value. After squaring, the value was once again standardized to fit the range of −1 to 1. This was necessary to eliminate the problem of squaring negative numbers, which always results in a positive value.

17.3.3   Creation of Probability Surfaces

Cable (1996) used a series of variables extracted from soil-drainage groups based on a GIS assessment done by Scurry (1989), including mean soil-drainage-rank diversity at the measurement point, soil-drainage-rank diversity at different search radii from the measurement point, and linear distances to each soil-drainage rank. In addition, he used a linear distance to nearest stream measure, along with linear distance to nearest known site. Through a series of correlation coefficients and a multiple regression evaluation, the statistically evaluated relationships to predict site potential were based on both site density and linear distance to the nearest known site. His initial choice of variables is conditioned by his familiarity with the region and the high interdependence between sites and soil types observed through years of fieldwork. But the application of those variables comes by evaluating their statistical relationship to a database of known sites and a collection of randomly placed "nonsite" control points.

Using a more data-independent approach, we assume that past discussions of prehistoric subsistence and settlement patterning made in previous studies are the basis upon which to model potential patterns of site placement. This means that we choose the variables to employ in the study on the basis of two factors: (1) how we perceive they might have contributed to the prehistoric or historic processes of site placement, and (2) how easily we can measure them today.

Although relatively few directed research projects have been carried out in the region around the Naval Weapons Station, a long sequence of work has existed for much of the Interior Coastal Plain of South Carolina. This work is summarized in reports by Anderson et al. (1982), Brockington et al. (1995), Brooks and Canouts (1984), Cabak et al. (1996), Cable (1993), Goodyear and Hanson (1989), Panamerican Consultants (1997), and Sassaman and Anderson (1995). Detailed examinations of past settlement and subsistence systems have coalesced into several definitive interpretations for specific time periods.

17.3.3.1   *Hunting-Gathering Adaptations*

Both the Paleo-Indian and Archaic periods are seen as intervals of intensive hunting and gathering dominance. Habitation locales may have been selected primarily for their low-cost access to preferred resources or as focal points for a hunting camp — base-camp seasonal strategy. Groups may have formed according to complex territorial arrangements that evolved early on. Such territories probably shrank considerably as populations increased or seasonal rounds developed based on smaller prey species (Anderson and Joseph 1988). The distribution of fluted points and diagnostic tools suggests to Anderson (1996) a highly mobile South Atlantic macroband (500–1500 people) consisting of at least eight parallel band territories encompassing the large watersheds. The average size of a band may have been approximately 50 to 100 people. Early Archaic exploitation of many animal species and ecotones is suggested by the scattering of low-density sites across a wide variety of landscapes contrasted with high-density sites in restricted stream terrace locales (Sassaman 1996).

It is often difficult to reconstruct the paleoenvironment, given the lack of available physical data and the confusion presented by historic disturbance (Cable 1993: 9). Some interpretations can be made to suggest that distributions of climax forest communities and hardwood species, in particular, were much more widespread in the distant past (Anderson et al. 1979: 15–18). With the ubiquitous presence of pine forest in the uplands today, the key ecological indicators of past plant and animal distributions are now limited largely to soil characteristics, especially structure and fertility.

Intensive Paleo-Indian and Archaic hunting and gathering reflected not only territorial arrangements (which are very difficult to project on a local level, especially with the lack of detailed site-location knowledge currently available), but local plant and animal resource availability. In the Interior Coastal Plain, emphasis may have been placed on the exploitation of both upland and marsh resources during the Archaic period (Anderson et al. 1982; Panamerican Consultants 1997; Mistovich and Clinton 1991). As more intensive use of local resources evolved, a larger reliance may have been placed on marsh or estuarine environments. Brooks and Canouts (1984) see this focus on marsh ecotones as increasing through the Woodland period and becoming most prominent during the Mississippian period.

17.3.3.2 Agricultural Economies

Similar to the gradual increased exploitation of marsh or estuarine resources during the transition from Paleo-Indian through Archaic into Woodland and Mississippian, agricultural intensification is reflected in the focus on well-drained, fertile soils beginning during the Early Woodland and continuing through Historic times. Though hunting and gathering never ceased to be an important activity, horticultural sophistication demanded a more sedentary lifestyle.

Brooks and Canouts (1984: 23) suggest a Middle-Late Woodland settlement pattern that reflects a "diffuse or generalized subsistence economy, involving exploitation of riverine and interriverine resources." The Mississippian pattern reflects a focal strategy involving intensive exploitation of the riverine resources (marsh ecotones). Anderson et al. (1979: 86–87) suggest that the multicomponent Cal Smoak site shows a long sequence of upland resource utilization (especially of deer and rabbits) and perhaps a later use of marsh resources (based on the presence of turtle remains and significant numbers of bird points). Although the site is located in a marginal area, it is adjacent to some of the most suitable microenvironmental zones for both plant and animal resources (Anderson et al. 1979: 23).

Anderson et al. (1982) suggest that Woodland and Mississippian occupation at the Mattasee Lake sites was very short term and geared toward exploitation of both marsh and upland resources. Little evidence exists to suggest long-term occupation or intensive agricultural use of this Interior Coastal Plain locale. The distribution of Mississippian period archaeological sites may have been keyed to well-defined sociocultural territories and trade networks as well. Because we have little definitive evidence for reconstructing prehistoric social territories, it is unlikely that we will be able to incorporate such variables into a probabilistic model anytime soon. For now, we can only focus on understanding the distribution of sites with respect to the more easily measured ecological variables.

17.3.3.3 Historic Settlement

Several sources have addressed the distribution of Historic period archaeological localities in and around the region (e.g., Drucker and Anthony 1979; Soil Systems 1982; South and Hartley 1985; Stine 1991; Stine et al. 1993; Tidewater Atlantic Research 1995; Zierden et al. 1986). They largely summarize historic settlement patterns as resulting from a combination of factors, including access to navigable waterways, suitable agricultural terrain (well-drained, fertile soils), and proximity to other travel arteries.

Although historic settlement may have been patterned on the dispensation of land grants and the ability of absent grantees either to settle their claim or to subdivide it and sell the parcels, other factors led to site-placement strategies within land units. Rivers, as the primary early travel arteries, assumed the most prominent role at first (Soil Systems 1982: 37). Later, roads allowed access to the interior, usually in the place of previously existing native trading paths. Some geophysical features contributed to the pattern of road networks, especially the distribution of marshes, but in some cases roads followed the patterns of settlement rather than vice versa.

Nevertheless, we should strongly suspect that historic sites are likely to occur in the near vicinity of roads (in the interior) and navigable waterways. Arable land and pasturage are clearly important, especially during times of most intensive cash cropping. Similarly, the distributions of specialty sites such as brick kilns may be a factor of available clay sources and access to navigable water. The soil surveys (Eppinette 1990; Long 1980) consistently rank soils by particle-size class, but not in sufficient detail to determine whether they were of use in brick manufacture.

17.3.4 Probabilistic Formulas

Based on observations of previous studies, it seems relevant that categorizing probabilistic formulas by time period can be limited to major settlement-subsistence strategies. A hunting-gathering primary subsistence mode is known for the Paleo-Indian through Early Woodland periods. Though the level of dependency on hunted or gathered resources decreases through time, a diffuse hunting-gathering economy is suggested. Weighting of open-land wildlife, woodland wildlife, wetland wildlife, and wild plants' soil-capacity values should rank fairly high.

With the progression of time and the increasing dependence on marsh ecotones, proximity to a marsh ecotone and the size of the nearest marsh ecotone contribute increasingly to the formulation. Proximity to navigable water may have played a much smaller role in comparison with later time periods. Soil drainage may have been a small consideration for comfortable habitation, along with the small regional variation in degree of slope. Distance to drinking water and soil seed/grain crop capacity would not likely have contributed to the process of Paleo-Indian–Early Woodland settlement, since water is nearly ubiquitous in the region and horticulture was not yet fully practiced.

For Middle Woodland through Mississippian periods, increased horticultural and agricultural sophistication supplanted hunting-gathering as the primary subsistence mode. Although dependence on open land, woodland wildlife, and wild plants' capacity values decreased, it is not suggested that wetland plant capacity, proximity to marsh ecotone, and the size of the nearest marsh ecotone was reduced. The regional analyses seem to indicate that marsh, estuarine, and riverine resources did not decrease in importance. Seed/grain crop capacity, however, would have played a larger role in the site-placement process, along with an increased dependence on access to navigable waterways. As sites became larger and more permanently settled, we would expect an increase in the role played by habitability concerns, namely slope and soil drainage.

Historic period sites are in part agricultural, and many would likely be keyed to soil seed/grain crop fertility values. Because we have period documentation in the form of historic maps (e.g., the 1825 Mills's Atlas), it is possible to include a high dependence on proximity to mapped historic resources, especially roads. Likewise, we know that the earliest historic travel arteries were navigable waterways, so we would expect to find sites in close proximity to them as well. Some level of minor importance may have been played by soil wildlife capacity values, but we would expect that unfamiliarity with the local conditions would have often led to settlement in unusually species-poor areas, at least in the beginning of the Historic period. Because historic sites represent the most intensive occupation and the sites least likely to have been destroyed, the greatest number of known sites date to the Historic period.

Although the total number of known archaeological resources in the study area is enough to evaluate the significance of the relationship between sites and probability formulas, we do not have a complete understanding of the nature of all previous survey areas and survey strategies. Without that information we have no way of determining if significance testing is valid. (We would have to make the shaky assumption that known sites represent a random sample of all sites.) Therefore, until that information is compiled and digitized, we can only use the information with respect to those survey areas where a known, uniform survey strategy has occurred; namely the survey tracts from the NWS since 1994. This affects the significance testing in three ways. First, all land units including archaeological sites are divided into 30×30-m squares, with a site value of either 1 or 0 (a single site may consist of one or many land units with the value 1). Second, temporal periods are divided only into Prehistoric and Historic, since we have too few archaeological land units from known survey areas in the NWS to create valid divisions between prehistoric subperiods. Third, because the longer Prehistoric period represents more-diverse settlement strategies, it is less likely that any one formula would be a very significant probabilistic tool. The one that fits the prehistoric data best will be more general than some that fit well with one settlement strategy.

Prior to the selection of weighting values for each independent variable, a multiple regression analysis was carried out to determine if the data were suitable for creating accurate beta values (statistically derived weighting values) and, therefore, an inferential probabilistic model. The values for each independent variable at each site were calculated based on the site centroid. In addition, three dependent variables (site size, site density within 1-km radius, and proximity to the nearest known site) were calculated. Twelve multiple regression analyses were carried out incorporating each temporal category (Prehistoric and Historic) — once with a straight linear relationship to the proximity and slope variables (cost distance to historic resources, cost distance to navigable waterways, cost distance to marsh ecotones, and degree of slope) and once using a logarithmic relationship to the same variables — for each dependent variable.

Though there was a total of 370 sites in the database (representing 429 occupations), the variation in the sample was such that in all 12 cases less than 20% of the variation in any dependent variable could be explained with a multiple regression analysis. The adjusted r^2 values could be increased by dropping independent variables, but it appears that site size, occupation density, and site clustering cannot be explained easily with the data on hand. Likely this is due to irregular survey coverage (plus the inability to accurately model most previous survey areas) and the inconsistency in assuming that site size or site density accurately represent intensity of occupation. A regression analysis incorporating artifact densities or more-detailed examination of site significance might fare a little better, but these data are not available.

Multiple regression analysis does not appear to be useful in this situation, and ultimately this is probably due to the homogeneous nature of the environment. Applying a wide range of variables with controlled weights allowed us to directly assess known or hypothesized settlement strategies with the archaeological data, rather than allowing the data to provide a single solution. Therefore, significance evaluation of the formulas with new and existing data forms the core of the hypothesis testing.

Fifteen formulas incorporating various combinations of the independent variables at different weights were used (Table 17.3). Using ArcView's internal map calculator routine, each formula created a 30-m grid theme

TABLE 17.3

Probabilistic Formulas Used in the Analysis

Label	Formula [a]
Formula 1	Y = (0.21)SWPC + (0.08)SWOWC + (0.08)SWTWC + (0.08)SOWC + (0.08)SNME + (0.08)CDME + (0.13)CDNW + (0.05)DOS + (0.08)SDR
Formula 2	Y = (0.21)SWPC + (0.08)SWOWC + (0.08)SWTWC + (0.08)SOWC + (0.08)SNME + (0.08)CDME2 + (0.13)CDNW2 + (0.05)DOS2 + (0.08)SDR
Formula 3	Y = (0.5)CDNW2 + (0.1)DOS2 + (0.4)SDR
Formula 4	Y = (0.5)CDNW + (0.1)DOS + (0.4)SDR
Formula 5	Y = (0.2)CDHR2 + (0.3)SNME + (0.4)CDME2 + (0.05)DOS2 + (0.05)SDR
Formula 6	Y = (0.11)SSCC + (0.07)SWPC + (0.07)SWOWC + (0.11)SWTWC + (0.07)SOWC + (0.11)SNME + (0.18)CDME + (0.18)CDNW + (0.03)DOS + (0.07)SDR
Formula 7	Y = (0.11)SSCC + (0.07)SWPC + (0.07)SWOWC + (0.11)SWTWC + (0.07)SOWC + (0.11)SNME + (0.18)CDME2 + (0.18)CDNW2 + (0.03)DOS2 + (0.07)SDR
Formula 8	Y = (0.4)CDME2 + (0.4)CDNW2 + (0.2)SDR
Formula 9	Y = (0.4)CDME2 + (0.2)CDNW2 + (0.4)SDR
Formula 10	Y = (0.2)SNME + (0.3)CDME2 + (0.3)CDNW2 + (0.1)DOS2 + (0.1)SDR
Formula 11	Y = (0.15)SSCC + (0.05)SWOWC + (0.05)SWTWC + (0.1)SOWC + (0.05)SNME + (0.05)CDME + (0.2)CDNW + (0.25)CDHR + (0.05)DOS + (0.05)SDR
Formula 12	Y = (0.15)SSCC + (0.05)SWOWC + (0.05)SWTWC + (0.1)SOWC + (0.05)SNME + (0.05)CDME2 + (0.2)CDNW2 + (0.25)CDHR2 + (0.05)DOS2 + (0.05)SDR
Formula 13	Y = (0.15)SSCC + (0.25)CDNW2 + (0.4)CDHR2 + (0.1)DOS2 + (0.1)SDR
Formula 14	Y = (0.5)CDHR2 + (0.1)DOS2 + (0.4)SDR
Formula 15	Y = (0.5)CDHR + (0.1)DOS + (0.4)SDR

[a] Variable abbreviations are given in Table 17.2 (Y = calculated probability value for each grid point).

(i.e., each grid unit covered an area equal to 900 m², or 30 × 30 m). For display over quad maps, the formula surfaces were resampled to a 5-m grid theme (25 m², or 5 × 5 m) after testing. Because the formulas always included proportional weighting of each independent variable, the final probabilistic values for each grid point ranged between –1 and 1. Categorization of the probability values was then based on ten discrete 0.2-range classes (–1 to –0.8, –0.8 to –0.6, –0.6 to –0.4, –0.4 to –0.2, –0.2 to 0, 0 to 0.2, 0.2 to 0.4, 0.4 to 0.6, 0.6 to 0.8, and 0.8 to 1).

17.3.5   Formula Evaluation

The 15 formulas were evaluated by χ^2 analysis of the numbers of observed archaeological land units and the expected number of archaeological land units (based on survey data from the NWS) by the probability value categories. There were 448 archaeological land units (i.e., 30 × 30-m grid units within the boundaries of previously recorded archaeological sites in the NWS) out of 15,976 total land units surveyed (Figure 17.2). Each site was categorized by two general time periods: Prehistoric and Historic (no smaller divisions were possible, since we were attempting to create a model that was most applicable to as many diverse sites as possible). The expected values were calculated by multiplying the proportion of each probability value class (of total land units surveyed) by the number of known site land units from each time period. The observed values were then tested against the expected values in a set of 30 χ^2 analyses. This χ^2 analysis was appropriate, given the categorical nature of the probability value classes. Significance of the χ^2 values is based on an assessment of a normal χ^2 significance table with 19 degrees of freedom at $p < 0.005$ (Rohlf and Sokal 1969: 165).

The χ^2 significance was the logical starting point to assess the potential for each formula to suggest archaeological site probability, but it was important to note where within the classes the positive and negative relationships occurred. A strong positive relationship between site locations and high-probability values was sought, along with a strong negative relationship between sites and low-probability values. It was also important, however, to ensure that the proportions of known sites in other areas outside the NWS reflected these relationships.

The most convenient way to examine the fit of the formula to the known archaeological site and survey data is to compare the histograms of observed minus expected values for each probability class. The ideal formula would combine a very high positive spike in the high-probability end of the histogram with a very low negative spike in the low-probability end. The closer the x-intercept falls toward the high end, the better is the potential for narrowing field efforts. A cutoff point can be assigned for high versus low archaeological site potential based on (a) assessing the histograms in conjunction with the total χ^2 value and (b) the comparison of surveyed land units with all land units in the study area.

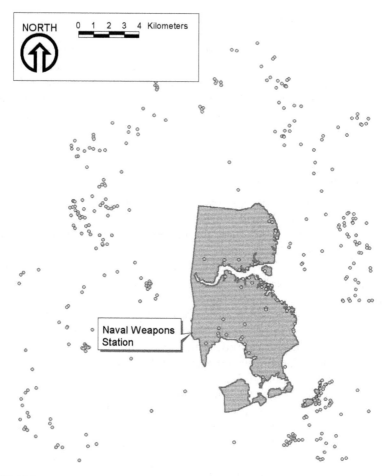

FIGURE 17.2
Known archaeological sites (centroids) in the study area.

When all of the site land units from all four quads (classed by Prehistoric and Historic periods) are compared against each formula, a mean probability value can be calculated. The mean probability value gives an indication of how well the formula would fit areas outside of the NWS, in the absence of knowledge concerning previous survey strategies. No statistical significance is assignable, since we do not know all of the survey strategies outside the NWS, but the mean probability values indicate how well the formulas fit the larger picture. The standard deviation shows an approximation of the spread of land units across all probability value classes.

A significant χ^2 in conjunction with a high x-intercept, a high probability mean, and a low standard deviation does not necessarily indicate a good formula. The formula must also show a good distribution of all land units across the probability value scale, preferably a normal curve. A good formula will not only show significant relationships, but also must be a good land-discrimination tool. If too many of the probability values are skewed to either end of the scale, it will not function well as a means to reduce survey areas or determine field strategies. In other words, we could easily create formulas that show significant relationships with archaeological sites, but if those formulas only produce mostly high-potential areas with a few scattered low-potential zones, then we might need to add more discriminatory variables.

17.3.5.1   Results

Fifteen probabilistic formulas were constructed based on a reading of the previous research in the region. Each formula was modeled in the study area and classified into probability classes. The χ^2 evaluations of known site data against expected site data by each formula were then calculated. Significance was set at $p < 0.005$ (which for 19 degrees of freedom falls at $\chi^2 \geq 38.582$).

Figure 17.3 and Figure 17.4 show the histograms of observed minus expected archaeological values for each formula (Prehistoric and Historic periods, respectively) by probability classes. Table 17.4 lists all of the significant χ^2 values by formula and time period, as well as the x-intercepts, the mean probability values, and their standard deviations, for all site land units in the study area. The highest χ^2 values in conjunction with strong positive spikes on the right of the histogram, strong negative spikes to the left, and a high-scoring x-intercept indicate the formulas that best fit the known pattern of sites. The formulas with the best compromise of these characteristics (with a good spread of all land units across probability value classes) show the most promise as land-management tools.

17.3.6   Prehistoric Site Probability

For the combined Prehistoric period (86 total land units in the NWS), significant χ^2 values were identified for only three formulas (Formulas 5, 10, and 12), of which only Formulas 5 and 10 show the required positive relationship between high-probability values and known archaeological site land units. The histogram for Formula 5 shows an x-intercept at 0.2, a high positive spike at 0.5, and a low negative spike at –0.1 (Figure 17.3). The χ^2 value for Formula 5 falls at 99.71. The mean probability value for all prehistoric sites is 0.4762, with a standard deviation of 0.2338. The minimum observed value at any archaeological site land unit is –0.29. These numbers indicate that Formula 5 has a significant positive relationship between high-probability values (those over 0.2) and prehistoric archaeological sites on the NWS. It also has a significant negative relationship between lower probability values (less than 0.2) and prehistoric archaeological sites on the NWS. The mean probability value for all archaeological sites in the four-quad study region indicates that 98% of the known prehistoric archaeological site land units have Formula 5 probability values greater than 0.0086 (two standard deviations less than the mean).

Prehistoric Period Sites

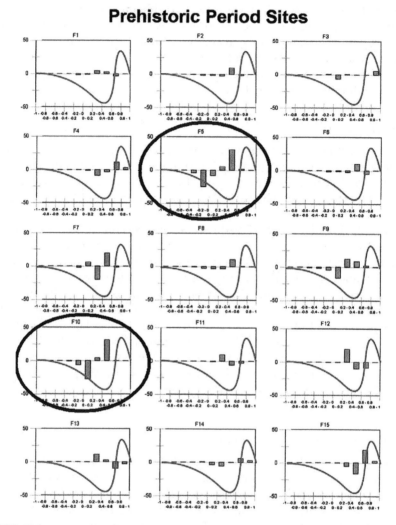

FIGURE 17.3
Histograms of numbers of observed minus expected prehistoric archaeological land units by probability value categories for each formula contrasted against idealized fit (curve).

Approximately 82% of the known prehistoric archaeological site land units have Formula 5 probability values greater than 0.2424 (one standard deviation less than the mean). No known prehistoric site falls in an area with less than −0.29 as a Formula 5 probability value.

Formula 10 has an x-intercept at 0.2, a high positive spike at 0.5, a low negative spike at 0.1 (Figure 17.3). The χ^2 value for Formula 10 is 85.91. The mean probability value for all prehistoric sites is 0.3245, with a standard deviation of 0.2323. The minimum observed value at any archaeological site land unit is −0.41. These numbers indicate that Formula 10 has a significant

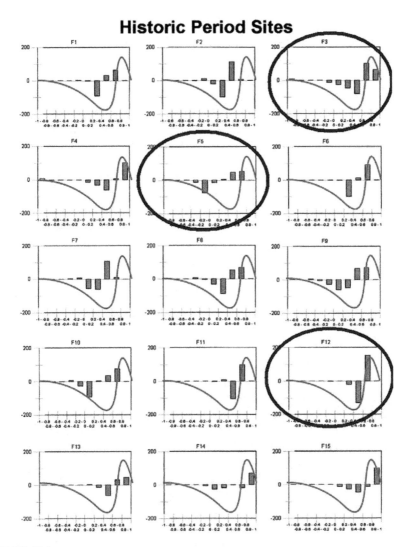

FIGURE 17.4
Histograms of numbers of observed minus expected historic archaeological land units by probability value categories for each formula contrasted against idealized fit (curve).

positive relationship between high-probability values (those over 0.2) and prehistoric archaeological sites on the NWS. It also has a significant negative relationship between lower probability values (less than 0.2) and prehistoric archaeological sites on the NWS. The mean probability value for all archaeological sites in the four-quad study region indicates that 98% of the known prehistoric archaeological site land units have Formula 10 probability values greater than −0.1401 (two standard deviations less than the mean). Approximately 82% of the known prehistoric archaeological site land units have

TABLE 17.4

Significant χ^2 Values, x-intercepts, Probability Means, and Standard Deviations for
Probabilistic Formulas by Time Period

Formula	Prehistoric Period				Historic Period			
	χ^2	x-Intercept	Mean	SD	χ^2	x-Intercept	Mean	SD
F1	—	—	—	—	172.07	0.4	0.3552	0.2496
F2	—	—	—	—	185.75	0.4	0.3088	0.2242
F3	—	—	—	—	559.24	0.6	0.1129	0.6265
F4	—	—	—	—	533.79	0.6	0.1866	0.6541
F5	99.71	0.2	0.4762	0.2338	170.27	0.2	0.4764	0.2572
F6	—	—	—	—	224.68	0.4	0.3567	0.2556
F7	—	—	—	—	198.39	0.4	0.2773	0.2351
F8	—	—	—	—	292.65	0.4	0.2156	0.2908
F9	—	—	—	—	411.06	0.4	0.0946	0.3704
F10	85.91	0.2	0.3245	0.2323	262.95	0.2	0.3048	0.2622
F11	—	—	—	—	135.91	0.6	0.4932	0.2536
F12	42.39	0.6	0.4564	0.2218	443.55	0.6	0.4302	0.2378
F13	—	—	—	—	202.99	0.6	0.4966	0.2740
F14	—	—	—	—	293.75	0.8	0.4545	0.3073
F15	—	—	—	—	413.44	0.8	0.5018	0.3039

Formula 10 probability values greater than 0.0922 (one standard deviation
less than the mean). No known prehistoric site falls in an area with less than
−0.41 as a Formula 10 probability value.

Formula 5 is indicative of a diffuse settlement pattern that is loosely
keyed to several different upland and marsh resources. The variables
used in the formula are size of and cost distance to the nearest marsh
ecotones, degree of slope, soil-drainage ranking, and cost distance to
known historic resources. Formula 10 differs only in the absence of a
dependence on historic resource locations in favor of cost distance to
navigable water. High Formula 5 probability values may, in fact, show a
positive relationship with prehistoric archaeological sites based on the
location of known historic resources only because historic resources themselves may be correlated with cost distance to navigable water, and
historic roads may reflect locations of prehistoric pathways. Therefore
although Formula 10 best fits what we would expect to see given the
previous research in the region, Formula 5 seems to be keyed slightly
more toward upland resource procurement through the association
between Historic period inland travel routes and possible prehistoric
interior pathways. Formula 5 appears to be most representative of a
general gloss of prehistoric settlement (in spite of the use of a Historic
period variable) and thus most applicable to land-management needs on
the NWS.

17.3.7 Historic Site Probability

All of the formulas had significant χ^2 values for the Historic period. Formulas 3, 4, and 12 had the most significant χ^2 values and the best spread of land units across probability value classes. The histogram for Formula 3 shows an x-intercept at 0.6, a high positive spike at 0.7, and a low negative spike at 0.5 (see Figure 17.4). The χ^2 value for Formula 3 falls at 559.24. The mean probability value for all historic sites is 0.1129, with a standard deviation of 0.6265. The minimum observed value at any archaeological site land unit is –0.96. These numbers indicate that Formula 3 has a significant positive relationship between high-probability values (those over 0.6) and historic archaeological sites on the NWS. It also has a significant negative relationship between lower probability values (less than 0.6) and historic archaeological sites on the NWS. The mean probability value and standard deviation for all archaeological sites in the four-quad study region indicates, however, that known historic archaeological site land units range widely in probability values. A small positive spike in the low-probability value end of the scale indicates that Formula 3 cannot accurately assess the probability of Historic period site land units in interior portions of the region.

The histogram for Formula 4 shows an x-intercept at 0.6, a high positive spike at 0.9, and a low negative spike at 0.5 (Figure 17.4). The χ^2 value for Formula 4 is very high at 533 .79. The mean probability value for all historic sites is 0.1866, with a standard deviation of 0.6541. The minimum observed value at any archaeological site land unit is –0.91. These numbers indicate that Formula 4 has a significant positive relationship between high-probability values (those over 0.6) and historic archaeological sites on the NWS. It also has a significant negative relationship between lower probability values (less than 0.6) and historic archaeological sites on the NWS. The mean probability value and standard deviation for all archaeological sites in the four-quad study region indicates, however, that known historic archaeological site land units range widely in probability values. A small positive spike in the low-probability value end of the scale indicates that Formula 4 cannot accurately assess the probability of Historic period site land units in interior portions of the region.

Formula 12 has an x-intercept at 0.6, a high positive spike at 0.7, and a low negative spike at 0.5 (Figure 17.4). The χ^2 value for Formula 12 is high at 443.55. The mean probability value for all historic sites is 0.4302, with a standard deviation of 0.2378. The minimum observed value at any archaeological site land unit is –0.27. These numbers indicate that Formula 12 has a significant positive relationship between high-probability values (those over 0.6) and historic archaeological sites on the NWS. It also has a significant negative relationship between lower probability values (less than 0.6) and historic archaeological sites on the NWS. The mean probability value for all archaeological sites in the four-quad study region indicates that 98% of the known historic archaeological site land units have Formula 12 probability values greater than –0.0454 (two standard deviations less than the mean). Approximately 82% of the known historic archaeological site land units have Formula 12 probability values greater than 0.1924 (one standard deviation less than the mean). No known historic site falls in an area with less than –0.27 as a Formula 12 probability value.

Between all of the formulas recognized as having a significant χ^2 values for the assessment of Historic period archaeological land units, Formula 12 shows the most promise as a land-management tool. Formula 12 indicates a historic settlement pattern that focused on exponential cost distance to navigable waterways and interior travel routes, but that also was keyed to seed crops and open-land wildlife capacities. A smaller emphasis was placed on marsh ecotones, wetland and woodland wildlife capacities, degree of slope, and soil-drainage ranking.

17.3.8 Field Testing the Model

The data provided by the best-fit probabilistic formulas can help determine field efforts and project planning in several ways:

1. By knowing what types of sites are likely to be encountered in any zone
2. By knowing the approximate frequency at which sites would be encountered
3. By being able to alter field methods to accommodate that information

By using the known site and survey data from the NWS, we were able to test the probability formulas against a uniform survey strategy and estimate their relationships with site areas outside of the NWS.

The next step in the process was to develop appropriate probability zonation to complete the assessment of unsurveyed lands in the NWS. Classes were created using the *x*-intercept as the cutoff point for high- versus moderate-probability zones, and using two standard deviations below the mean value as the cutoff between moderate- and low-probability zones. The following equations show the transformation of probability values into probability classes:

Low probability ≥ -1, and $\leq$ mean – 2·SD (from Table 17.4)
Moderate probability > mean – 2*SD, and $\leq$ *x*-intercept (from Table 17.4)
High probability > *x*-intercept (from Table 17.4)

For the Prehistoric period (Formula 5), the probability classes are defined as:

Low probability ≥ -1 and ≤ 0.01
Moderate probability > 0.01 and ≤ 0.2
High probability > 0.2

TABLE 17.5

Potential for Finding Cultural Remains by Probability Class

	Probability Class (%)					
	Low		Moderate		High	
Period: Formula	Observed	Projected	Observed	Projected	Observed	Projected
Prehistoric: Formula 5	0.5	0.8	0.4	2.5	5.1	4.0
Historic: Formula 12	0	0	1.5	1.1	2.4	7.6

For the Historic period (Formula 12), the probability classes are defined as:

Low probability ≥ -1 and ≤ -0.04

Moderate probability > -0.04 and ≤ 0.6

High probability > 0.6

A total of 15,967 land units have been surveyed on the NWS (3592 acres) out of a possible 61,047 land units (13,736 acres). This constitutes 26% of the total NWS acreage. Another 15,704 land units (3,533 acres) have been field-verified to be disturbed beyond the potential to contain intact archaeological resources. The remaining 29,376 land units (6,610 acres) include a mix of surveyable and potentially disturbed land that was not field-verified prior to formula development.

Table 17.5 shows a projection of the likelihood of encountering cultural resources by best-fit formula for each time period. The likelihood is calculated by dividing the number of observed site land units by the total land units in each class. Using the best-fit formulas (Figure 17.5 and Figure 17.6) in high-probability areas, 4 out of every 100 shovel tests placed at 30-m intervals will contain prehistoric artifacts; about 8 out of every 100 will contain historic materials.

In moderate-probability areas, though, the efficiency of shovel testing at 30-m intervals drops to almost half for prehistoric sites and less than one-fifth for historic sites. In low-probability zones, 30-m shovel testing produces positive prehistoric or historic results very rarely and never identifies significant sites (based on known site-eligibility determinations).

After the model was created, an additional 8889 land units (about 2000 acres) were surveyed at the NWS, including at least 3168 land units (about 713 acres) located within Formula 5 and 12 low-probability zones. Survey of these areas was carried out as a uniform 30-m shovel-testing regime (including those areas previously deemed unsurveyable, i.e., wetlands). The goal was to field-verify the model formulas. The results are shown in Table 17.5 as the observed values.

For Formula 5 we projected 1 positive shovel test for every 125 excavated in the low-potential areas, and we encountered only 1 positive for every 200

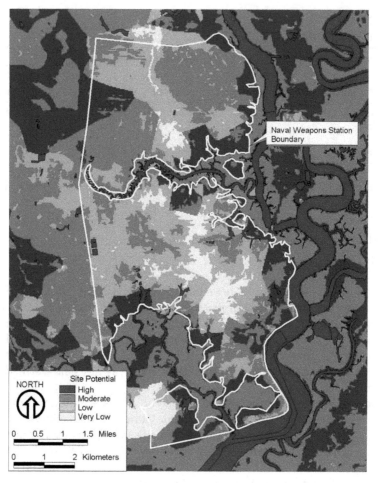

FIGURE 17.5
Probability zones at the NWS based on Formula 5 (for prehistoric sites).

excavated. We projected 1 positive shovel test for every 40 excavated in the
moderate zones, and we encountered only 1 positive for every 278 excavated.
We projected 1 positive for every 25 excavated in the high-potential areas,
and we encountered 1 positive for every 19 excavated.

For Formula 12 we projected no positive shovel tests in the low-potential
areas, and we encountered none. We projected 1 positive shovel test for every
91 excavated in the moderate zones, and we encountered 1 positive for every
67 excavated. We projected 1 positive for every 13 excavated in the high-
potential areas, and we encountered only 1 positive for every 43 excavated.
After the results indicated general concurrence with the projections of the
modeled formulas, the remaining 20,487 land units (about 4600 acres) at the

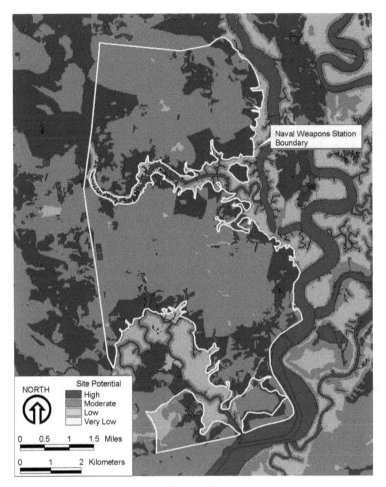

FIGURE 17.6
Probability zones at the NWS based on Formula 12 (for historic sites).

NWS were surveyed with differential survey strategies, including pedestrian reconnaissance of approximately 2200 disturbed acres.

17.4 Conclusions

The development of a GIS-based data-independent archaeological probabilistic model at the Charleston Naval Weapons Station provided several important observations about the region and probabilistic modeling in general:

1. A better grasp of the cognitive nature of site placement and overall settlement patterns in any area results from evaluating environmental variables in a data-independent framework. Archaeological data dependency creates an additional level of potential error and provides much less incentive to develop truly explanatory settlement hypotheses.

2. In a GIS framework, data-independent probabilistic models are not more difficult to create than inferential ones. They are more intuitive and potentially more likely to be employed by nonspecialists as well.

3. Homogeneous environments are difficult settings within which to create accurate probabilistic models. Without examining causal factors from the perspective of cognitive decision making, it becomes virtually impossible to create a model that is both accurate (encompasses most known sites) and precise (has a high site-to-probability-area ratio).

4. Although the model focuses strictly on a general categorization of sites as "prehistoric" or "historic" (due to the limitations of the available archaeological data, and the need to recommend differential survey strategies), the greatest potential for data-independent models might be realized in a situation where specific site types or characteristics are more easily determined.

References

Adams, N.P., Early African-American Domestic Architecture from Berkeley County, South Carolina, Master's thesis, Department of Anthropology, University of South Carolina, Columbia, 1987.

Anderson, D.G., Models of paleo-Indian and Early Archaic settlement in the lower southeast, in The Paleo-Indian and Early Archaic Southeast, Anderson, D. and Sassaman, K., Eds., University of Alabama Press, Tuscaloosa, 1996, pp. 29–57.

Anderson, D.G. and Joseph, J.W., Prehistory and History along the Upper Savannah River, Interagency Archaeological Services, National Park Service, Atlanta, 1988.

Anderson, D.G. and Logan, P.A., Francis Marion National Forest Cultural Resources Overview, USDA Forest Service, Columbia, SC, 1981.

Anderson, D.G., Lee, S.T., and Parler, A.R., Cal Smoak: Archaeological Investigations along the Edisto River in the Coastal Plain of South Carolina, Occasional Paper 1, Archaeological Society of South Carolina, 1979.

Anderson, D.G., Cantley, C.E., and Novick, A.L., The Mattassee Lake Sites: Archaeological Investigations along the Lower Santee River in the Coastal Plain of South Carolina, Special Bulletin 1, Archaeological Services Branch, National Park Service, Tallahassee, Florida, 1982.

Brockington, P.E., Jr., Markham, M.V., Butler, C.S., and Jones, D.C., Cultural Resources Survey of the Charleston Naval Weapons Station, Berkeley and Charleston Counties, South Carolina, report prepared for U.S. Army Corps of Engineers, Savannah District, Brockington and Associates, Atlanta, 1995.

Brooks, M.J. and Canouts, V., Modeling Subsistence Change in the Late Prehistoric Period in the Interior Lower Coastal Plain of South Carolina, Anthropological Studies 6, occasional papers of the South Carolina Institute of Archaeology and Anthropology, University of South Carolina, Columbia, 1984.

Cabak, M.A., Sassaman, K., and Gillam, J.C., Distributional Archaeology in the Aiken Plateau, Savannah River Archaeological Research Papers 8, occasional papers of the South Carolina Institute of Archaeology and Anthropology, University of South Carolina, Columbia, 1996.

Cable, J.S., Prehistoric Adaptations to the Interior Uplands Environment of the Central South Carolina Coast II, report prepared for the USDA Forest Service, New South Associates, Irmo, SC, 1993.

Cable, J.S., A Study of Archaeological Predictive Modeling in the Charleston Harbor Watershed, South Carolina, Technical Report 414, New South Associates, Irmo, SC, 1996.

Dalla Bona, L., Cultural Heritage Resources Predictive Modelling Project, Vol. 3, Methodological Considerations, report prepared for the Ontario Ministry of Natural Resources, Center for Archaeological Resource Prediction, Lakehead University, Thunder Bay, ON, 1994.

Drucker, L.M. and Anthony, R.W., The Spiers Landing Site: Archaeological Investigations in Berkeley County, South Carolina, Carolina Archaeological Services, Columbia, 1979.

Dunnell, R.C., The notion site, in *Space, Time, and Archaeological Landscapes*, Rossignol, J. and Wandsnider, L., Eds., Plenum Press, New York, 1992.

Ebert, J.I., The state of the art in "inductive" predictive modeling: seven big mistakes (and lots of smaller ones), in *Practical Applications of GIS for Archaeologists: A Predictive Modeling Kit*, Wescott, K. and Brandon, R.J., Eds., Taylor and Francis, London, 2000, pp. 129–134.

Eppinette, R.T., Soil Survey of Dorchester County, South Carolina, USDA Soil Conservation Service, South Carolina Agricultural Experiment Station, and South Carolina Land Resources Conservation Commission, Columbia, 1990.

Goodyear, A.C., III and Hanson, G.T., Eds., Studies in South Carolina Archaeology: Essays in Honor of Robert L. Stephenson, Anthropological Studies 9, occasional papers of the South Carolina Institute of Archaeology and Anthropology, University of South Carolina, Columbia, 1989.

Goodyear, A.C., III, House, J.H., and Ackerly, N.W., Laurens-Anderson: An Archaeological Study of the Inter-Riverine Piedmont, Anthropological Studies 4, occasional papers of the South Carolina Institute of Archaeology and Anthropology, University of South Carolina, Columbia, 1979.

Hempel, C.G., *Aspects of Scientific Explanation*, Free Press, New York, 1965.

Jeffrey, R.C., Statistical explanation versus statistical inference, in *Statistical Explanation and Statistical Relevance*, Salmon, W., Ed., University of Pittsburgh Press, Pittsburgh, 1971, pp. 19–28.

Judge, W.J. and Sebastian, L., Eds., *Quantifying the Present and Predicting the Past: Theory, Method, and Application of Archaeological Predictive Modeling*, Bureau of Land Management, Denver, 1988.

Kohler, T.A. and Parker, S.C., Predictive models for archaeological resource location, in *Advances in Archaeological Method and Theory*, Vol. 9, Schiffer, M.B., Ed., University of Arizona Press, Tucson, 1986.

Kvamme, K.L., Geographic information systems in regional archaeological research and data management, *Archaeological Method and Theory*, 1, 139–203, 1990a.

Kvamme, K.L., The fundamental principles and practice of predictive modelling, in *Mathematics and Information Science in Archaeology: A Flexible Framework*, Voorrips, A., Ed., Vol. 3 in *Studies in Modern Archaeology*, Holos-Verlag, Berlin, 1990b, pp. 257–295.

Kvamme, K.L., A view from across the water: the North American experience in archaeological GIS, in *Archaeology and Geographic Information Systems: A European Perspective*, Lock, G. and Stančič, Z., Eds., Taylor and Francis, London, 1995, pp. 1–14.

Long, B.M., Soil Survey of Berkeley County, South Carolina, USDA Soil Conservation Service, South Carolina Agricultural Experiment Station, and South Carolina Land Resources Conservation Commission, Columbia, 1980.

Mills, R., *Atlas of the State of South Carolina*, 1825, reprint, Sandlapper Co., Lexington, SC, 1979.

Mistovich, T.S. and Clinton, C.E., Archaeological Data Recovery at the Track Site, 38BU927, Marine Corps Air Station, Beaufort, South Carolina, report of Investigations 60, Division of Archaeology, Alabama Museum of Natural History, University of Alabama, Tuscaloosa, 1991.

Panamerican Consultants, Historic and Archaeological Resources Protection Plan for Naval Weapons Station Charleston, Charleston, South Carolina, report prepared for U.S. Army Corps of Engineers, Savannah District, Panamerican Consultants, Tuscaloosa, AL, 1997.

Pearl, J., *Causality: Models, Reasoning and Inference*, Cambridge University Press, Cambridge, 2000.

Rohlf, F.J. and Sokal, R.R., *Statistical Tables*, W.H. Freeman, San Francisco, 1969.

Rust, T., Cultural Resources Survey of the Proposed Commissary Site, Naval Weapons Station Charleston, Berkeley County, South Carolina, report prepared for General Engineering Laboratories, Brockington and Associates, Charleston, SC, 1997.

Salmon, W.C., *Statistical Explanation and Statistical Relevance*, University of Pittsburgh Press, Pittsburgh, 1971.

Salmon, W.C., *Scientific Explanation and the Causal Structure of the World*, Princeton University Press, Princeton, NJ, 1984.

Salmon, W.C., *Causality and Explanation*, Oxford University Press, Oxford, 1998.

Sassaman, K.E., Early archaic settlement in the South Carolina coastal plain, in *The Paleo-Indian and Early Archaic Southeast*, Anderson, D. and Sassaman, K., Eds., University of Alabama Press, Tuscaloosa, 1996, pp. 58–83.

Sassaman, K.E. and Anderson, D.G., Middle and Late Archaic Archaeological Records of South Carolina, Savannah River Archaeological Research Papers 6, occasional papers of the South Carolina Institute of Archaeology and Anthropology, University of South Carolina, Columbia, 1995.

Scurry, J.D., The Design and Testing of a Geographic-Based Information System for Archaeological Resources in South Carolina, Masters thesis, Department of Anthropology, University of South Carolina, Columbia, 1989.

Soil Systems, Cooper River Rediversion Canal Historic Sites Archaeology, draft report prepared for Interagency Archaeological Services, National Park Service, Atlanta, Soil Systems, Marietta, GA, 1982.

South, S.A. and Hartley, M., Deep water and high ground: seventeenth century low country settlement, in *Structure and Process in Southeastern Archaeology*, Dickens, R.S. and Ward, H.T., Eds., University of Alabama Press, Tuscaloosa, 1985, pp. 263–286.

Stančič, Z. and Kvamme, K.L., Settlement Patterns Modelling through Boolean Overlays of Social and Environmental Variables, Publications of the Centre for Spatial Studies, Scientific Research Centre of the Slovene Academy of Sciences and Arts, Predictive Models, 2000; available on-line at http://www.zrc-sazu.si/pic/pub/predictive/predictive.html.

Stančič, Z., Veljanovski, T., Ostir, K., and Podobnikar, T., Archaeological predictive modeling for highway construction planning, in *Computing Archaeology for Understanding the Past*, CAA 2000 proceedings, Stančič, Z. and Veljanovski, T., Eds., BAR International Series, 931, Archaeopress, Oxford, U.K., 2001, pp. 233–238.

Stephenson, K., Cultural Resources Survey of the Ordnance Railroad Modifications, Naval Weapons Station Charleston, Berkeley County, South Carolina, report prepared for Tuner, Collie and Braden, Inc., Brockington and Associates, Inc., Charleston, SC, 1998.

Stine, L.F., Revealing Historic Landscapes in Charleston County: Archaeological Inventory, Contexts, and Management, report prepared for County of Charleston Planning Department and the South Carolina Department of Archives and History, South Carolina Institute of Archaeology and Anthropology, University of South Carolina, Columbia, 1991.

Stine, L.F., Drucker, L.M., Zierden, M., and Judge, C., Eds., Historic Landscapes in South Carolina: Historical Archaeological Perspectives of the Land and Its People, Council of South Carolina Professional Archaeologists and the South Carolina Institute of Archaeology and Anthropology, University of South Carolina, Columbia, 1993.

Tidewater Atlantic Research, A Submerged Cultural Resource Management Document and GIS Database for the Charleston Harbor Project Study Area, Charleston, South Carolina, report prepared for the South Carolina Coastal Council, Tidewater Atlantic Research, Washington, NC, 1995.

Van Dalen, J., Probability modelling: a Bayesian and geometric example, in *Geographical Information Systems and Landscape Archaeology*, Gillings, M., Mattingly, D., and van Dalen, J., Eds., Vol. 3, *The Archaeology of Mediterranean Landscapes*, Alden Press, Oxford, U.K., 1999, pp. 117–124.

Wescott, K.L. and Brandon, R.J., Eds., *Practical Applications of GIS for Archaeologists: A Predictive Modeling Kit*, Taylor and Francis, London, 2000.

Whitley, T.G., Archaeological Probabilistic Modeling of the Charleston Naval Weapons Station, Berkeley and Charleston Counties, South Carolina, report prepared for U.S. Navy Facilities Engineering Command, Southern Division, Charleston, SC, Brockington and Associates, Atlanta, 1999.

Zierden, M.A., Drucker, L.M., and Calhoun, J., Home Upriver: Rural Life on Daniel's Island, Berkeley County, South Carolina, report prepared for South Carolina Department of Highways and Public Transportation, Carolina Archaeological Services, Charleston Museum, Columbia, SC, 1986.

18

Predictive Modeling in Archaeological Location Analysis and Archaeological Resource Management: Principles and Applications

Tatjana Veljanovski and Zoran Stančič

CONTENTS

ABSTRACT The main objective of this chapter is to illustrate the various aspects of academic and practical applications of predictive modeling and to discuss the need to assess the reliability of the results and the implications of reliability assessment on the further development of predictive models. To accomplish this, some focus is directed to reconsideration of error propagation. The chapter begins with a short introduction to prediction and the predictive-modeling approach in archaeology. This is followed by a discussion of a number of methodological issues and concerns, with emphasis on the presentation of several case studies. The first set of case studies can be treated as academic applications. They focus on the examination of a possible mental space of the Bronze Age culture on the island of Brač, in Croatia, by modeling the potential distribution of Bronze Age hill-fort locations and barrows on the island. Next, some results of the application of multivariate statistical techniques and the Dempster -Shafer method of land-use modeling (used for the analysis of Roman settlement patterns) are presented as an anticipated enhancement to the Boolean overlay approach. Finally, a case study of a project in the Pomurje region, Slovenia, is used to show how potential archaeological locations were predicted in a highway construction project. This final case study demonstrates the extent to which the predictive-modeling approach can be employed in a useful and practical application. The chapter concludes with general remarks based on the experience gained with these models while emphasizing the need to develop new approaches to predictive modeling, to test model accuracy, and to evaluate the predictive power of archaeological models.

18.1 Background to Principles and Applications

18.1.1   Prediction

Prediction is a conventional procedure used in countless situations. It is used particularly in cases where the main endeavor is oriented toward the observation of a possible continuation (yet unknown) of manners, actions, or some behavior, or where prediction (or, more likely, interpolation) is used as a tool to supplement missing data. Predictions are made in weather forecasting, and a similar principle can also be found in risk analysis, animal behavioral studies in biology, material behavior in chemistry, etc. The predictive techniques and technological environments in which predictions are made differ from field to field. Technological environments can range from a specific computer-based environment such as a geographic information system (GIS) to complex expert systems joining several specific environments especially tailored for artificial knowledge-based applications. Within a GIS environment, the concept of prediction is more restricted than the generally understood meaning of the comprehensive term "prediction." A GIS-based prediction yields spatially focused predictive information about specific locations that are likely to be of archaeological significance. The combination of

GIS with the principles of advanced modeling techniques has been successfully used to tease out archaeological implications in the landscape (Lock
and Stančič 1995; Exon et al. 2000; Gillings et al. 1999).

18.1.2 Modeling and Prediction in Archaeology

Although predictive modeling is something archaeologists always keep in
mind when attempting to locate yet unknown sites, the early beginnings of
computer-based predictive-modeling techniques (i.e., GIS) date back to the
late 1970s, when geographical information systems were introduced into
archaeology. Predictive models have a fairly long tradition in American
archaeology (Judge and Sebastian 1988), where they are intensively used for
cultural-resource management. During the past decade, such use has become
common throughout the world. There are two branches of archaeological
predictive models (van Leusen 1996; Wescott and Brandon 2000): (a) those
developing in relation to the pragmatic aspect of use (i.e., mapping sensitive
areas within a landscape to protect the current archaeological heritage) and
(b) the academic models, where an attempt to reconstruct past societies leads
toward models that would increase our understanding of settlement logic
on the basis of location determinants. The reasons for these developments
have been discussed in numerous publications, but it is useful to briefly note
the role of conceptual distinction between the two principles.

The main distinction between the two approaches is the evident inclination
of the academic approach to establish the determinants that influenced the
settlement behavior in the past. Following this, a given archaeological site
can be understood as a reflection of certain decisions representing the final
choice within a certain cultural system. For example, the selection of a
dwelling place within a given landscape is a complex reflection of such
decisions which may also be influenced by personal preference, in accordance with the potential offered by the natural environment. The incorporation of the human component within location modeling acknowledges the
role of subjective judgment. However, making the models more cognitive
may also introduce a hardly manageable complexity. Despite this, the analysis of the social environment might include those variables that are understood as descriptive of humans' cultural, religious, ideological, or economic
relationships with the landscape, as they can offer us additional information.
Still, the fact remains that they are often omitted from the modeling procedure because they are more difficult to obtain and because it is nearly impossible to reliably transform them into a meaningful GIS layer.

Besides, human knowledge of the past is neither perfectly certain nor
absolutely uncertain. Predictive modeling should therefore be viewed as
a vital attempt to draw recognized behavior further, i.e., an extension of
an otherwise unobservable phenomenon. One possible way of dealing
with limited knowledge certainty is to think in terms of belief reflected
through probabilities. Another way that has also been adopted in the field

of archaeology (and even more so in cultural-resource management) is to think in terms of the potential that a given location may possess. However, both methodological approaches rely heavily on statistics. Thus, when speaking about an archaeological potential map within a given landscape (as a result of a prediction), we should be well aware that it can only play a role of a valuable output to the main interest of environmental planning, i.e., on a landscape-management level, and rarely support archaeological fieldwork planning on a local scale. Together with the nature and qualities of modeling techniques, the methodology process itself should be reserved to stretch more deeply into the scope of reconstructing the past, archaeological explanation, and anticipated interpretation.

Within archaeology, the study of settlement patterns is important when debating any degree of knowledge as regards the landscape and past human occupation of the area. Through this point, the expression "archaeological predictive modeling" is generally and hereby introduced. Predictive modeling is a procedure used to predict site locations within a region. This is performed on the basis of observed patterns or on assumptions about human behavior (Kohler and Parker 1986; Judge and Sebastian 1988). Archaeological predictive models are essentially based on the fundamental assumption that our knowledge of the known archaeological sites allows us to establish which factors influenced their location in the landscape and to use this data in empirical testing. Prediction is merely the elucidation of settlement "rules" in a form that allows us to map locations that conform to the "conditions" predicted by the model for settlement. To achieve this, we analyze the relationship between the natural and social environment and archaeological site location. The idea itself is very simple and good, but what should never be neglected is the fact that we can never be certain to what degree the available (thus far representative) archaeological data sample can represent the context of past settlement.

18.1.3   Model Verification and the Methodology Transparency Problem

The step open to discussion in developing predictive models is how to test the model accuracy and evaluate its predictive power. It is important to emphasize this particular problem, since the quality of archaeological predictive models is mainly based on the ability to test their efficiency. An archaeological sample of representative sites or other past features within a region is not unlimited, and we may be forced to include all or at least the majority of the existing data into the model-developing process. Because most techniques that we employ in the predicting procedure still rely upon a statistical basis, a sample size must turn out to be a key issue. However, if this is the case, we soon realize that no real independent data is left with which to test the model or the modeling procedure. Luckily this is not applicable on areas where archaeology is dense.

The predictive modeling procedure's result map frequently shows information that is too general and thus may be hard to interpret in a sober archaeological context. Consequently, the interpretation component usually depends on the techniques involved in model development and the principle for verifying the model's results. In most cases, this approach is acceptable, even though it can hardly be used for regional planning purposes. On the other hand, the archaeological context is seldom discussed in publications, yet it should encourage a greater interest.

There are three important objectives related to quality and the transparency problem, all of which need special attention in predictive modeling.

1. Concern about the quality of the data and its reliability within the past landscape context
2. Concern about the quality of techniques implemented in the predictive-modeling procedure within specific environments (both natural and computer)
3. Concern about the possibility of misleading or opposite interference with the settlement "rules" that might occur if such rules were derived solely empirically

However, all this remains within the analyst's frame of decision making, modeling care, and aims. Almost two decades have passed since the adoption of predictive modeling within the archaeological community, and there is an urgent need for further innovation in acquiring qualitative interpretation and evaluation of modeled results and modeling procedures. A possible approach may be offered by the so-called tracing methods known within data tracing and the error-propagation field familiar to those who use GIS to model spatial data.

Data tracing is a process used to gather information on data behavior (for example, changes in attribute value, coordinate, or path). In general, tracing tools are able to calculate and visualize the execution (behavior) of a program or procedure. However, because data tracing in computer and software sciences usually means inserting a special code at the entry and exit points, data tracing introduces the cunning problem of correctly implementing the code (ASKALON 2002), and at the same time, it requires comprehensive programming efforts. The basic idea is that the code collects information on the code region, which is comparable with the processing stage in the modeling. This information is stored in trace files that can then be used to describe the dynamic behavior of the code region, providing a means of obtaining information on data behavior at a particular processing stage. The major problem with data tracing occurs when data with a spatial dimension is introduced. Tracing the attributes of spatially referenced data should not cause problems, for this can be taken into account by the analyses of error propagation (see Heuvelink 1998; Burrough and McDonnell 1998). However, tracing two-dimensional spatial accuracy in the modeling process may prove to be more problematic, mainly because of the exposed problem that arises from coding.

Nevertheless, it seems that the described approach will be applicable to the geographical computer environment in the near future, and this could consequently facilitate an advanced evaluation of the results of predictive-modeling procedures. Indeed, intelligent GIS might be a solution for all GIS users (Longley et al. 2001).

18.1.4   Predictive-Modeling Techniques

The purpose of this chapter is not to advance the very first technology tools, but to stimulate thoughts on what could be improved and to advocate archaeological predictive-modeling efforts. It is also not our purpose to discuss in detail the theoretical approaches to predictive modeling, but rather to comment on the various modeling techniques that are used in predictive modeling today. We have defined two theoretical approaches to archaeological prediction: inductive and deductive. The truth is that both approaches overlap in practice, as do the modeling techniques. The technological limits when applying any of these approaches are also worthy of consideration. Generally speaking, one can apply a number of alternative procedures, each having its own advantages and disadvantages. The Boolean overlay of variables, which has been proved to have influenced the location patterns, can be utilized alongside multivariate statistical techniques (linear regression, logistic regression, discriminant analyses) or some other decision-support methods (i.e., Dempster-Shafer theory, a belief method). Knowledge-based approaches (neural-network applications, agent systems) show promise as a viable alternative to the established predictive methods. At this point in the development of predictive models, the appropriateness of the chosen technique is primarily subject to the modeling aim and the available data range.

To overcome the statistical requirements for scarce archaeological data, a variety of techniques have been developed based on probability theory, fuzzy theory, and agent theory, although less rigorous modeling of complicated (location) systems may also be used. These techniques, which are born from the recent explosion of interest in chaos theory, dynamic modeling, and complex systems theory, have been applied to a variety of problems and may well lead to fresh insights. Hopefully, some of them could become relevant to archaeological explanation or hypothesis testing, while the vast majority could improve interpretation of modeled results. However, within this sphere of sophisticated approaches, it is imperative that we recognize the axioms underlying past modeling efforts and the subsequent approximations resulting from modeling simplification. We must be aware of the lack of the absolute knowledge and its involvement in the success and failure within archaeological computer modeling.

18.2 Case Studies Perspective

This section examines some results and experience that the authors obtained using different predictive-modeling techniques. The first set of case studies focuses on the application of multiple overlays of thematic layers for modeling potential Bronze Age hill-fort locations. Some results of the multivariate statistics applied to Bronze Age barrow locations and to the analysis of Roman settlement patterns are also shown. These are then followed by the implementation of the Dempster-Shafer belief method, and the results are briefly compared. All of these approaches were performed to test the behavior of three different techniques and to observe their response to the interpretation of location patterns, and they are therefore considered as "academic applications" of predictive modeling. Finally, a case study that can be considered to be a useful "pragmatic application" of predictive-modeling efforts is presented. This case study shows how archaeological potential was predicted in a Slovenian national highway construction project.

18.2.1 "Academic" Applications on the Island of Brač, Croatia

Applications on the island of Brač relate to the work carried out in the Central Adriatic, where an international team has been studying the past colonization of Central Dalmatian islands. The Adriatic Island project has carried out work within a transect of islands. Currently, a synthesis of the project is in progress. During the field campaign in 1994, the island of Brač was surveyed using an extensive sampling strategy. A total of over 600 sites were recorded within the database (Stančič et al. 1999). Next, a natural-environment database was produced. Thematic data on soil, geology, and contour lines were acquired from different data sources. The alternative, vegetation data, had been collected by satellite images using a Landsat thematic mapper (TM). More-sophisticated models, combining data for variables from the natural database (e.g., the vegetation index, erosion model, etc.), were produced.

Note that the aim of the following sections is focused on the observed behavior of the various techniques that we tested. A detailed description of the response of these techniques in the context of archaeological interpretation is beyond the scope of this text.

18.2.1.1 *Predicting Bronze Age Hill-Fort Locations with the Boolean Overlay Technique*

The main objective was to test the extent to which predictive modeling can be based on a small number of sites. The model was developed on an area measuring approximately 120 km², which is approximately one-quarter of the total surface of Brač selected area also greatly overlaps with one of the three physiographic regions on the island where almost half of all Bronze

Age barrows and hill forts were located. The area is commonly considered as a center of Bronze Age activities on the island of Brač.

Because the hill-fort locations could be predicted using a very small data set (eight hill forts), the goal of this study was to predict locations using a number of variables describing both the natural and the social environment. The relationship between the hill-fort locations and the following variables was analyzed: relief (elevation, slope, ridge/drainage index, rim index, and relief-below index), cumulative viewshed index, distance from the coast, distance between the hill forts, and cumulative distance from barrows and other points (see Stančič and Kvamme 1999) that were considered to be of potential importance. All the correlations were examined by various quantitative and descriptive statistical approaches.

A threshold value on each variable was defined based on the insights provided by the relationships. The threshold values could be defined in different ways, but in this case a simple criterion was employed, such that the threshold would capture all known sites. Finally, a predictive model was made as a sum of those grid cells that proved to have influenced hill-fort locations (Figure 18.1).

Besides testing the technique's predictive power, we also wished to evaluate its transparency. The results show that the predictive model of the Bronze Age hill-fort locations is simple yet powerful. The model efficiently excludes the reliably potential areas from the background. Given its simplicity and procedural transparency, the Boolean overlay technique appears to be suitable for use with smaller (but homogeneous) archaeological data samples in a nonmonotonous landscape. The main inconvenience of the technique lies in the uniformity of the predictors' arrangement in a settlement pattern, but this can be overcome by incorporating weighting into the process.

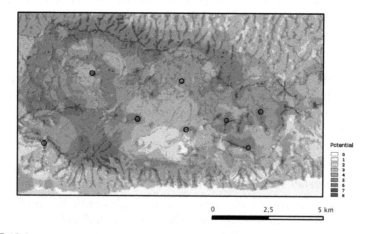

FIGURE 18.1

Archaeological predictive model based on Boolean overlays for the Bronze Age hill-fort locations on the island of Brač and the distribution of hill-fort locations.

18.2.1.2   Predicting Roman Settlement Sites with the Linear Regression Technique

The second case study was aimed at examining the potential of multivariate statistical techniques in the development of predictive models. Special emphasis was paid to the location analysis of Roman settlement sites dating between the second century B.C. and the second century A.D. A total of 29 sites out of approximately 90 Roman sites that were recorded could have been interpreted as settlements (Stančič et al. 1999). Because the number of these Roman settlement sites is not great (concerning multivariate statistic requirements), and considering the specifics of site distribution across the entire island of Brač, we decided that the model should be developed for the island as a whole.

An important step in analyzing settlement patterns is to show that the characteristics of site locations significantly differ from those of other locations across the landscape. The first step in variable selection is based both on previous experience and empirical measurement of data. Therefore, the initial data set incorporated eight variables: elevation; degree of the surface slope; the surface aspect; quality of soil; land use; proneness to erosion; distance from the coast; and the proximity to Sennone limestone, which is viewed as an extension of surface geology (see Stančič and Veljanovski 2000). Finally, four variables were accepted as the possible predictors for Roman settlement locations on the island of Brač: aspect (southwest-facing slopes), distance from Sennone limestone (proximity to the thin limestone zone), elevation (avoidance of elevated areas), and slope (places that are less steep). Promising variables were subjected to a linear regression analysis, regarded as a linear-trend extrapolation technique.

Given the fairly large number of sites, it was hoped that a regression analysis might be applied. However, regression solutions and the resulting equation relating site potential on 29 observations were not very promising in any variable combination we made. (The adjusted r^2, interpreted as a part of explained variance, did not exceed 48% in any of these cases.)

In general, it was learned that the linear-regression-based predictive technique makes it possible to predict the potential that a site possesses at a given location only when substantial care in handling data is introduced (data ordering and correlation observation, data normalization, etc). Despite the fact that the results were not as good as anticipated, it should be emphasized that this technique is good for providing a more detailed insight into the importance of individual variables and their relative contribution to the settlement pattern. Illustrating this on the Roman settlement case study, it was confirmed that the thin layer of Brač "marble" might have gravely influenced the decision as regards the location on the island during the Roman period. According to the regression analysis results, the other three important factors forming common sense for Roman settlements on Brač also seem to be aspect (southwest-facing terrain), followed by elevation, and finally the slope.

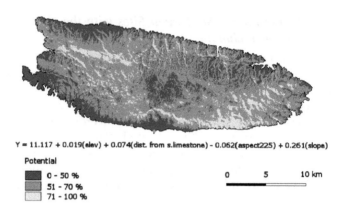

Y = 11.117 + 0.019(elev) + 0.074(dist. from s.limestone) - 0.062(aspect225) + 0.261(slope)

Potential

■ 0 - 50 %
▨ 51 - 70 %
☐ 71 - 100 %

0   5   10 km

FIGURE 18.2
Archaeological predictive model based on linear regression for the Roman settlement sites on
the island of Brač and the distribution of sites.

When the results (Figure 18.2) were analyzed in greater detail, it became
apparent that the poor statistical performance might be seated in the archaeological data. What we have interpreted as homogeneous Roman settlement
sites could have been a mixture of several types of Roman settlements.
Therefore, it was decided to carry out a more refined analysis to provide
new ideas on the possible clusters of sites. The K-means cluster analysis
identified three distinguished clusters representing the proneness to three
different strategies of control over natural resources. We provisionally
denoted them as seaport type, quarry type, and agricultural type of Roman
sites (Stančič and Veljanovski 2000).

Predictive models derived with the linear regression technique (also predictive models of identified subtypes) are not satisfactory because the predictive power is quite poor. Besides the lack of any statistical significance,
the main deficiency of the models is that the insight into the impact of single
variables, observed as spatial extrapolation in a predictive-model result (a
map), is somewhat blurred. This is mainly due to the mixture of the positive
and negative influences of variables (a direct consequence of data ordering),
therefore possibly nullifying the identified trends as well as the potential
exaggeration at the edges belonging to the linear regression equation (and
linear modeling nature). However, the linear regression technique (applied
to archaeological data) remains promising and powerful as a method to
explore the relative relationships between variables affecting site location.
Other case studies — the linear regression technique applied to 88 barrow
locations and even the technique applied to eight hill forts and Roman
settlement subtypes — also proved the above statement about linear regression behavior on spatially referenced problems (see Figure 18.3). For mapping the archaeological potential, it would be more appropriate to use at
least a polynomial regression form and a sufficiently large sample size of
archaeological data.

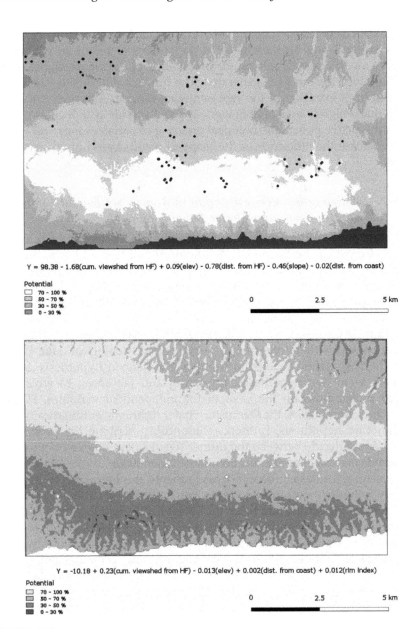

Y = 98.38 - 1.68(cum. viewshed from HF) + 0.09(elev) - 0.78(dist. from HF) - 0.46(slope) - 0.02(dist. from coast)

Potential
☐ 70 - 100 %
▨ 50 - 70 %
▨ 30 - 50 %
▨ 0 - 30 %

0   2.5   5 km

Y = -10.18 + 0.23(cum. viewshed from HF) - 0.013(elev) + 0.002(dist. from coast) + 0.012(rim index)

Potential
☐ 70 - 100 %
▨ 50 - 70 %
▨ 30 - 50 %
▨ 0 - 30 %

0   2.5   5 km

FIGURE 18.3
Archaeological predictive model based on linear regression for the Bronze Age barrows (upper) and hill forts (lower) locations on the island of Brač with corresponding regression equations.

While the transparency of the regression technique can be obtained in the intermediate stage (the given regression equation), this cannot always be said for the final stage: mapped potential. The potential map directly reflects a quite generalized trend surface, which is due to the nature of this technique's underlying premise that the surface is most often regularly continuous, with no significant changes (defragmentation) on small distances. However, with the case study it was also learned that the success or failure in producing predictive models heavily rely on the preparation of the archaeological sample upon which the model is generated. The functionality of sites should be envisaged in preparation of the representative sample. Although it may be claimed that a higher level of homogeneity would result in an increase of the model's predictive power, the nonideal ordering of variables may still significantly decrease the efficiency of the mapped results based on a regression method.

18.2.1.3   Modeling the Reconstruction of Roman Settlement Land Use with the Dempster-Shafer Belief Method

This section presents the results of a small joint project between the Scientific Research Center of the Slovenian Academy of Sciences and Arts and the Department of Prehistoric Archaeology of the University of Aarhus. The main objective was to compare different predictive-modeling techniques and to develop modeling examples based on the Dempster-Shafer theory.

The Dempster-Shafer theory is a variant of the Bayesian probability theory, which explicitly recognizes the existence of ignorance due to incomplete information and is therefore directly suited for archaeological data. Within this method we define a decision frame of the defined hypothesis. In our case there were two hypotheses, [presence] and [absence], to which we assigned probabilities on the basis of the independent variables. Having defined these hypotheses, the Dempster-Shafer technique automatically generates another hypothesis, [presence, absence], describing the probability that we cannot decide between the former two, i.e., our degree of ignorance (Bo Ejstrud, personal communication, November 2000).

When setting the basic probability assignments, it is important to consider which hypothesis is supported by the given evidence or variable. There are no strict rules for setting the basic probability assignments. The results of the Dempster-Shafer method are divided into three. The first result, *belief*, is a calculation of the minimal conditional probabilities in favor of our hypotheses [presence]. Secondly, there is a calculation, *plausibility*, indicating how little speaks against it. And finally, by measuring the difference between these two, we get a calculation, *belief interval*, representing the degree of uncertainty. High values in the belief interval point out areas where we need additional or better data.

Three belief models were performed for three subtypes of Roman settlement sites on the island of Brač. Then the highest belief values for each were isolated and consequently merged into what can be interpreted as a prediction of the most likely land use on Brač during the Roman period (Figure 18.4). The same location hypothesis as gained with the regression efforts was incorporated into the development of the belief models.

When comparing the results of the Dempster-Shafer method with the results gained from the regression procedure, it can be said that the slightly

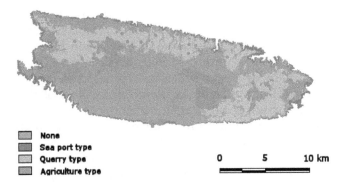

None
Sea port type
Quarry type
Agriculture type

0   5   10 km

FIGURE 18.4
Reconstruction of the Roman settlement land use based on the Dempster-Shafer belief method and the distribution of Roman settlement sites.

more fragmented landscape of the Dempster-Shafer method (compared with the regression method) seems more consistent with the archaeological interpretation of the landscape. However, this is a far more complex technique, and a great deal of previous knowledge on location hypotheses is required. Making predictions with the Dempster-Shafer belief method is an iterative procedure, and the above results present only the first iteration. Based on the significant behavior of variables, the belief method gives us the ability to incorporate partially deductive variables in the otherwise inductive models. However, a better differentiation — or a stronger fragmentation — of the landscape (belief intervals) was expected.

18.2.2 Practical Application in Archaeological Resource Management in Pomurje, Slovenia

It is important to stress that within the context of the Pomurje study, the goal was not oriented toward the development of a model that could improve our understanding of the known distribution of sites, as it was in the Dalmatian case studies. Our aim was clearly designed: to develop a consistent model that would prove helpful during the process of highway planning (Stančič et al. 2001). The Slovenian national plan seeks to complete and link highways in a network across the entire country. Concern over the costs of archaeological fieldwork led to an invitation to design a methodology for predicting archaeological sites within a test section of a highway corridor in the Pomurje region, in northeastern Slovenia. The corridor measured approximately 11.2 km in length, and archaeological research had already been carried out in the area. This provided an opportunity to use fieldwork results for a true evaluation of the predictive model's efficiency.

The project's specifics guided us in the choice of the approach we finally implemented. Considerable problems can occur when the samples of sites analyzed are small and we intend to integrate multivariate statistics. Another problematic issue was that most of the variables playing a key role in location studies are naturally based on the topographic characteristics of the area. Surface topography is important due to its association with past land use and shelter. Because the Pomurje case study area is almost flat, it would be difficult to extract any delineating rules or to predict potential based solely on the relief-based distribution of the sites. Given this set of circumstances, complex methods based on multivariate statistics or belief failed to suit our data. Therefore we decided to apply a simple traditional method of Boolean overlay to create a predictive model of different site locations within the region. Because the deficient number of variables was a potential pitfall in performing location analysis for the purpose of prediction, we additionally investigated and substantially implemented the potential of remotely sensed data.

Along a high-resolution and accurate DEM (digital elevation model), a remotely sensed land-use layer was created in addition to geological structure and soil data. Two layers, the proximity to the nearest stream and to the main river within the region, were calculated. We were concerned with the variables that could provide a description of differences in the lowland, and we felt that this information was likely to be obtained from remotely sensed data. Besides land use and a vegetation index, original information relating to surface soil type was obtained through the generation of mineral delineation layers. These indices are used extensively in mineral exploitation and vegetation analyses, essentially because they can indicate small differences between various rock types and vegetation classes. Therefore, three layers suitable for the GIS analysis were produced: clay minerals (correlation to pottery manufacture), ferrous minerals, and ferric minerals (iron oxide), interpreted as pointers of the soil quality. In calculating these indices, we hoped to obtain evidence for significant variation in the lowland that could not have been anticipated from the geomorphic variables.

Exploration and statistical analyses were carried out for five archaeological data samples. All records were acquired from the National Archaeological Database ARKAS (Tecco-Hvala 1997): prehistoric settlements (9), Roman and Early Medieval settlements (11), barrows (29), undated barrows (18), and isolated finds (mainly prehistoric axes, 15). Univariate descriptive statistical tests (χ^2 and Student t-test) were used to test the correlation between site locations and each of the variables. The results indicated that different site types in the study area tended to occur in different environmental settings; however, from a statistical point of view, the correlation was quite tentative. Nevertheless, all variables that were initially examined for significance were later included in the model, and the appropriate "threshold" conditions were applied (for details, see Stančič et al. 2001). As a result, we obtained five models for five different archaeological data types, based on 12 variables. For the purposes of project fulfillment, merging the five (weighted) model outputs created a predictive model of the region.

The weights are used to represent the relative relations between location determinants within the past settlement system and to calibrate the model.

The criterion for weighting the five models was based upon the assumption that a good predictor isolates an explicit area and therefore acts as a distinct pointer in comparison with the background situation. Consequently, weights were applied in proportion to the identified power of the variable. Three mineral indices and the vegetation index seemed to be highly correlated (complemented information), and it was decided to give them additional weights. Each of their powers was diminished by one-fourth, and ultimately they represented, in a combined form, the role of a single variable.

The validity of the calibrated model was tested by measuring the predicted potential accuracy using two different test samples. The first was a sample of the archaeological sites that were used to develop the model. This type of testing is interpreted as internal testing and was carried out as a control measure. Its results indicated that the model's prediction of potential was correct in 81% of the sites, where "high potential" incorporates areas with potentials ranging from 70 to 100 (Figure 18.5). The remaining 19% of sites fell within the limits of "medium potential" (40 to 70); no sites fell into the "low potential" range. The data used for external testing was obtained through a recent archaeological survey in the highway corridor, and was therefore considered as an actual independent test sample. It additionally proved that the model's predictions are highly accurate and efficient within the corridor area; all identified locations were captured within zones representing the potential of over 85%.

What should be highlighted in relation to the Pomurje case study is that the modeling procedure succeeded in distinguishing between locations where sites are most likely to be present and locations where they are not, even in geomorphologically disagreeable conditions. Although this process has deficient substance issues from the archaeological interpretation point

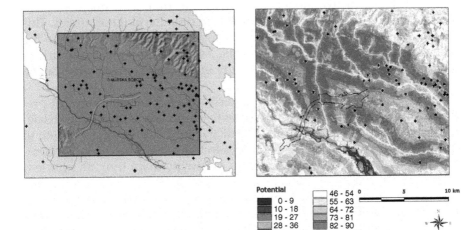

FIGURE 18.5
Archaeological predictive model for the Pomurje region: the situation (left) and weighted model (right) with the distribution of sites upon which the model was developed.

of view (the model performance is poor when observing an archaeological context for a region), the predictive power of such a model may still be regarded as useful for highway network planners.

18.2.3 What Can Be Learned from These Case Studies?

The strength of predictive modeling is that the methodology itself enables exploration and evaluation of location factors or location specifics deriving from social spaces. Due to the fact that this is mainly important in archaeological settlement-pattern research, it remains a specific segment of past-location understanding. The knowledge of possible site-location determinants can be tested and quantified, and therefore they might prove useful in assisting archaeological interpretation. Despite this, we must be aware that each of the predictive-modeling techniques has its advantages as well as disadvantages, and the choice of a suitable technique should always be guided by a clear purpose of modeling aims and the availability and quality of the data. It is often the case that the modeling procedure is where new insights for archaeological explanation are accumulated, and not in the results.

The case studies performed on the island of Brač in Croatia encourage predictive-modeling efforts, since some interesting correlations of natural conditions versus site locations have been identified, which seems an improvement in content from what was previously thought. The regression technique and the Dempster-Shafer belief method have both proved promising in location hypothesis testing and in implementing the corpus of location determinants into the modeling procedure. Unfortunately, differentiation or a stronger fragmentation of the results (regression-based potential, belief intervals) was not as good as anticipated. Finally, the relatively poor performance of the predictive models may also result from inadequate archaeological data and the nonexpressive landscape situation of these studies. This seems to remain the main problem of mapping potential locations. On the other hand, even with a poor landscape situation and empirically driven location determinants, the Pomurje study in Slovenia showed what can be achieved.

The highway case study encourages us to reach the conclusion that predictive modeling has great practical potential, especially for planning archaeological activity. The integration of predictive models offers great benefit to the early stages of planning (e.g., while selecting a highway corridor); therefore the results of predictive models (a sensitivity map) provide sufficient information for damage estimates over archaeological sites. Secondly, they can also be used during the construction stage, their main objective being to optimize the archaeological fieldwork methodologies in accordance with the construction work. Thirdly, a predictive model can be used as a planning tool to define the required fieldwork methodologies. Each of these propositions can facilitate time efficiency as well as promote a cost-efficient practice in the event that the model is trustworthy.

Predictive modeling in a GIS environment has the potential for a wide range of further practical applications. The advantage of using computers is, of course, the ease with which large databases are processed. However, by using GIS, we also have the additional bonus that we are able to graphically show the results of an analysis on maps, and thus give landscape or planning managers something visible and concrete to base their decisions upon.

18.3 Conclusions

The range of applications and techniques for location-modeling ideas is enormous. Location modeling with GIS technology deserved extensive attention mainly because the environmental variables such as slope, elevation, and soil type are easy to quantify digitally. However, it soon became apparent that this was not the case with the concepts of social space. According to its definition and underlying theoretical concepts, modeling archaeological potential is a form of location modeling. However, if archaeological prediction is to be successful, social and cultural concepts should not be mistreated during the modeling procedure. Therefore, it is perhaps worth reconsidering what has already been achieved, what is still left to do, and what are the limits of predictive-modeling efforts.

On the basis of the established methodological approaches to location-prediction problems, a mass of more or less meaningful maps of potential, or probability, of archaeological proneness or sensitivity were produced. The success of different approaches and techniques has most often been evaluated and stressed on the basis of the fulfillment of results into acceptable levels of meaningful interpretation and simple testing of the model's apparent efficiency. Failures, such as evidently false areas of potential and the problem of archaeological context assessment, are discussed on a few occasions. A disappointing aspect of the previous publications is the manner in which they present the quality of the results. We do know that the quality of results (map accuracy) is reflected through the following components: the quality of data used (spatial and attribute accuracy), the appropriateness of the applied technique (in relation to the used data), and finally the quality of algorithms used in the modeling procedure. To evaluate the results of predictive-modeling efforts, the value of these components (which also contribute to error generation) also needs to be estimated. Total error estimation is a necessity in any modeling with GIS technology, and therefore it is essential for a reliable evaluation of archaeological location modeling.

The presented predictive techniques may satisfy cultural-resource management needs, and the resulting trustworthy models may be used for planning purposes. The success of predictive-modeling techniques as an exploratory tool is thereby mainly related to a positive coincidence of the data used in the modeling procedure and the modeling skills and aims. For a better understanding of and eventually the ability to predict the complex settlement systems or culture–location interaction during the past times, major emphasis should be placed on demanding the involvement of artificial knowledge or numerical modeling, and certainly the work should move from the location itself (empirical measurements on site locations) toward landscape perception. The modeling efforts have to determine the variability or similarity of the observed locations, whether one component/variable of the system derives variability in another, and whether complex modes of location variability/similarity, once identified, are predictable. The predictive-modeling procedure therefore has to solve a complex causality — an interactive task beyond and betraying only location.

The location predictive-modeling procedure is particularly crucial due to the lack of observational records of sufficient length for landscape variables, combined with the need to synthesize sparse archaeological data. With hope to overcome this critical point and to step ahead from sensitivity maps, other techniques with predictive capabilities need to be developed and demonstrated. However, the value of artificial-knowledge-based applications and numerical modeling for archaeological prediction tasks will be impossible to evaluate until feedback from any resulting models and predictive-modeling procedures becomes available. A new depth of understanding the archaeological prediction phenomenon will hopefully be explored at a later stage.

References

ASKALON 2002, Askalon: A programming Environment and Tool Set for Cluster and Grid Computing; available on-line at http://www.par.univie.ac.at/project/askalon/.

Burrough, P.A. and McDonnell, R.A., *Principles of Geographical Information Systems*, Oxford University Press, Oxford, 1998.

Exon, S., Gaffney, V., Woodward, A., and Yorston, R., *Stonehenge Landscapes: Journeys through Real-and-Imagined Worlds*, Archaeopress, Oxford, U.K., 2000.

Gillings, M., Mattingly, D., and van Dalen, J., Eds., *Geographical Information Systems and Landscape Archaeology*, Vol. 3, *The Archaeology of Mediterranean Landscapes*, Oxbow Books, Oxford, U.K., 1999.

Heuvelink, G.B.M., *Error Propagation in Environmental Modelling with GIS*, Taylor and Francis, London, 1998.

Judge, W.J. and Sebastian, L., Eds., *Quantifying the Present and Predicting the Past: Theory, Method, and Application of Archaeological Predictive Modeling*, U.S. Government Printing Office, Washington, DC, 1988.

Kohler, T.A. and Parker, S.C., Predictive models for archaeological resource location, in *Advances in Archaeological Method and Theory*, Vol. 9, Schiffer, M.B., Ed., Academic Press, Orlando, FL, 1986, pp. 397–452.

Lock, G. and Stančič, Z., *Archaeology and Geographical Information Systems: A European Perspective*, Taylor and Francis, London, 1995.

Longley, P., Goodchild, M.F., Maguire, D., and Rhind, D., *Geographical Information Systems and Science*, John Wiley and Sons, New York, 2001.

Stančič, Z. and Kvamme, K., Settlement pattern modelling through Boolean overlays of social and environmental variables, in *New Techniques for Old Times*, CAA 98 proceedings, Barcelo, J.A., Briz, I., and Vila, A., Eds., BAR International Series, 757, Archaeopress, Oxford, U.K., 1999, pp. 231–237.

Stančič, Z. and Veljanovski, T., Understanding Roman settlement patterns through multivariate statistics and predictive modelling, in *Beyond the Map: Archaeology and Spatial Technologies*, Lock, G., Ed., NATO Science Series, IOS Press, Amsterdam, 2000, pp. 147–157.

Stančič, Z., Veljanovski, T., Oštir, K., and Podobnikar, T., Archaeological predictive modelling for highway construction planning, in *Computing Archaeology for Understanding the Past*, CAA 2000 proceedings, Stančič, Z. and Veljanovski, T., Eds., BAR International Series, S931, Archaeopress, Oxford, U.K., 2001.

Stančič, Z., Vujnovič, N., Čače, S., Podobnikar, T., and Burmaz, J., The archaeological heritage of the island of Brač, Croatia, in *The Adriatic Project*, Vol. 2, BAR International Series, 803, Archaeopress, Oxford, U.K., 1999.

Tecco-Hvala, S., Towards a standardisation of archaeological data, *Journal for the Protection of Monuments*, 37, 1997, pp. 176–183.

Van Leusen, P.M., GIS and locational modeling in Dutch archaeology: a review of current approaches, in *New Methods, Old Problems: Geographical Information Systems in Modern Archaeological Research*, Maschner, H.D.G., Ed., Occasional Paper 23, Center for Archaeological Investigations, Southern Illinois University, Carbondale, 1996, pp. 177–197.

Wescott, K.L. and Brandon, R.J., Eds., *Practical Applications of GIS for Archaeologists: A Predictive Modelling Kit*, Taylor and Francis, London, 2000.

19

The Changing Mesopotamian Landscape as Seen from Spot and Corona Images

Carrie Ann Hritz

CONTENTS

19.1 Introduction

Predictive modeling as a tool for archaeological research is not new. The use of different predictive models to understand historical questions such as shifts in settlement patterns can be seen in archaeological projects from as early as Thorkild Jacobsen's work in Mesopotamia (preserved in unpublished notes with basic methodologies in 1936 [1958]), Braidwood's work in Anatolia (1937), and V. Gordon Willey's (1953) Peruvian research, to the modern studies of the Roman European countryside (Robert 1996). Given the importance and the characteristics of the Mesopotamian alluvium, it is not surprising to find that many researchers have used these data in predictive-modeling efforts to enhance historical understanding.

The alluvial plain of southern Mesopotamia saw the rise of the earliest urban settlements in the ancient world. To understand the rise of these urban agglomerations and utilize analytic tools such as predictive modeling, the harsh environment that characterizes southern Mesopotamia must be understood. Choices that shape social, economic, and political history are based, in part, upon environmental constraints such as natural changes and human subsistence needs, and also result in a conglomeration of natural and artificial transformations of this complex landscape.

In southern Mesopotamia, prediction of irrigation systems from settlement patterns were attempted by Thorkild Jacobsen as early as 1936 (1958). Robert Adams (1965, 1972, 1981) continued this analysis of settlement in a systematic program. Subsequently, a Belgian team led by Leon De Meyer and Herman Gasche attempted an integrated and updated approach that employs a wider range of archaeological, historical, and geoarchaeological data. These two approaches and their resulting seminal case studies provide a basis for understanding the way in which settlement, and in effect history, in southern Mesopotamia developed. While these two approaches are often contrasting, it seems that an integration of approaches proves more appropriate when analyzing data for southern Mesopotamia. A major component of such an integrative approach is the use of GIS as a tool for the organization, classification, and presentation of vast amounts of different types of data from southern Mesopotamia. The present chapter results from a period spent working on a GIS system developed by the Ghent team, specifically Kris Verhoeven, and the continued development of a methodological protocol to be used in combination with remote sensing and archaeological data. Case studies of several small areas (Nippur, Nahrwan Canal, and Shatt al-Gharraf) provide a first stage in the analysis, which establishes the basic model for the larger analysis of the Mesopotamian alluvium as a whole.

19.2 The Environmental and Archaeological Context of Mesopotamian Irrigation

The environment of Mesopotamia is diverse. Northern Mesopotamia is characterized by semiarid plains with seasonal *wadis* that enable rain-fed agriculture to be the main mode of food production. This fairly easily cultivated northern landscape contrasts with southern Mesopotamia, which is arid and generally harsh, with a mean annual rainfall of less than 150 mm. Vegetation is sparse throughout the annual cycle, and irrigation is necessary for successful cultivation.

The southern landscape can be described as consisting of four primary topographic components. They are (a) lines of dunes, (b) remnants of canals, (c) river channels, and (d) mounds of former settlements (Adams 1981: xviii).

These prominent physical features overlie the primarily flat plain. This physical landscape is a direct product of the two major watercourses, the Tigris and Euphrates.

The natural flatness of the plain renders it susceptible to river changes over time because there are few topographic restraints. In general, rivers tend to shift their channels abruptly and move to lower elevations, a process known as avulsion (Schumm 1977). Despite being conducive to irrigation works, the gentle gradient of the plain is a negative factor in that it hinders proper drainage, with the result that the plain experiences repeated cycles of salt and water table problems (Buringh 1969: 90–100)

As a result of this environment and the struggle to maintain balanced conditions, the southern alluvium exhibits complex patterns of topographic features (Figure 19.1). Abandoned canal patterns crisscross the entire plain, as do relict meanders from the shifting Tigris and Euphrates courses. Mounds standing above the southern Mesopotamian plain signify areas of former habitation that include important historical cities (Figure 19.2). It is likely that erosion and alluvial deposition have destroyed or obscured prehistoric and historical settlement remains, leaving a somewhat unclear picture of settlement pattern.

19.3 The Adams Approach

Robert Adams, following Jacobsen, approached settlement and watercourse patterns in southern Mesopotamia by applying archaeological principles to a theoretically predictive framework. Adams used basic archaeological survey to recover "the river and canal network of successive periods, and therewith their settlement patterns" (Jacobsen 1982: xiii). His basic assumption is that in a harsh semiarid environment, such as that of the alluvial plain, settlement was only possible where water was available along rivers and canals (Adams 1965: app. I). From this, he assumed that where the settlements of a given period showed linear patterns, it indicated that the lines were a reflection of the watercourses upon which the settlements depended. The concept is that settlement accompanies agriculture, and that both are dependent upon assured supplies of irrigation water. From this analysis, he extrapolated the delineation of settled areas in a region and contrasted these areas with those of swamps or desert. He was also able to suggest where large-scale abandonment had occurred. In this approach, Adams tied the stability and flux of settlement of the Mesopotamian plain to the availability of irrigation agriculture.

He concluded that mounds and watercourses can be brought together into an integrated and constantly changing system or sequence of systems (Adams 1981: 28). Thus, if site habitation is dependent on channel success and use, then site abandonment is linked to channel abandonment. These

FIGURE 19.1
Corona image 1968. Photo courtesy of U.S. Geological Survey.

phenomena are made more complex for multimounded sites in which occupation moves from place to place within a confined area. The watercourse development and human settlement are wholly interdependent. Using archaeological survey collections, Adams employed ceramic chronological indicators to date the sites as well as their dependent courses.

Adams checked where his framework and model placed courses on both aerial photographs and on the ground. "Unfolding awareness of the pattern formed by sites occupied during a particular period led to hypotheses about where the associated canal routes should be sought" (Adams 1981: 28). In other words, canals on aerial photographs provided a cross-check on the ground of visible site alignments.

FIGURE 19.2
Spot image 1992–1995. Photo courtesy of National Geo-spatial Intelligence Agency.

The methodology of the survey of Adams must be considered in light of the norms of the time in which it was conducted. It excluded large areas of the plain, which could not be reached for ground truthing for numerous reasons.

19.4 The Belgian Approach

The University of Ghent uses microtopographic data to reconstruct the picture of river networks and main canals. The team employed this approach

on the northern Mesopotamian plain, concentrating on the area around Tell
ed-Der. This area was specifically chosen because of the amount of ground-
truthing data available in excavation reports and surveys (Cole and Gasche
1998).

In contrast to the approach of Adams, the Ghent team suggests that it is
"no longer adequate to draw a line between two settlements because the
textual sources situate them on the same watercourse, nor is it any longer
sufficient to reconstruct watercourses connecting sites which supposedly
belong to the same period" (Cole and Gasche 1998: 2). Their method builds
on the basic predictive idea put forth by Adams and attempts to include
additional variables such as geomorphology to explain the disparities in
Adams's data.

This approach essentially posits that the history of watercourses and settlement patterns can be interpreted on the basis of topographic features of
the plain using geomorphological features as the primary variables for settlement. The Ghent group begins with a classification of the plain based on
its geomorphological attributes. The most prominent geomorphological features of the plain are levees and channel avulsion scars.

Levees are long, slightly swelling features that form very low ridges
throughout the plain. In Mesopotamia, these features can be up to several
kilometers in width and are rarely higher than 4 m, making their visibility
on the ground extremely difficult. From the perspective of satellite imagery
and aerial photography, these features contrast sharply with the natural
flatness of the plain.

Levees are built up over millennia.

> With each inundation the terrain close to the channel receives most of
> the sediment load, with the rest being transported to more distant areas.
> The perennial rhythm of seasonal inundations occasions a continuous
> heightening of the river's approaches, with the water flowing more and
> more on its own alluvial material. Finally, the river channel is elevated
> well above the plain, with basins forming progressively on either side.
> The practice of irrigation along fluviatile arteries — where most favorable
> soils for agriculture are encountered — has also significantly contributed
> to their construction (Cole and Gasche 1998: 6).

Large and long-used levees leave an impressive and clear mark upon the
landscape. They encourage long-term human use and provide relatively
well-drained soils (Hill and Rapp 1998: 61–63). Thus, if the river shifted,
humans had to either refurbish the old bed or replace it by a canal, which
then could have taken advantage of the topography. The replacement canal
could have flowed along the slopes, rather than directly on top of the levee,
depending on the level of water in the river from which the canal originated.
This action, over time, would have broadened the levee and helps explain
why relict meanders and alignments of sites could be at some distance from
the levees. Thus, by the identification of levee remnants, both on images and

on the ground, ancient watercourse patterns could be recognized as well as indications of associated sites. It also supplied an understanding of the area under cultivation.

River shifts leave clear traces upon the landscape in the form of meander scars and avulsions. The plain is virtually flat and, as topographic barriers are created over time through the creation of levees, the landscape is changed. When the river course becomes barred from its natural path over time, the course moves invariably to a lower elevation area in the plain. While river discharge and barriers are constantly changing as a result of human agency and natural alluvial buildup, the river's courses are in a constantly dynamic state. These changes and their causes leave topographic traces upon the landscape, which can be related temporally to one another.

19.5 Our Theoretical Approach

If these two approaches can be combined, a synchronic view of the present Mesopotamian landscape can be presented. The following types of data can be used for such landscape analysis.

1. Irrigation channels
2. Archaeological sites
3. Topography of the plain
4. Data derived from ancient texts

Shortcomings are inherent in the Adams model in part due to limitations in the available technology and data sources. For example, Adams, at times, connected watercourses to sites without regard for the topographic reality of the plain because he lacked the appropriate data, such as detailed contour maps. In other words, he did not consider barriers in the landscape when positing a river course based on site alignments. Our approach does not discount the theoretical concept that settlements are tied to water availability and that channel patterns can be inferred from settlement pattern as shown by Adams, but we also recognize the significance of the topography of the plain itself. We place emphasis on the visible features of the landscape, which can be traced readily. This approach sees an interaction between settlement location and water availability, so that the process can be seen as interrelated with both settlement locations being created by water availability and water availability created by settlement processes.

By emphasizing the present landscape, we are taking a retrogressive approach to interpreting the complex settlement and watercourse patterns on the southern alluvium. From textual, satellite imagery, and survey evidence, it has become apparent that the most visible network of irrigation

patterns are of the first millennium B.C., i.e., Parthian, Achaemenian, Sasanian, and Islamic periods (Adams 1981; Le Strange 1905). For those periods,
a relatively clear picture of intensive settlement and canalization appears.
Sites are focused on the use of the Euphrates, and the canals form a crisscrossing pattern across the plain to reach the Tigris. Identification of these
patterns and subsequent topographic units of settlement, as well as associated canalization, is the first step in the testing of the predictive methodologies, protocols, and understanding of the landscape.

19.6   The Combined GIS Methodology

In our approach, traditional cartography and historical geography are
applied to a combined data set of survey maps and satellite imagery. The
Ghent team created a grid using latitude and longitude as well as a UTM
projection. The grid, created in Microstation, covers all of Mesopotamia
proper (excepting Northeast Syria and the extreme southern boundaries of
the plain). This basic GIS framework was then integrated into the more
capable and more easily manipulated ArcGIS program.

The benefit of a multicoordinate grid is enormous. The use of latitude and
longitude allows for the referencing of important early maps. For example,
the British 1918 series maps, which can be used to locate sites in areas that
our archaeological teams have not yet visited, can be incorporated by latitude
and longitude coordinates. The use of UTM coordinates allows for the incorporation (Figure 19.3 and Figure 19.4) of more recent material without the
loss of the older basic and invaluable sources. UTM coordinates also allow
for a usable framework for future survey.

The grid was then overlaid with 1/400,000 Spot negatives. These negatives
provided the basic map for tracing both ancient and modern sites and
hydrology. By using larger resolution Spots and the basic GIS grid, a base
map GIS was created to act as a framework for more-detailed mapping of
the landscape features of the plain.

At this point, a series of Spot images, overlaid with a UTM grid, were
incorporated. These images cover a large portion of the central and southern
plain, but do not extend as far south as the area immediately surrounding
the head of the Persian Gulf. These images are panchromatic Spot images
from mission 2 February 1990 covering an area of 80×80 km per image with
a resolution of roughly 10 m. They were scanned from the negatives at 508
dpi so that each pixel equals 11 m. The images were incorporated into the
digital vector grid using the UTM grid coordinates for each image. Once
these images were referenced, survey maps covering numerous parts of the
plain were referenced in both UTM and latitude and longitude. These surveys are: M. Gibson's Kish survey (Gibson 1972); Adams's *The Land behind
Baghdad* (1965), *Heartland of Cities* (1981), and *Uruk Countryside* (Adams and

FIGURE 19.3
Modern river channels and archaeological sites from the Heartland of Cities (Adams 1981) survey vectorized in ArcGIS.

Nissen 1972) surveys; and the Belgian mission to Tell ed-Der survey of the Sippar area. For the first time, all of these surveys are in a digital format and a UTM coordinate system so that, theoretically, all site locations can be referenced on the ground with a GPS.

The survey data are compared with the visible landscape of the Spot imagery. The perimeters of relict settlements and canals are based on Spot tonality interpretation and are not as accurate as those based on aerial photographs or larger scaled maps such as those used by the field surveyors. It is necessary to mention that what we map as relict settlements, after comparison with the original survey data, are only those settlements and watercourses clearly visible on the imagery.

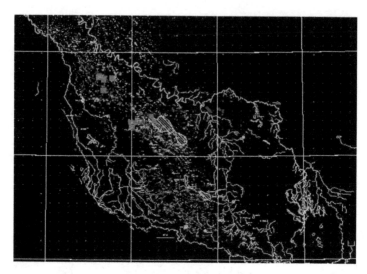

FIGURE 19.4
Microstation UTM Grid vectors. Photo courtesy of Kris Verhoeven of University of Ghent, Belgium.

Once this joint preliminary identification is completed, the less extensive but far more relevant and detailed Corona satellite photographs covering areas of roughly 17 × 180 km can be added and analyzed. The Spot imagery, because of the date of acquisition, presents problems in analysis and comparison. The effects of development projects of the 1980s and 1990s are included on the Spot images, whereas the Coronas (Figure 19.5), dating from December 1967, show these same areas without major disturbances. The Coronas were scanned at 800 to 1200 dpi. The coverage available at the time of the study covered an area of 50 km N/S and 240 km E/W. This forced a focus on the southern and central portions of the plain. The resolution of these images, while on the whole superior to the Spot images, is variable throughout each individual strip, decreasing from the center toward the edges as a result of satellite rotation. Error is possible, and resolution is lost when these positives are scanned. When scanned, one pixel equals roughly 9 m. Corona images are the most useful because they were acquired at nearly the same time as many of the ground-mapping surveys. Thus, the data from the ground surveys are more easily referenced, compared, and placed into context with the Corona images than with the Spot images. A comparison of the two sets of images forms an invaluable tool in assessing modern landscape changes and how those changes affect ancient remnants of topographic features.

19.7 Analysis

By using Spot and Corona imagery and comparing these with ground-survey data, it becomes clear that the predictive approaches based on the concept

FIGURE 19.5
Spot image 1992–1995. Photo courtesy of National Geo-spatial Intelligence Agency.

of settlement and watercourse interaction as well as topographic features can and must be integrated. The images present a new way to approach these interactions based on what is visible. We have been able to confirm visually some of the long-held theories of watercourses on the southern alluvium. For example, Adams mentions two major courses visible in part on the aerial photographs and in part on the interpretation of settlement locations. One of these courses is clearly identifiable in the images (Figure 19.6).

FIGURE 19.6
Corona image 1968. Photo courtesy of U.S. Geological Survey.

Adams mapped a historically important course, in sections, running from the site of Adab to Umma. At the time of the survey, and on the corresponding Coronas, this course had a large dune field covering extensive parts of it and which continued in a southeasterly direction. Adams posits that the dunes were part of a major levee running alongside and over parts of the actual watercourse. Using a combination of both sets of data, the course clearly appears to approach the site of Adab from the northeast and to continue past the site of Tell Jidr in a southeast direction.

The Corona images clearly show a linear watercourse feature with a dark tonality signature as a result of continually held moisture (Lillesand and Kiefer 2000: 240). This feature is obscured, at times, and runs parallel to an ever-widening band of dune fields. On the Coronas, the dunes cast off a shadow not unlike tells, but their tendency toward overlap and a "star"

FIGURE 19.7
Archaeological sites and relict channels within sand dunes. Corona image 1968. Photo courtesy of U.S. Geological Survey.

shape (Cooke and Warren 1973: 283), together with their clustering over great tracts of land, clearly differentiate them from tells (Figure 19.7). Their distinct morphology identifies them as classic Barchan-type sand dunes, which are ever shifting in location and shape as part of an alluvial desert environment.

The Spot imagery, however, presents a different signature for this course (Figure 19.8). During the 30 years between the images, the dunes have shifted slightly east, and the line of the course from Adab to Umma is more distinct and shows a greater contrast with the background topography. The Spot signature shows the dunes as washed-out light features, not nearly as distinct in morphology. The neighboring pixel signatures have become averaged in the processing, and the contrast between the dunes and moist areas has been

FIGURE 19.8
Dunes and channels near the site of Nippur. Corona image 1968. Photo courtesy of U.S. Geological Survey.

obscured. Nonetheless, the line is still clearly identifiable. Topographically, the Spot imagery clearly shows an ever-broadening levee running parallel to the line in this area. The watercourse and levee lines are accompanied by two parallel lines of settlement, one following the levee and one farther out on the levee. The settlements farther out can be assumed to be following the levee as it broadens over time. This is a way in which inhabitants of the plain could take advantage of the natural features of the landscape and shift their settlement placements in response to a migrating channel. In this particular case, the combination of imagery allows for a more comprehensive

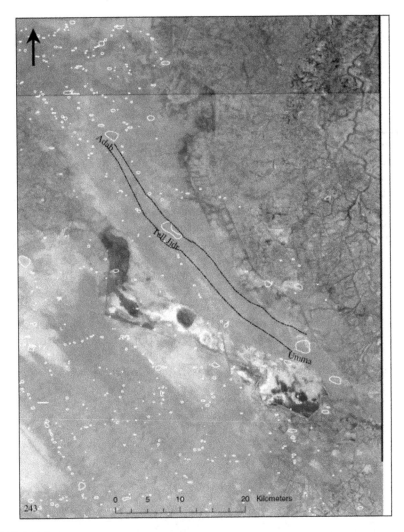

FIGURE 19.9
Dunes, large relict channels, and archaeological sites. Photo courtesy of National Geo-spatial Intelligence Agency.

understanding of the relationship of three important topographic features: watercourse, levee, and sites.

The junction of the Tigris and the Shatt al-Gharraf is another area where the modern and ancient topography appear together (Figure 19.9). Because Corona imagery was not available for this area at the time of the study, this area provides a clear example of the utility of the Spot imagery. The Shatt al-Gharraf, which is a major right-bank effluent at Kut, is needed to irrigate fields from the Tigris.

In this area, the Spot images show clear meander scars to the west and south of the modern river course. These meanders appear in the modern cultivated fields with clearly recent topographic features overlying them. The meander traces are within 6 km of one another but exhibit different characteristics. Two meanders average 2 km in width, while a third is less than 1-km wide.

The geometry of the meanders can provide a basis for determining the hydraulic characteristics of the river at a certain period in time. One of the two ancient courses appears to decrease in width as it moves to the east. The character and tonality signature visible on the Spot images changes as the line moves to the east. This may be an indication that there are two temporally different river shifts. The wider meander loops may represent an older shift of the river over a considerable period of time, which is then bisected by a more recent movement of the river in a dynamic state or the partial abandonment of one riverbed during an avulsion event (Verhoeven 1999). This course and its dating would not have been identified by the line-of-site method, since no sites were identified in this area by the Adams survey. However, sites appear clearly on the imagery both in the surveyed area and just outside the survey boundaries.

I have identified several sites in this area that appear just off the line of this relict course and on the associated levees (Figure 19.10). Further north of the levee, linear patterns associated with a meander abandoned by an eastward movement of the Tigris crosscut the modern cultivation. Firm

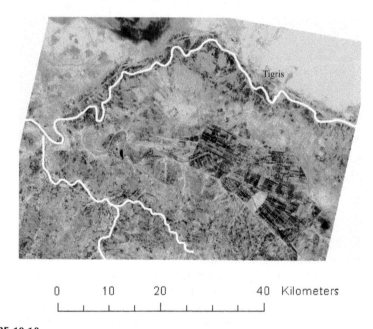

0   10   20   40   Kilometers

FIGURE 19.10
Modern Tigris River outlined in white. A large relict channel appears in the cultivated fields. Spot image 1992–5. Photo courtesy of National Geo-spatial Intelligence Agency.

dating is not possible, but written sources do give an indication of historical periods that are relevant. Texts (Le Strange 1905) tell us that this kind of gridlike crosscutting linear pattern is representative of Sasanian-Islamic period watercourse and field patterns. We know that branches of the Tigris were under intensive irrigation in these periods.

19.8   Abu Salabikh: A Case Study

The area around Abu Salabikh, surveyed and mapped by Wilkinson (1990), provides our best data for ground truthing. This area presents a clear example of the integration of approaches that can be used to analyze the watercourse and settlement patterns. The area was mapped for geomorphology and archaeology field survey, and is covered by both the Spot and Corona images. The site and its immediate environs represent the westernmost limits of Adams's survey data. Wilkinson's survey objective was to "determine if an early branch of the Euphrates ran through the site, and if so, to attempt to trace its course.... It has been assumed that such a channel probably ran north-south between the main and west mounds of Abu Salabikh and its presence must have determined or limited the growth and shape of the Uruk and Early Dynastic towns" (Wilkinson 1990: 75).

The area immediately to the west of the mounds of Abu Salabikh appears moist and under cultivation on the Corona images. The later Spot imagery shows the area remaining under cultivation but in a season of less moisture. Both sets of images retain enough contrast in tonality to illuminate ancient topographic features. These features can be understood when compared with the field survey and geomorphological data collected by Wilkinson (1990).

For example, a relatively recent but abandoned river called the Shatt al-Hawa is clear at a distance of less than 2 km from the main mounds of the site and running through the surrounding cultivated fields (Figure 19.11). More important, the Corona of this area shows clearly a remnant of a sinuous canal that runs from northwest to southeast between the eastern and western mounds of the site. This course, identified by Adams as third millennium, was traced by Wilkinson through the mounds. The images show that this course can be traced farther to the northwest, outside of both Adams's and Wilkinson's survey boundaries, and closer to the Euphrates branch. In this case, the imagery has refined and clarified the conclusions presented by the surveyors.

Turning to the east of the site, another landscape picture appears from the imagery. Dunes appear just to the east of the main mounds, obscuring many of the sites mapped by the ground surveyors and providing a boundary for the Wilkinson survey. Nonetheless, within these dune fields, a linear canal feature system appears. These channels come off a main channel that leads to the site (Figure 19.12). A small meander scar and traces of canals from the

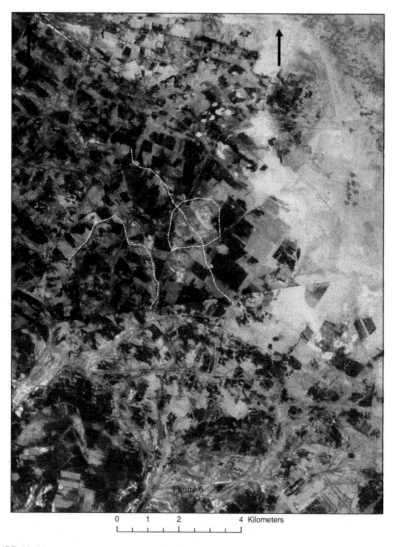

FIGURE 19.11
Shatt al-Hawa and the site of Abu Salabikh. Corona image 1968. Photo courtesy of U.S. Geological Survey.

scar may indicate that (1) the system followed a more westerly course just south of Abu Salabikh and (2) sinuous canal traces linked the Abu Salabikh area to a larger regional system.

This larger system is represented by the third functional unit identifiable around the site of Abu Salabikh (Figure 19.13 and Figure 19.14). While the system appears primarily on the Corona images, it is also traceable on parts of tracts on the Spot imagery. This system cuts down the center of the plain just east of the main mounds of Abu Salabikh. The imagery shows the Shatt

FIGURE 19.12
Channels in dune fields around the site of Abu Salabikh. Corona image 1968. Photo courtesy of U.S. Geological Survey.

al-Nil clearly, and its feeder canals to the surrounding area take off from the area of the site of Ishan al-Jihariz. A comparison of the Spot and Corona images shows that in the later Spot imagery, the Shatt al-Nil line is almost totally covered by dunes, while the linear remains on the Coronas are unmistakable. This is a result of the eastward movement of the dunes and the drying of the marshes due to large-scale drainage projects. Off the right bank of this line, a clear pattern of canals emerges. A large canal from this system can be seen running from Ishan al-Jihariz to just past Abu Salabikh. North

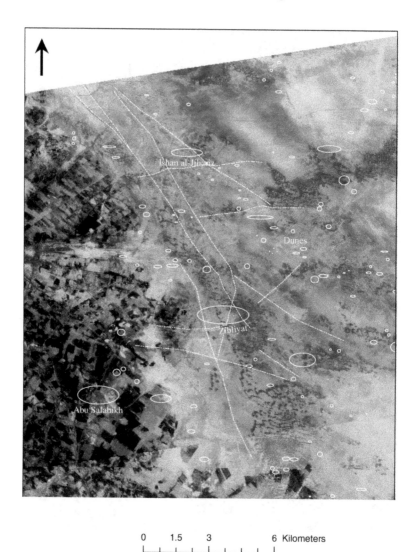

0   1.5   3   6 Kilometers

FIGURE 19.13

Channels and canals in the central portion of the Mesopotamian alluvium appearing on the 1968 Corona image. Photo courtesy of U.S. Geological Survey.

of this canal, the landscape appears to show meander scars and a levee. This system of levee and Shatt al-Nil line date to the post-Early Dynastic period (Adams 1981). The traces of a sizable canal appear to be coming into contact with the Shatt al-Nil from the northeast. While this system is periodically lost on the Spot images, it is clear on the Coronas. When it enters the vicinity of Tell Zibliyat, a complex pattern of takeoff canals emerges. This pattern appeared on a smaller scale in the Ur III period (Adams 1981). In the first millennium, this pattern is intensified, and the area is maximized in terms

0   1.5   3   6 Kilometers

FIGURE 19.14
Channels and canals in the central portion of the Mesopotamian alluvium appearing on 1992–1995 spot image. Photo courtesy of National Geo-spatial Intelligence Agency.

of cultivation. The pattern of maximization of agricultural production is described by Adams (1981: 188) as:

> An interlocking, much more artificial grid of watercourses that broke large contiguous areas of cultivation into polygons of fairly uniform size and shape. In the older pattern, one must assume that a fairly broad uncultivated band was left between parallel branches, while the new,

more tightly arrayed polygons suggest that fairly broad expanses of continuous cultivation had begun to be introduced.

These canals run from the Shatt al-Nil to Sassanian-Islamic sites such as 702, 726, 728, and so on.

Islamic geographers tell us that these patterns were common in the bureaucratic land management programs and that this particular line was in use during this period (Le Strange 1895). These canals and channels were used and reused through the Sasanian and Islamic periods. They present a network of irrigation works that fan out over the landscape in an east-west direction off the main north-south lines to cultivate the now dune-covered fields and their associated sites. The Coronas show this most clearly in a system of canals that run from the vicinity south of Zibliyat across a dunelike area, joining Sasanian and early Islamic sites and rejoining the Shatt al-Nil near Abu Sarifa. Adams maps a few canals and channels as well as two unconnected meanders in this area. The Coronas show four canals as well as clear meander scars that are related to the main north-south line and its associated field systems.

This detailed study of the area around Abu Salabikh has provided three new insights into the ancient landscape. First, it is clear that the Shatt al-Hawa and canal running through the main mounds mapped by ground surveyors can be traced beyond the previous survey boundaries. These systems can then be related to a larger regional system and placed in their proper historical context. The imagery in conjunction with the survey data allows for a more comprehensive understanding of this system and its relation to the site of Abu Salabikh. Second, it is clear that the Sasanian-Islamic field pattern east of the site is linked to the Shatt al-Nil and continues west. Third, a larger functional system of the Shatt al-Nil has been traced over an extensive expanse of the landscape, and in areas such as Tell Zibliyat, the relationship between this system and field systems as well as their associated sites can be traced. This presents a more clearly identifiable picture of irrigation works and settlement in this area in the Sasanian-Islamic periods.

Methodologically, the area provides an example of the integration of Spot and Corona imagery, as a combination of both data sets was needed to trace a functional landscape unit. The textual, archaeological, topographic, and geomorphological data were combined into the GIS system to illustrate the landscape. The combination of these interdisciplinary methods and data sets allow this system to be placed into its historical and sociocultural context.

19.9 Conclusions

It is clear from the case study around Abu Salabikh that a combined and integrated methodology is preferred to a predictive framework when analyzing settlement and watercourse patterns. The Spot and Corona images proved to be valuable tools in analysis of the topographic features of the plain. It follows that while the concept of settlement alignments forming along channels remains valid, the topography, archaeology, and geomorphology of the plain is equally important and a key component to the development of these patterns. The use of remote sensing and GIS methodologies allow past assumptions to be tested in a systematic manner. These technological tools enhance organization of the multitudes of data and add a new level of physical realities of the plain to the research analysis.

Precise dating of features remains a problem whose only solution would involve ground collection and stratigraphic analysis of canal sections to produce precise dates. Relative dating of irrigation patterns is possible through analysis of levee and topographical formations such as dunes. Through analysis of topographic relationships, we are able to expand the "picture" of settlements and watercourse patterns by including areas that were historically neglected for one reason or another by ground surveyors. Through this integration of GIS, predictive modeling, and textual sources — using a theoretical framework that emphasizes principles of archaeology, geography, and geomorphology and places emphasis on landscape features such as irrigation channels, levees, and sites — we can begin to analyze the complex landscape of southern Mesopotamia and subsequently understand the sociohistorical context of settlement in the area. This methodology allows for a more inclusive and broader picture of the landscape as a whole, traces its successive changes, and produces a visually understandable catalog of those changes and their historical significance.

References

Adams, R., *Heartland of Cities*, University of Chicago Press, Chicago, 1981.
Adams, R. and Nissen, H., *Uruk Countryside*, University of Chicago Press, Chicago, 1972.
Adams, R., *The Land Behind Baghdad*, University of Chicago Press, Chicago, 1965.
Braidwood, R., *Mounds in the Plain of Antioch: An Archaeological Survey*, University of Chicago Press, Chicago, 1937.
Buringh, P., *Soils and Soil Conditions in Iraq*, Republic of Iraq Ministry of Agriculture, Baghdad, 1960.
Campbell, B., *Human Ecology*, Aldine De Gruyter, New York, 1983.
Cole, S.W. and Gasche, H., *Changing Watercourses in Babylonia*, Mesopotamian History and Environment Series II: Memoirs V/1, University of Ghent and Oriental Institute of Chicago, Chicago, 1998.
Cooke, R. and Warren, A., *Geomorphology in Deserts*, University of California Press, Berkley, 1973.
Gibson, M., *The City and Area of Kish*, Field Research Publications, Miami, 1972.

Hill, C. and Rapp, G., *Geoarchaeology: The Earth-Science Approach to Archaeological Interpretation*, Yale University Press, New Haven, CT, 1998.

Jacobsen, T., Salinity and irrigation agriculture in antiquity; report on essential results June 1, 1957 to June 1, 1958. Diyala Basin Archaeological Project.

Jacobsen, T., *Salinity and Irrigation Agriculture in Antiquity: The Diyala Basin Archaeological Project*, Biblotheca Mesopotamia 14, Undena Publications, Malibu, CA, 1982.

Le Strange, G., Description of Mesopotamia and Baghdad, written about the year 900 A.D. by Ibn Serapion, *Journal of Royal Asiatic Society*, 2, 1–76, 1895.

Le Strange, G., *The Lands of the Eastern Caliphate*, Cambridge University Press, Cambridge, 1905.

Lillesand, T. and Kiefer, R., *Remote Sensing and Image Interpretation*, 4th ed., John Wiley and Sons, New York, 2000.

Robert, B., *Landscapes of Settlement*, Routledge, New York, 1996.

Schumm, S., *The Fluvial System*, Wiley, New York, 1977.

Verhoeven, K., Geomorphological research in the Mesopotamian flood plain, In *Changing Watercourses in Babylonia: Towards a reconstruction of the ancient environment in lower Mesopotamia*, Eds., Gasche, H. and Tanret, M., Series MHE V/1, Oriental Institute, University of Chicago and University of Ghent Press, 1999.

Wilkinson, T., Early channels and landscape development around Abu Salabikh, preliminary report, *In Iraq*, 52, 75–83, 1990.

Willey, G., Prehistoric Settlement Patterns in the Viru Valley, Peru, Bureau of American Ethnology Bulletin 155, Washington, DC, 1953.

20

Quantitative Methods in Archaeological Prediction: From Binary to Fuzzy Logic

Eugenia G. Hatzinikolaou

CONTENTS

ABSTRACT   Archaeological prediction has its origins in the late 1970s. Since then, this new and interesting field has evolved significantly. Technological achievements and the evolution of software have played a key role in this growth. Many models have been generated, some more empirical than others. This chapter discusses the quantitative methods that have been used to predict archaeological site locations. Beginning with the most traditional binary method, regression, and ending with the most extravagant, fuzzy logic, an evaluation of the methods is presented.

20.1 Introduction

Prediction of archaeological site locations has become very popular during the past decades. It started on an experimental basis and developed through the years as a new science. The interest of a variety of scientists — geologists, computer engineers, surveyor engineers, geographers — in this domain has contributed to its rapid expansion. In many countries, predictive modeling is used as a powerful tool in cultural-resource management.

The basic assumption in predicting archaeological locations is that the choice of a site's location can be explained by the presence of certain environmental parameters. Consequently, knowledge from sites that have already been excavated, or site survey, forms the basis for predicting where else those predefined patterns can be met within a region with certain geographical characteristics.

Archaeological prediction consists of three main phases:

1. *Data collection*, a difficult task, as data must be accurate and suitable for its intended purpose. Moreover, data sets in digitized format are not always available in all countries. In many applications, instead of making any decisions on data quality, the methodological frame is defined by what coverages are available.

2. *Data processing*, suitable for integration into a predictive methodology. A geographic information system (GIS) is the ideal environment for this task, being a box of tools for inserting, handling, and mapping geographical data.

3. *Application* of a methodological tool to predict site locations.

The aim of this chapter is to discuss methodological issues of quantitative methods in archaeological prediction. The use of GIS is fundamental for this type of applications. The topic of interest in this chapter, however, is what comes next to GIS analysis: how the processed information can be transformed into a predictive output.

After each method is presented, there is a reference to its use in archaeological prediction. A final evaluation of the methods at the end of the chapter attempts to group their common features or to point out their differences in an effort to set a methodological frame for future applications.

20.2 Quantitative Methods and Their Applications in the Prediction of Archaeological Site Locations

Quantitative methods are those that use a mathematical frame to collect, analyze, and classify data in a systematical manner. The most important quantitative methods used to solve geographical problems are:

Regression analysis

Multicriteria analysis

Expert systems

Supervised classification

Unsupervised classification (or cluster analysis)

Fuzzy logic

Supervised and unsupervised classification can be based either on statistical logic, fuzzy logic, neural networks, or any combination of these (Hatzichristos 1999).

Prediction of archaeological site locations could be considered as a geographical problem of classifying space in the landscape as sites or nonsites, or in specific site categories, according to certain characteristics (environmental, topographical, historical, social). Therefore, archaeological prediction could be considered as a geographical problem that can be "solved" using the previously mentioned methods.

Logistic regression has been the most widely used method. The term "predictive modeling" has been generated through this method and has prevailed as a term used to describe archaeological prediction. Recently, other methods have been applied in an effort to predict site locations. In the following sections, the six quantitative methods identified here are presented along with a brief description of their application in archaeological prediction.

20.2.1 Regression Analysis: Logistic Regression

Logistic regression, or logit analysis, belongs to a family of statistical procedures called probability models. These models were developed for special regression problems in which the dependent variable is a categorical measure rather than an interval or ratio-scale measure (Warren 1990). Logistic regression is similar to discriminant function analysis, but is less constrained by statistical assumptions. The basic concept of the method is that it uses a sample area of sites and nonsites and environmental variables, and through a formula it is able to predict site-presence probabilities in unsampled areas.

The standard logistic regression formula is:

$$p(b) = \frac{\exp(a + \beta_1 X_{1i} + \beta_2 X_{2i} + \dots + \beta_n X_{ni})}{1 + \exp(a + \beta_1 X_{1i} + \beta_2 X_{2i} + \dots + \beta_n X_{ni})}$$

or, calibrated, is:

$$p(b) = \frac{1}{1 + \exp[-(a + \beta_1 X_{1i} + \beta_2 X_{2i} + \dots + \beta_n X_{ni})]}$$

where

p(b) is the probability that case i is a member of group B
exp is the function that raises e (Euler's number) to a parenthetical value
is a y-intercept constant
$\beta_2,\ldots,\beta_n$ are regression coefficients for variables $X_1,\ldots, X_n$
$X_{1i},\ldots, X_{ni}$ are values of variables $1,\ldots, n$ for the ith case

Logistic regression is undoubtedly the most well-known and commonly applied method for the prediction of archaeological site locations. This method was initially applied by Kenneth L. Kvamme in several models in western and southeastern Colorado. Kvamme's models are almost unparalleled in terms of their innovative definition of variables, their persistent methodological rigor, and their exposition of the vast research potential of predictive modeling (Warren 1990). The basic idea behind the methodology involves the examination of known archaeological sites or settlements for statistical associations in a region with various environmental conditions, such as ground steepness, elevation, aspect, soil, or distance-to-water preferences (Kvamme 1995). In the early 1980s, Scholtz-Parker and Hasenstab developed their own models (Kvamme 1995), similar to Kvamme's models.

Logistic regression has also been applied by Carmichael (1990) in central Montana; by Warren and Asch (2000) to predict prehistoric site locations in the Eastern Prairie Peninsula of central Illinois; and by Duncan and Beckman (2000), who formulated four predictive models to predict archaeological site locations in Pennsylvania and West Virginia.

20.2.2 Multicriteria Analysis

Multicriteria analysis is a multidimensional decision and evaluation tool. The basic characteristic of the method is that all possible and predictable impacts of any decisions to be taken are assessed.

The fundamental part of the multicriteria analysis method is the construction of a matrix of effectiveness, which consists of the criteria (according to which the specific problem is evaluated) and the alternatives, which may be either discriminated solutions of the problem or steps before the final solution.

Suppose that matrix consists of k alternatives and j criteria; then the element describes the contribution of alternative k to the criterion j:

$$\Sigma = \begin{bmatrix} \Sigma_{11} & \Sigma_{12} & & \Sigma_{1j} \\ \Sigma_{21} & \Sigma_{22} & & \Sigma_{2j} \\ \cdot & \cdot & & \cdot \\ \Sigma_{k1} & \Sigma_{k2} & & \Sigma_{kj} \end{bmatrix}$$

Multicriteria analysis has recently been applied in the prediction of archaeological locations (see Krist's and Verhagen's chapters [Chapters 16 and 9, respectively], in this volume).

20.2.3   Expert Systems

The main principle in expert systems is that an expert (i.e., a person who has a deep and thorough understanding of a problem area) makes his knowledge available to a computer program (Burrough 1986). The user who wants to acquire this knowledge enters queries into a program called an "inference engine."

The main drawback of the method is the high level of knowledge that must be available in a specific field. Moreover, an expert system takes a significant amount of time to form its rules. Another characteristic of these systems is that if the problem to be resolved is even slightly different from the one expected, then the performance is reduced. It would be interesting to apply expert systems in archaeological prediction.

20.2.4   Supervised Classification

Classification is a procedure that classifies different objects or incidents in classes according to the specified criteria. Classification can be either supervised or unsupervised.

Supervised classification is a classification in which one or more experts has *a priori* defined the groups in which an object and a training set belong. In supervised classification, there are two main phases: the training phase, during which the training set is used to define the weights of the parameters and their best combinations so that the classes can be discriminated effectively, and the realization phase, during which the defined weights are used to specify the classes in which the objects belong. Supervised classification can be realized with statistical logic, neural networks, or fuzzy logic.

Supervised classification has been applied in archaeology with neural networks (Druhot 1993). Neural networks, or more accurately, artificial neural networks, represent the human brain at the most elementary level of process. Specifically, neural networks retain as primary features two characteristics of the brain: the ability to "learn" and the ability to generalize from limited information (Hewitson and Crane 1994). There are many types of neural networks, depending on their topology and the use of supervised or unsupervised learning.

In Druhot's application, a back-propagation network was chosen. Back propagation is a feed-forward network, meaning that there are no direct feedbacks to previous processing elements. Unfortunately, Druhot's attempt did not manage to train back-propagation networks. This is due to a software restriction such that, when he combined grids, a limit of values forced him to reduce the number of variables. The predicted output was expected to be a value between 0.0 and 1.0, and every value equal or greater than 0.5 could be called a site or a site presence. However, for any attempt he made, the output values were less than 0.5. Druhot's application may be considered as a challenge, as it has proved that the methodology works.

20.2.5   Unsupervised Classification (Cluster Analysis)

Unsupervised classification, or cluster analysis, is a classification method that tries to create "natural" data groups. It is a data-driven method in which the user need only specify the number of classes. Unsupervised classification can be realized with statistical logic, neural networks, or fuzzy logic (or any combination of these).

In archaeology, this method has been applied recently with neural networks for the prediction of site locations in Brandenburg in northeastern Germany (Ducke 2003). Unsupervised classification overcomes the problem of determining "sites" and "nonsites," since there is no need to define classes. According to Openshaw, there are four types of unsupervised learning architectures that are of greatest relevance to neurospatial classification: the competitive learning nets, the self-organizing map, the adaptive resonance theory, and the associative memory nets (Openshaw 1994). Ducke's application uses Kohonen's self-organizing map, one of the most interesting of all the competitive neural nets. Among the more important advantages of the method, according to Ducke, is that information is processed more efficiently and economically, and a PC with 128 MB of main memory is completely sufficient. It would be interesting to see more results for this ongoing application in the future.

20.2.6   Fuzzy Logic

Fuzzy logic is a new science with a strong potential to simulate real-world conditions. In fuzzy logic, a proposition, apart from being true, may be anything from almost true to hardly true (Kosko 1994). Fuzziness is the modern term that stands for the scientific term "multivalence." It means three, more, or infinite options instead of just two extremes. While classic Boolean logic is binary, meaning that a certain element is true or false or an object either belongs to a set or it does not, fuzzy logic permits the notion of nuance: an element can be true to a degree and false to a degree, while an object can, to a degree, belong in more than one set.

Fuzzy logic was introduced by Zadeh in 1965. The key to Zadeh's idea is to represent the similarity a point shares with each group with a function (membership function) whose values (called memberships) are between $0 < m < 1$. Each point will have a membership in every group, with memberships close to unity signifying a high degree of similarity between the point and a group, while memberships close to zero imply little similarity between the point and that group. Additionally, the sum of the memberships for each point must be unity.

In Boolean logic the complement of A is notA and both A and notA are unique. In fuzzy logic the fuzzy complement applied to the fuzzy set A with the membership function $\mu_A(x)$ is

$$\mu_{notA}(x) = 1 - \mu_A(x)$$

To solve a problem with a knowledge-based fuzzy system, there is a four-step process that has to be followed:

1. *Fuzzification*: the assignment of a membership function to every variable of the problem. During this process, crisp sets are transformed to linguistic subsets (for example, short or long distance).
2. *Knowledge base*: the definition of the rules. Rules follow the format "if ..., then ..." and express logical assumptions. Experts with general knowledge on the specific field usually accomplish the task of rules definition.
3. *Processing of the rules*: also called inference. All the Boolean algebra operations (like intersection, union, negation, etc.) can be easily extended to fuzzy-set operations (Kandel 1986), and they can be used in this stage.
4. *Defuzzification*: the procedure of transforming the result of rules processing back into a crisp value. There are several methods to achieve defuzzification (Bezdek 1981).

Fuzzy logic has been used in archaeology to predict prehistoric sites on the island of Melos in Greece (Hatzinikolaou et al. 2003). The most important advantage of the method is that it reflects realism through linguistic variables. The final output is gradual and consists of more alternatives, which are lost in statistical analysis. Moreover, in this application there was no need to have data for sites and nonsites. The final output consisted of two subsets: the existence of settlements and the existence of special-purpose sites (agricultural units, mining units, and observatories). Membership functions were specified for nine criteria, and 15 rules were generated by a group of four experts to obtain the predictive output. The final output was compared with the results of a site survey to prove its accuracy.

20.3 An Evaluation of the Methods: From Binary to Fuzzy Logic

Binary logic dominates our modern world. Most of our technological achievements are based on it, since computer technology is binary. On the other hand, fuzzy logic can describe real-world conditions, as most life events are not just black or white boxes, but a gray-scale fluctuation.

Applied in archaeology, binary logic seems to be an ideal choice because, theoretically, a location either is a site, or it is not. From the predictive point of view, however, there is a great degree of uncertainty: a location has a potential to be a site to some degree and not to be a site to another degree. Going even deeper, at the heart of the problem, a hypothetical site has a potential to be a settlement and, at the same time, a potential to be a cemetery according to a set of hypothetical rules. Only by excavating can one be sure of a site's identity.

The use of quantitative methods for the prediction of archaeological site locations is a powerful tool with a potential for future development. To date, most predictive models have been constructed based on the logistic regression method. Those models are binary, and the final predictive output is a probability map showing a certain area's classification according to the probability (high, medium, low) of the existence of sites.

The use of classification methods (supervised and unsupervised) has been limited in archaeological applications. Both classification methods use neural networks with supervised and unsupervised learning.

Druhot's application with supervised learning did not work out as expected due to a software restriction. However, the method has potential and may be a stronger predictor than regression due to the great abilities of neural networks. In both regression and supervised classification with neural networks, a sample area of "sites" and "nonsites" is required. The output is consequently binary, 1 or 0, i.e., sites or nonsites.

Unsupervised classification overcomes the restriction of the training sample. This can be quite important because data on site absence are not always available. This method takes an "unbiased" look at the archaeological record by determining data structure exclusively from the data itself (Ducke 2003).

The above-mentioned methods are binary. Fuzzy logic presents a totally different approach to the prediction problem.

Using fuzzy logic, the final output is a predictive map for each of the sites' categories specified in the survey, showing the potential of each unity of the study area to belong in the specified category. For example, if the output is settlements, mines, ancient ports, and observatories, the final output will consist of four maps showing in color scale the potential of each pixel of the study area to belong in any of the specified groups. Areas not likely to be of archaeological interest will appear in the lowest category of the chosen color scale.

Fuzzy logic has a great potential to be applied in countries with a rich archaeological background, especially in areas that have not yet been excavated. The fact that the output result is not based in any training site-survey sample gives the method the freedom to be applied in any area with similar geographical characteristics. The generation of rules by an ideal group of experts might seem to be vague, but it should actually be considered stronger than the statistical analysis of dense artifact locations. The human mind is a most powerful tool when it comes to making decisions, and prediction with fuzzy logic is based on the same capacity.

In conclusion, it is important to mention that all of the methods described in this chapter are scientifically accurate, and because their use is substantiated, there must not be any doubt as to their validity. Classifications with neural networks are a tempting alternative to traditional regression methods. On the other hand, fuzzy logic is a new perspective: it has the potential to specify a site's origin; it is realistic; and it uses the abilities of human expertise in combination with a strong scientific theory.

It would be interesting to see additional archaeological applications using the quantitative methods described here. Prediction in archaeological survey is a new scientific field, and it deserves to be supported by all methodological tools.

References

Bezdek, C.J., *Pattern Recognition with Fuzzy Objective Function Algorithms*, Plenum Press, New York, 1981.

Burrough, P., *Principles of Geographical Information Systems for Land Resources Assessment*, Clarendon Press, Oxford, U.K., 1986.

Carmichael, D., GIS predictive modeling of prehistoric site distributions in central Montana, in *Interpreting Space: GIS and Archaeology*, Allen, K.M.S., Green, S.W., and Zubrow, E.B.W., Eds., Taylor and Francis, London, 1990, pp. 216–225.

Ducke, B., Archaeological predictive modelling in intelligent network structures, in *The Digital Heritage of Archaeology: Computer Applications and Quantitative Methods in Archaeology*, Archive of Monuments and Publications, Hellenic Ministry of Culture, Greece, Doerr, M. and Sarris, A. Eds., CAA 2002 proceedings, 2003, pp. 267–273.

Duncan, R. and Beckman, K., The application of GIS predictive site location models in Pennsylvania and West Virginia, in *Practical Applications of GIS for Archaeologists: A Predictive Modeling Kit*, Wescott, K. and Brandon, R., Eds., Taylor and Francis, London, 2000, pp. 33–58.

Druhot, R., Neural networks and spatial modeling, in *Proceedings of Thirteenth Annual ESRI User Conference*, ESRI, USA, 1993.

Hatzichristos, T., Delineation of Ecoregions Using GIS and Computational Intelligence (in Greek), Ph.D. thesis, National Technical University of Athens, Athens, 1999.

Hatzinikolaou, E., Hatzichristos, T., Siolas, A., and Mantzourani, E., Predicting archaeological site locations using GIS and fuzzy logic, in *The Digital Heritage of Archaeology: Computer Applications and Quantitative Methods in Archaeology*, Archive of Monuments and Publications, Hellenic Ministry of Culture, Greece, Doerr, M. and Sarris, A., Eds., CAA 2002 proceedings, 2003, pp. 169–177.

Hewitson, B. and Crane, R., Looks and uses, in *Neural Nets: Applications in Geography*, Hewitson, B. and Crane, R., Eds., Kluwer Academic, Netherlands, 1994, pp. 1–9.

Kandel, A., *Fuzzy Mathematical Techniques with Applications*, Addison-Wesley, New York, 1986.

Kosko, B., *Fuzzy Thinking: The New Science of Fuzzy Logic*, Flamingo Press, London, 1994.

Kvamme, K.L., A view across the water: the North American experience in archaeological GIS, in *Archaeology and Geographical Information Systems: A European Perspective*, Lock, G. and Stančič, Z., Eds., Taylor and Francis, London, 1995, pp. 1–14.

Openshaw, S., Neuroclassification of spatial data, in *Neural Nets: Applications in Geography*, Hewitson, B. and Crane, R., Eds., Kluwer Academic, Netherlands, 1994, pp. 53–70.

Warren, R., Predictive modeling in archaeology: a primer, in *Interpreting Space: GIS and Archaeology*, Allen, K.M.S., Green, S.W., and Zubrow, E.B.W., Eds., Taylor and Francis, London, 1990, pp. 90–111.

Warren, R. and Asch, D., A predictive model of archaeological site location in the eastern prairie peninsula, in *Practical Applications for GIS for Archaeologists: A Predictive Modeling Kit*, Wescott, K. and Brandon, R., Eds., Taylor and Francis, London, 2000, pp. 5–32.

Zadeh, L.A., Fuzzy sets, *Information and Control*, 8, 338–353, 1965.

21

The Use of Predictive Modeling for Guiding the Archaeological Survey of Roman Pottery Kilns in the Argonne Region (Northeastern France)

Philip Verhagen and Michiel Gazenbeek

CONTENTS

ABSTRACT   The Argonne survey project (carried out from 1996 to 1998 in the northeast of France) was aimed at obtaining a reliable overview of the occurrence of Roman pottery kilns in the area. Many pottery kiln sites in the area are under threat of erosion by changing land use, and as the area is an important center of Roman pottery production in northwestern Europe, a protection program was to be developed. The survey project used a combination of techniques to make an inventory that was as complete as possible (field walking, geophysical survey, and augering), and it relied heavily on predictive mapping as a tool to guide the survey. This chapter focuses on the predictive model developed and the consequences of using such a model as the basis for surveying the area.

21.1 Introduction

The Argonne region, situated in the northeast of France (Figure 21.1), was an important center of Roman pottery production in northwestern Europe. Today it is a quiet area with abundant forest (covering about 50% of the region), but during the First World War it was the stage for fierce frontline fighting between the German and French forces. Numerous trenches are still present in the area, and the remains of ammunition, barbed wire, and weapons can be found on many fields.

   The area is currently experiencing rapid land-use changes that are potentially damaging to the archaeological remains. The most important of these is the conversion of grassland to agricultural land, a development that will increase erosion of the topsoil. Furthermore, new infrastructure in the area is being developed by the French government as part of a revitalization campaign in the area. One of the aims of this campaign is to draw tourists to visit the World War I relics. Also, the new TGV Est high-speed railway, connecting Strasbourg to Paris in 2005–2006, will be running straight through the area.

   Given these developments, the Service Régional d'Archéologie (SRA) of the Lorraine region decided in 1996 to launch a project (Gazenbeek et al. 1996) to make an inventory of the distribution and state of conservation of pottery kiln sites. The SRA of the region Champagne-Ardennes decided to join the project in 1997, which brought the total study area to 725.62 km² in 51 municipalities.

FIGURE 21.1
Location of the study area in France.

The aims of the project, as expressed by the SRA Lorraine, were to:

1. Establish survey methods appropriate to the region (field walking, geophysical survey, and augering)
2. Decide which sites were to be excavated
3. Acquire land and establish archaeological reserves
4. Elaborate protection measures in negotiation with land owners and users

The survey was to focus on both the known sites as well as sites still undiscovered. As it was clear from the beginning that not all the area could be surveyed, it was necessary to start the project by preparing a predictive map. The survey was carried out in three consecutive years (1996–1998), and before the start of the campaigns a predictive map was prepared to guide the survey; a revised map was produced before the last field campaign, after which a final map was produced and presented to the SRA. The survey consisted of field walking to discover kiln sites, augering to establish their extent and precise location, and in selected cases, geophysical survey to obtain an impression of the remains still present underground. The field-walking survey and excavations were carried out by students and staff from the Université de Paris I Panthéon-Sorbonne; the geophysical survey, augering, and predictive mapping were performed by RAAP Archeologisch Adviesbureau.

21.2 Archaeological Context

The Argonne area has exported industrial quantities of ceramics all over northwestern Europe during nearly all of the Antiquity, from the 1st century AD until at least the 5th century. These products, especially the fine slip ware, are very important for the dating of consumer sites. However, of the production centers themselves, their range of products, their production techniques, and their life span, little was known, even though research had been going on for more than a century. The data collected in the Argonne project helped significantly to better understand the economical and environmental background that ensured the success of the Argonne pottery production, and to understand its place in the regional occupation network.

The Roman settlement pattern of the Argonne area appears as a patchwork of villages surrounded by extensive workshop areas (including pottery kilns), of isolated ceramic and glass production centers (that often mask the associated dwellings), and of villages and modest farms spread out over the countryside. Except for some large production centers that were active during the whole period, the workshops were mostly short lived, and moved rapidly from one place to another. This model is evident as early as the 1st century AD when the initial Belgian wares were produced. From this period onward, the geographical expansion of the pottery kilns shows only minor changes until well into the 4th century, even though the products themselves changed completely, moving from Belgian wares to red slip wares and covering a whole range of black slip beakers and common coarse wares as well.

The success of the Argonne wares has often been explained by the privileged position of the region close to several major Roman roads that connected it to the main communication network of this part of the Roman Empire, putting it at a relatively short distance of several of the larger cities of northern Gaul, such as Reims and Metz, and providing many markets and redistribution centers for its products. Export over the Marne and Meuse rivers certainly also was very important for the spread of the ceramics, and the nearby town of Verdun on the Meuse probably played a key role in the river transport. However, the importance of production for the regional domestic market has long been underestimated. The fact that the slip wares, the main export product, have been studied in more detail, means that the local markets and the associated potteries have been neglected in research. As it is, coarse wares and roof tiles were produced in large quantities by the various workshops simultaneously with the rest. Our fieldwork also demonstrated the importance of glass and iron workshops in the area. Especially the 3rd and 4th centuries appear to be an important period for the local glass industry, with rows of deep shafts dug into the sandstone that was used as raw material. Its products, such as glass cubes for mosaics, were strongly oriented towards export.

These activities show that the ceramics were far from being the only product manufactured in the region. All together, they testify to a flourishing economy during most of the Antiquity, with a very active industry maintaining itself over a long period of time. The importance in this of the natural factors that are characteristic of the Argonne, cannot be underestimated: the geological context is responsible for the environmental conditions that could supply at the same time the basic raw materials needed for different products, and the fuel (wood) necessary to produce the (semi-)finished products. The large area covered by the geological formations involved guaranteed the long life span and the profitability of these industries on such a large scale. The road network that crosses the region, allowing access to the markets, is therefore only a factor that made this particular economical development easier, but it did not command it.

21.3 Area Description

The Argonne region is an area of undulating hills that form part of the French Ardennes. The area has a general slope toward the northwest, and is dissected by several watercourses. Elevation generally ranges between 175 and 300 m above sea level. The Aire River divides the area into two distinct parts: the Forêt d'Argonne to the west, and the Forêt de Hesse to the east. The World War I front line runs from west to east through the area (Figure 21.2).

The geology of the area consists mainly of Jurassic and Cretaceous sedimentary rocks. The oldest formations are found in the east, the youngest in the west of the area. The most important rock types to be found in the area, from old to young, are:

Kimmeridgian clays and marls

Portlandian limestones (Calcaires du Barrois)

Lower Albian greenish clayey sands (*sables verts*)

Upper Albian yellowish-brown sandy clays (Argiles du Gault)

Cenomanian calcaric sandstones (Gaize)

It is assumed that the pottery in the area was produced using both the *sables verts* and Argiles du Gault; however, no definite answer to that question has been found.

Apart from the Jurassic/Cretaceous deposits, colluvial and alluvial deposits are found in the valleys. Along the valleys of the larger watercourses (Aire, Biesme, and Aisne), older alluvial deposits can be found in terraces located 10 to 15 m above the current valley bottom (Figure 21.3).

The geomorphology of the Forêt d'Argonne is dominated by the Gaize sandstones and consists of a strongly dissected plateau with numerous springs where the Gaize rests on the Argiles du Gault. The Forêt de Hesse is more varied, with Calcaires du Barrois in the valleys, followed by *sables verts* and Argiles du Gault uphill, and often capped by a Gaize sandstone butte. Slopes are more gentle in this area. Springs are not as numerous and can also be found on the transition between *sables verts* and Calcaires du Barrois. To the south of the river Vadelaincourt, this geomorphology is replaced by a plateau of Calcaires du Barrois. This means that the area where pottery clay can be found is primarily concentrated in the Forêt de Hesse.

Soils in the area are not very well developed. The *sables verts* will weather to a yellowish-brown sandy clay that is sometimes difficult to distinguish from the Argiles du Gault. The Calcaires du Barrois weathers to a brown clay, but it may be strongly eroded on steep slopes. Weathered Gaize has not been found in the area; the sandstones are usually covered by a shallow, dark brown, sandy topsoil. Soil erosion is an important phenomenon in the area. The augering campaigns revealed downhill accumulations of colluvial deposits in many places, and often these could be dated to post-Roman age. Especially the *sables verts* and Argiles du Gault are highly susceptible to erosion. Currently, however, the rate of erosion is not very high, as most of the area is protected by a vegetative cover of grassland and forest (Timmerman et al. 1998).

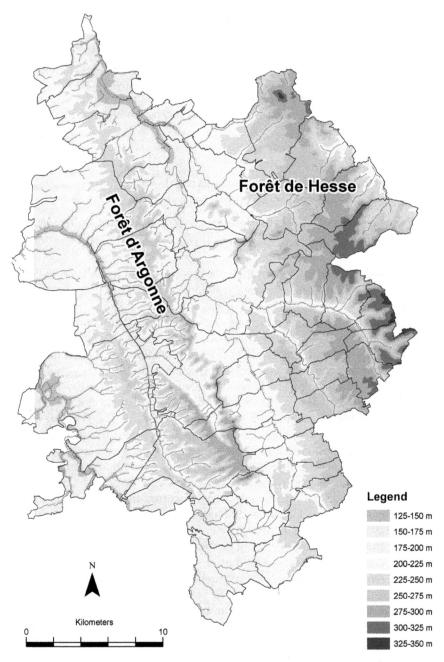

FIGURE 21.2
Topographical map of the Argonne study areas. *Source*: Institut Géographique National.

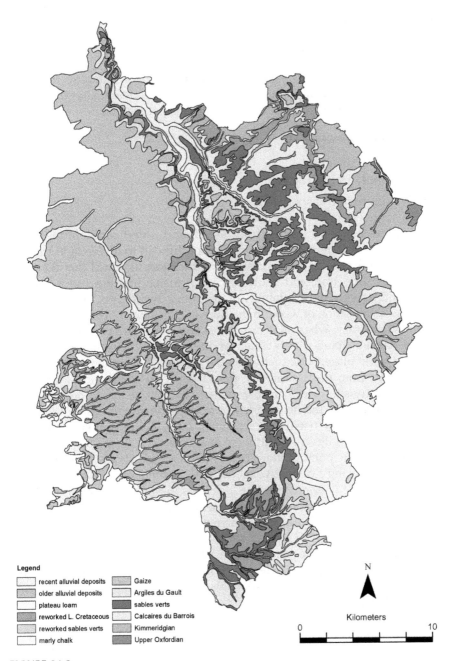

FIGURE 21.3
Geological map of the Argonne study area. *Source*: Bureau des Recherches Géoloques et Miniéres.

21.4 The First Predictive Model

When confronted with the question of where pottery kiln sites might be located, it is not difficult to understand that the selection of sites for pottery production must have depended on four principal factors:

1. Proximity to pottery clay
2. Proximity to water
3. Proximity to firewood
4. Proximity to transport routes

The source materials needed (clay, wood, and water) could, in theory, be transported to a different place, but it is not difficult to understand that the pottery is more readily transportable. It is therefore assumed that the proximity to existing transport routes is not a major limiting factor for site placement, whereas the availability of the source materials is. Of these source materials, the location of wood in the Roman period cannot be reconstructed with any accuracy, whereas it can be assumed that the position of watercourses has not changed very much, and the geological formations will still be in the same place. A deductive model of kiln site location will therefore have to start from the assumption that distance to water and pottery clay are the primary site-placement parameters that can be operationalized.

The first results of the fieldwork seemed to confirm this assumption. The kiln sites found were usually located near valley bottoms or springs where *sables verts* and Argiles du Gault were available.

From a scientific point of view, the model building should probably have started by preparing a deductive model. The first model built, however, was an inferential one (Gazenbeek et al. 1996). As in many inferential modeling exercises before, the available environmental information (elevation, slope, aspect, geology, and distances to watercourses and springs) was subjected to a χ^2 test on the basis of a small and biased data set. However, the χ^2 test was not used to select the significant variables; all available map layers were reclassified according to site density and then averaged. Even though the reliability of this model was questionable, it did serve to highlight the weaknesses of the existing archaeological data set.

Until the start of the Argonne Project, the knowledge of the distribution and state of pottery kiln sites in the region was scant. A total of 30 kiln sites had been reported to the French national archaeological database DRACAR, almost exclusively found in the Forêt de Hesse area. These sites were used to construct the first model. Even with this small number of sites, it was clear that the *sables verts* were very important for kiln site location. It also suggested that the known data set was biased, as very few sites were reported in forested areas. This provided two strong objectives for the survey campaigns: to increase the number of known kiln sites to a level where

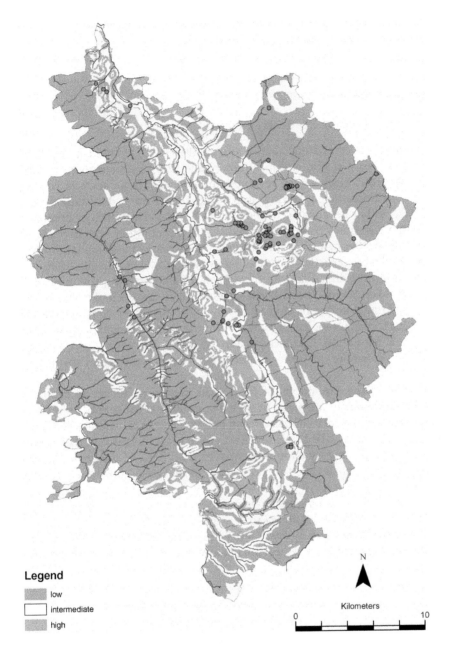

FIGURE 21.4
The final predictive map. The dots indicate kiln sites.

statistical analysis could be done in a meaningful way, and to improve the representativeness of the known site sample.

21.5   The Second Predictive Model

The three field campaigns carried out in 1996–1997 resulted in a dramatic increase of known kiln site locations (Table 21.1). Furthermore, the previously reported 30 sites were revisited and checked. In most cases, the registered coordinates were wrong, and some sites were withdrawn from the database, either because no traces of the site could be found, or because the site had erroneously been interpreted as a Roman pottery kiln. In September 1997, the number of sites inside the surveyed area had become large enough to justify the construction of a new inferential model of kiln site location. As it was assumed that kiln site location is related to the proximity of key geological formations (notably the *sables verts*), distances to these formations were calculated and subjected to a χ^2 test. The distances were calculated as distances to the geological formation boundary, both outside and inside the geological formation. Distances inside the formation were given a negative value. Furthermore, distances to permanent watercourses and springs were used.

Because of the limitations of the χ^2 test, it was necessary to reduce the number of distance categories in such a manner that the number of expected sites per map category should not fall below 5 (see, e.g., Thomas 1976). With a total of 56 sites available inside the surveyed area, this meant that preferably no single map category should be smaller than 8.9% of the area. Therefore, the distance categories used are not equal-interval zones, but equal-area zones that can be produced in Arc/Info Grid by using the Slice command. The resulting distance maps were analyzed for autocorrelation. It turned out that distance to Argiles du Gault is rather strongly correlated to distance to *sables verts* and to Gaize ($r = 0.59$ and $r = 0.56$, respectively). This can be explained by the fact that Argiles du Gault are usually found as a narrow band between

TABLE 21.1.

Area surveyed during the four consecutive field campaigns. Before the start of the survey, only 30 kiln sites were known in the area.

survey period	area surveyed	inside study region	# kiln sites	in surveyed area
Nov 96	1037.42	1037.42	42	15
Feb 97	1349.03	1308.34	70	47
Sep 97	1381.00	1114.39	74	56
Mar 98	2750.59	1991.19	91	83
TOTAL (ha):	6518.04	5451.34		

Gaize and *sables verts*. Furthermore, the distance to Calcaires du Barrois is negatively correlated to the distance to Gaize ($r = -0.51$). This is due to the fact that the Gaize is predominantly found in the west, and Calcaires du Barrois in the east of the area. Other, weaker correlations were found between springs and permanent watercourses ($r = 0.49$) and between permanent watercourses and recent alluvial deposits ($r = 0.44$).

After χ^2 analysis of all variables, the model was based on those variables that were statistically significant at the 95% probability level and not strongly autocorrelated: slope, distance to *sables verts*, distance to Gaize, and distance to recent alluvial deposits. Although the initial intention was to use only the data from inside the surveyed area, it turned out that the "kiln hunt" had been successful in the number of sites found, but biased in terms of the area visited. For example, the mainly forested Gaize and steep slopes had been avoided by the field walkers, for understandable reasons. To compensate for this effect, the full data set was used, which meant using a possibly biased data set instead of a certainly biased one. The resulting model showed high probabilities for kiln site location in the Forêt de Hesse and along the valley of the Aire, whereas low probabilities were found in the Forêt d'Argonne. However, some outliers were found close to the major rivers and the presumed location of Roman roads, possibly indicating a preference for location close to transport routes instead of close to the source materials needed.

21.6 The Final Model

Although the second map was useful in indicating the important zones for kiln site location, it was less well-suited to predict low-probability zones. A large area was still designated medium probability, and this was primarily a consequence of the survey bias. The last field campaign in 1998 was therefore dedicated to extending the surveyed area into the Gaize and steep-sloped zones. The last campaign included over 50% of Gaize. This made it possible to produce a model with optimal reliability, from a statistical point of view.

Apart from that, it was decided to perform a field check to see if the geological maps used were accurate enough for the predictive modeling. The field check (Timmerman et al. 1998) confirmed that the quality of the geological maps is adequate for most of the area, with two notable exceptions. Firstly, where outcrops of geological formations have a limited extent, they are not always mapped. In one particular case, a kiln site was found close to a pocket of *sables verts* that was not depicted on the map, and it can be expected that similar locations exist in the area, especially near valley bottoms. Secondly, south of the Forêt d'Argonne, a relatively large area of *sables verts* shown on the geological map was not found at the surface, but only at considerable depth. The high probability assigned to this area on the predictive map is therefore incorrect.

The final model (Figure 21.4) was satisfying from the point of view of cultural resource management: the area of medium probability had been substantially reduced, and the high-probability zone showed a gain of 55.5%, (Table 21.2). It should, however, be noted that the model can be this specific because the location of kiln sites considered is primarily linked to a very specific location factor, the availability of pottery clay.

21.7 Conclusions

The Argonne project succeeded in producing a predictive map that is useful for locating the principal areas of pottery kiln sites with high accuracy. Given the lack of information at the start of the project, this is an impressive achievement.

Nevertheless, it seems that the model can only be this successful because the site type aimed at is highly predictable from a deductive point of view. The availability of pottery clay is the most important location factor to be taken into account, and therefore highly limits the area were kiln sites can be found. In general, this means that predictive models will be most successful when they aim at predicting specific functional site types. This in turn implies that the site sample used for the modeling should be analyzed on specific site types, and the variables used should reflect the possible limiting and attracting factors for locating these sites. This is not a very surprising conclusion, it seems, but one that is frequently overlooked in "commercial" predictive modeling.

Furthermore, the project made eminently clear that field testing of the predictive maps, both archaeologically as well as geologically, was very useful and, in fact, necessary to obtain a reliable model. It is also clear that the amount of field testing done should be substantial, and it should be guided by the questions that arise from the predictive mapping itself. In practice, this may be a very difficult point to get across to contractors, as

TABLE 21.2.

Comparison of the three predictive models made. In the third model, the gain of the high probability zone is 55.5 %, and the area of intermediate probability is considerably smaller than for the previous models.

	1996		1997		1998	
probability	%area	%sites	%area	%sites	%area	%sites
low	27.5	2.8	29.9	0.0	63.2	6.6
intermediate	58.6	61.1	52.9	32.4	22.0	23.1
high	13.9	36.1	16.0	67.6	14.8	70.3

predictive modeling is often seen as a means to prevent costly field campaigns. The costs for arriving at the Argonne predictive map were high: a total of 7.5% (54.5 km²) of the area had to be surveyed to obtain the representative site sample needed for the final predictive model. Even with a field campaign primarily looking for sites as easily detectable as pottery kilns, this represents a considerable amount of work to be done. In the case of the Argonne project, four months of field walking were done, using 20 students. However, when commercial prices have to be paid, this would represent an investment (at current rates) of approximately 600,000 euros or dollars. In the context of European public archaeology, this is a very high price for a field-survey project.

Acknowledgments

The author would like to thank the following people collaborating in the project:

Professor Sander van der Leeuw (Université de Paris I/Panthéon-Sorbonne, UFR03, Histoire de l'Art et Archéologie), and Roel Brandt (former director of RAAP Archeologisch Advies Bureau) coordinators of the Argonne Project.

The colleagues from RAAP Archeologisch Adviesbureau involved in the project, especially Joep Orbons (geophysical survey) and Jan Roymans (field survey).

The students from the Vrije Universiteit Amsterdam who did the geological fieldwork: Saskia Gietema, Nicole Rosenbrand, and Rinke Timmerman.

References

Exaltus , R., Orbons, J., Papamarinopoulos, S., van der Leeuw, S., and Verhagen, P., Les ateliers céramiques gallo-romains et médiévaux de l'Argonne, rapport triennal (1996–1998), Vol. 3, Carrottages, prospections géophysiques, Université de Paris I, Paris, 1998.

Gazenbeek, M., Orbons, J., Spruijt, T., and Verhagen, P., Les ateliers céramiques gallo-romains et médiévaux de l'Argonne: bilan, recherche et gestion patrimoniale, rapport soumis aux Services Régionaux de l'Archéologie de Lorraine et de Champagne-Ardennes, Université de Paris I, Paris, 1996.

Thomas, D.H., *Figuring Anthropology: First Principles of Probability and Statistics*, Holt, Rinehart and Winston, New York, 1976.

Timmerman, R., Rosenbrand, N., and Gietema, S., Verslag van het bodemkundig/geologisch onderzoek in het kader van het Argonne Project, unpublished report, Vrije Universiteit, Amsterdam, 1998.

Index

Milton Keynes UK
Ingram Content Group UK Ltd.
UKHW021908071024
449327UK00022B/1642